The Genie out of the Bottle

The Genie out of the Bottle

World Oil since 1970

M. A. Adelman

The MIT Press
Cambridge, Massachusetts
London, England

Second printing, 1996

This book was set in Palatino by Asco Trade Typesetting Ltd., Hong Kong and was printed and bound in the United States of America.

Library of Congress Cataloging-in-Publication Data

Adelman, Morris Albert.
The genie out of the bottle : world oil since 1970 / M. A. Adelman.
p. cm.
Includes bibliographical references and index.
ISBN 0-262-01151-4
1. Petroleum products—Prices—History. 2. Petroleum industry and trade—History. I. Title.
HD9560.4.A28 1995
338.2′3282—dc20 95-8271
CIP

To MIL, whose love and devotion made it possible

Contents

Detailed Contents

Figures and Tables

Preface

In March 1971, I said, "The genie is out of the bottle. The oil producing countries had a great success using the weapon of a threatened concerted stoppage and they cannot be expected to put it away." [NYT 3-29-71:49] A year later: "I would expect prices to describe a parabola in the 1970s —first rising, then falling." [NYT 4-10-72:53] The parabola took much longer than I expected. The world price of oil, inflation adjusted, rose to fourteen times the 1970 level, then declined by 1993 to somewhat over three times.

Huge oil revenues in the Middle East were the means and the motive for war. In 1961, Iraq had claimed Kuwait and mobilized on the border. A British brigade flew in to end the threat. In the next thirty years, over $2.5 trillion (at 1991 prices) flowed into the area. To thwart Iraq in 1991 took 100 times as many troops as in 1961, massive armament, and much destruction. The Iran-Iraq war had already left a million dead between 1980 and 1988, and weapons continue to accumulate in this area.

Among the oil-importing countries, the poorest suffered the most, often by devastating their forests to provide fuel. In every industrial country there was a sharp down deflection in economic growth after 1973. Energy consumption was reduced, but from investment that instead would have gone to expand output. Economic forecasts are routinely qualified, "provided there is no oil shock".

Stagnant productivity has meant stagnant per capita incomes. Each social group has tried to hold its own at the expense of others, with wearisome deadlock over budget deficits or reducing barriers to international trade. We do not know how much of the long stagnation can be traced to the oil shocks and their interaction with other events and government policies.

The oil-exporting nations used their new revenues largely for subsidies, consumption, and weapons, with very little nonoil development.

Considering inflation, the raising of false expectations, swollen then decaying public works and services, and the ruin of agriculture and other traditional industries, many (some think most) of their inhabitants are worse off than they would have been with constant oil prices.

But our business in this book is with causes, not effects. How and why did the price of oil explode and fluctuate? Oil is known as the glittering prize; the source of world power and empire; the blood of the people; the family jewels. But what is precious also may be precarious: "limits to growth," "the strait of stringency," "energy crisis," "the wolf is here," "*ce monde affamé d'énergie*," the "moral equivalent of war," "wells drying up," and much more.

Glamour robs people of their common sense. As Samuel Johnson warned: "The mind which has feasted on the luxurious wonders of fiction has no taste for the insipidity of truth." World oil shortage is a fiction, but belief in this fiction is a fact.

Some small, undeveloped nations overcame the world's greatest industrial corporations. They were free to seize oil properties and fix oil prices because colonialism was gone. I share the conventional American opinion: good riddance. But were I a *pukka sahib* mourning past imperial glories, there would be no reason to change a line of how it all happened.

The documentary record is sparse. Historians of recent events, wrote Marc Bloch, are handicapped; they lack involuntary confidential disclosures. Although they hear supposed inside dope, "it is hard to tell information from gossip. A nice cataclysm would often do us more good." Just so: the 1973 events did give us some documents. All along, a free press has forced companies and governments to disclose more than they wished. They may distort facts, but they rarely can invent them.

Chapter 1 is a road map to the book. Then chapter 2 lays out the general economic ideas and measures used in it. This book is written to accommodate general readers, who can move past the few equations or jargon used as a bridge to the companion volume, *The Economics of Petroleum Supply* (MIT Press, 1993). Nevertheless, the story unfolded here is a history of price change, not a collection of tales adorned with talk of deep tides running. To follow the history, and to avoid seeing what was not there, readers must be able to distinguish changes in scarcity from changes in control of the market.

Oil production is the daily drawdown of inventories, or proved reserves, so scarcity is defined by the investment needed to add reserves and capacity. The competitive price is the return necessary to induce the investment. Unhampered competition offers carrots: unusual profits on new and better products, processes, and sources of supply.

Competition also has a stick, which is coercion. The "invisible hand" is not gentle. It slaps and cuffs sellers to stay in line and keep moving where they do not want to go, "to promote an end which was no part of his intention" (Adam Smith). Sometimes it hits hard enough to kill. Sellers think they are entitled to a "fair" price. Entitled or not, they can get it only by some kind of arrangement to evade or reduce competition.

The narrative theme of chapters 3 through 8 is how the owners of most of the world's oil resources have restrained production from an abundant oil supply to raise the price. To restrain output, they must first restrain one another. They feel the threat of a war of all against all. Readers might prefer fewer meetings of OPEC oil ministers. So would I. More important, so would the ministers, but they had no choice. What I once called "the clumsy cartel" tried to fine-tune with coarse instruments. Hence the "energy crises" we examine.

The winter of their discontent has lasted for over a decade now, and I see no spring ahead. They will continue as best they can to dam up oil abundance. They do not intentionally seek disorder and violence, but sometimes they cannot avoid it.

The nations that raised oil prices despite abundance and used overflowing revenues to prepare and wage war received great help from the U.S. and other governments that believed oil was running out and sought "peace in the Middle East." There may be more such ironies. If this history, as Thucydides wrote of his, "is useful to those who want an accurate account of the past to help understand the future, that is enough."

Acknowledgments

The work for this book and for much that appeared in *The Economics of Petroleum Supply: Papers by M. A. Adelman 1962–1993* was supported by the National Science Foundation, grant SES-8412791, and by the Center for Energy and Environmental Policy Research at the MIT Energy Laboratory. I am grateful for their support.

It is heart-warming to recall the long-time collaboration at MIT with Henry Jacoby, Gordon Kaufman, and most of all Michael Lynch. I wish James Paddock, David Wood, and Zenon Zannetos could hear my thanks. Fortunately, many others can. Richard Gordon put me deeper in debt by reading the whole manuscript, as did Carol Dahl and Campbell Watkins. I thank as well Paul Bradley, Robert Deam, Paul Eckbo, James Griffin, Mariano Gurfinkel, Charles Kindleberger, Walter Levy, John Lohrenz, Richard Mancke, Stephen McDonald, John Moroney, Francisco Parra, James Smith, Michael Telson, Philip Verleger, and Martin Zimmerman. This book embodies a dialogue of over thirty years with Jack Hartshorn (*Oil Trade*, Cambridge University Press, 1993).

Therese Henderson was a constant coworker, as was Kate Schoninger in the later stages. The help of my undergraduate assistants Rachel Obstler, Chris Roberts, Manoj Shahi, and Kathleen Liew Kie Song is visible in many places. None of those mentioned here are responsible for errors, only for there not being more.

1 A Road Map

Basic Economics

A price of any commodity may change because of a change in *scarcity*—the supply/demand balance—or a change in *market control*—the degree of monopoly. We will examine the scarcity side of oil first.

Oil and other minerals will never be exhausted. If and when consumers will not pay enough to induce investment in new reserves and capacity, the producing industry will dwindle and disappear. Nobody will ever know, or even want to know, how much is still in the ground. Only cost and price matter.

Over the long run, most mineral prices have declined, along with costs. But for any given mineral at any given time, the tug of war can go either way: growing knowledge versus diminishing returns.

The cost of oil (figure 1.1) consists mostly of investment to find and delineate reservoirs and create capacity. Oil is produced from a reservoir at a diminishing rate. The pool's estimated total production over time is a *reserve*. Reserves are inventories, constantly used up and replaced.

Supply and demand are equated when the price makes it worthwhile to replenish reserves and capacity. A persistent rise in the investment needed per unit is the warning signal of greater scarcity and higher prices ahead.

From 1945 to 1970, the "low" price of oil induced enough investment to expand world capacity and output sixfold. Outside the United States, it was fifteen-fold. At the all-time price low in 1970, the Persian Gulf price was $1.21 (at 1993 price levels, about $4.27). A producing company kept only 33 cents, but even that minor fraction provided an annual return on new investment of over 100 percent. Naturally enough, companies were planning on continued growth. If there has been an increase in investment requirements since then, it is too small to explain any price change.

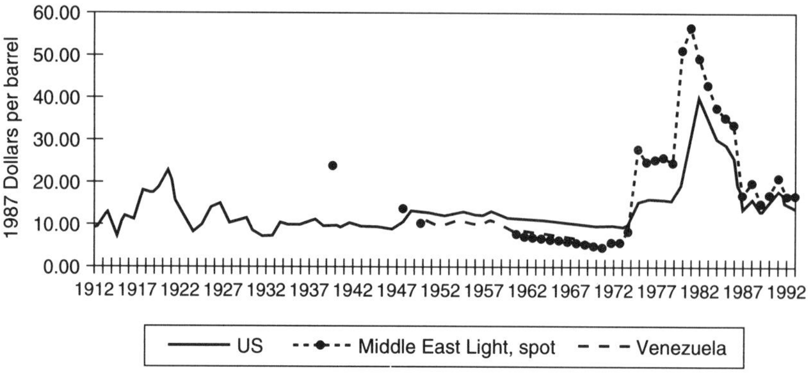

Figure 1.1
Crude oil prices, United States, 1912–1992 and Middle East, 1947–1992. Sources: Price deflators: 1912–1939, from Bureau of the Census, *Long Term Economic Growth* (Washington: GPO 1066). Thereafter from annual *Economic Report of the President.* Oil prices: Middle East, before 1960, from table 3.2, therafter from DOE:AER and DOE:MER (ultimate source PIW); Venezuela, Adelman (1972), p. 342; United Startes, 1912–1945, from [API 1959], p. 374, thereafter from DOE:AER.

Low investment cost has been waved off because of the belief that oil is a limited stock, created once and for all. Hence it is thought to have an inherent value separate and apart from any investment, but since there is no fixed initial stock, there is no inherent value. In the real world, there are many alternative ways to invest to create reserves. Old fields are constantly extended, and new fields are occasionally found; all must be developed. At the margin, the many methods of reserve creation are all substitutes, and their costs all change in the same direction.

The industry in the United States is the prime historical example. The wellhead price of a barrel, its market value in-ground, its development cost, and the value of an undiscovered barrel were constant to declining for many years before 1973. Then they rose.

Elsewhere, development costs declined, and they have not changed much since. Where development is very cheap, the value of newly found oil is low, since the two are substitutes. In 1976, when the wellhead price was $12, an undiscovered barrel in Saudi Arabia was valued at 1 or 2 cents. Such a value does not explain a price of even $1.20, let alone $12.00 or $34.00.

Once the vague notion of "limited resources" is reduced to an observable market value, we can say (with apologies to Winston Churchill), rarely have so many made so much of so little.

Reversal in the 1970s: Water Runs Uphill

From World War II to 1973, growth in new low-cost oil-producing areas greatly outstripped growth in the high-cost areas. That is how a competitive system operates to save resources. Then came the reversal. The lowest-cost producers (members of OPEC, the Organization of Petroleum Exporting Countries) cut back investment and output, producing only what they could sell at current prices. But non-OPEC producers sell all they can produce, so expansion has been in these higher-cost areas.

If reserve owners were keeping an asset in the ground for future higher prices, the higher-cost producers would gain more by retaining it, and the lower-cost producers would profit by promptly selling it. Since the price explosion, both groups have done just the contrary. By the rules of competitive economics, water began to run uphill about twenty-five years ago. If we cannot make sense of this reversal, we have not much to say.

If the nation-owners of oil deposits had a very long view and preferred to keep oil in the ground rather than sell it, each OPEC nation would have reduced output on its own. In fact, each urged the others to cut back more.

Economic theory suggests that these nations' horizons would be not long but very short—that they would be avid for more money and quick to spend it. Income from a single source is more risky than is an income flow from diverse sources and is to be discounted at a higher rate. The producing countries had and have little but oil income. Moreover, most are unstable societies, where power is transferred by force. They reside in unsafe neighborhoods and fear attack. It was said of some in 1935, "The future leaves them cold. They want money now." Fifty years later, commentators noted "OPEC's seeming inability to do without instant gratification." OPEC nations acted by the book: spending soon caught up with receipts and then exceeded them. Saudi Arabia's foreign assets, once $160 billion, were nearly gone even before the 1991 war.

There is no way to explain the price upheavals by higher demand, deficient supply, changes in discounting, or political objectives. The only story that makes sense is that the sellers achieved some degree of market control: monopoly.

Challenge and Response

History is full of examples of challenge and response. In the world market, the challenge was and still is oil abundance. The owners of world oil resources have restrained production to keep up the price. Cooperation went smoothly for many years when the owners were private companies; their aims were limited and their methods well adapted. They did not keep prices artificially low; that would have been very odd behavior. Lower prices meant lower profits and political dangers, in both the producing and the consuming countries, where they were attacked for cheap oil imports.

For decades before World War II, the world price was the U.S. price plus freight. It was far above the cost of adding and using reserves in the new lightly tapped fields outside the United States. Investment in much greater output to displace older areas would have been highly profitable, but it would have lowered the world price and on balance would have injured the companies. Hence, they expanded output only to match the growth of consumption.

But after World War II, while increasing competition brought prices far down, host governments increased their tax take. Thus the company margin kept shrinking. That was a danger signal to the governments because the margin was the buffer between price and government take. Price erosion became a growing threat to the governments.

In the 1960s, governments learned to act in concert, to stabilize their revenue per barrel. Between 1970 and 1973, they actually began to raise it—slowly at first, then faster, and then they made it soar.

In 1973, a two-month output cutback panicked the market and drove prices far up. But volatility was permitted only up, not down. When panic abated, OPEC nations worked a ratchet: they put a firm floor under the new high price level by raising their excise taxes. In 1974, demand fell, and installed capacity was in excess, yet they kept raising excise taxes and prices in concert.

The Cartel Unbound

In a little over a decade, OPEC became a cartel unbound. In an industrial society, no group of private companies would dare to raise the price of a commodity like oil fourteen-fold. It would injure those much more numerous and powerful than themselves. Every country would have its own methods of bringing them to heel.

But there is no such conflict in a small, undeveloped country. More oil revenues are good for everyone; the burden is all on foreigners. Moreover, OPEC members were not private companies; they were sovereigns, accountable to no outside power and free to take all the market allowed.

How much that "all" amounted to was for them to find out. Freedom to act was freedom to make mistakes. Like soldiers peering into the fog of war, sellers do their best with what information they have. Even before a situation has worked itself out, it is already changing.

Even harder than deciding what is best for the group is sharing the gains and the burdens of restricting output, or deciding who holds back how much. It is a zero-sum game. Every solution invites cheating, because every member would profit by producing more at the expense of others. Individual self-interest must constantly be repressed.

Economic analysis can show the limits within which the group worked and some principles of burden sharing, but within those limits, many things were possible, and some of them actually happened.

Supplanting the Oil Companies

By 1970, the oil companies were in the saddle only as well-paid jockeys; the oil exporter nations owned the horses and collected the winnings. They could have continued to set excise taxes as a price floor and left the companies in place to invest and produce efficiently and to compete on the narrow margins left to them. But the OPEC nations were prisoners of current opinion and of their own past, resentful at having been exploited and despised as poor and backward. To expel the companies was a moment of bliss. But they soon began to pay for it, and the bill keeps growing.

The vertically integrated companies had competed mostly in selling refined products in many local markets. Sellers at each point of sale were few. The value of a purchase (mostly gasoline at the pump) was very small in relation to buyers' incomes, so they had no incentives to hunt for bargains. Hence competition was sluggish. Reactions to price changes were limited and slow and provided little incentive for sellers to discount prices. Moreover, because of joint ventures, each major company knew the others' investment and sales plans for years ahead. Actual collusion had ceased long before 1970, but the long price retreat had been masterfully slow.

Seizure of the companies' producing assets "upstream" divorced production from the refining and marketing "downstream". It created worldwide open markets in crude oil and products. OPEC members now sold to refiners, which had thin profit margins and were strongly affected by even small price differences. Hence, futures, swaps, and other stabilizers developed. A buyer looked for better offers all over, like Argos in the Greek myth, who had a thousand eyes. A small discount could attract many buyers and more volume; being too slow was costly.

Most companies had produced in various countries. They could tolerate quality and location differentials on crude oils that were out of sync with product prices. Benefits in some places offset penalties in others. There was a brake on overlifting more desirable crudes in some countries and underlifting the less profitable in others: companies wished to keep their valuable concessions in all countries. But when the OPEC nations became sellers, price differentials by quality and location of crude oil became their problem. Differentials were a constant irritant. When genuine, they could divert sales. They were also a convenient screen for discounts.

In the new order, the cartel members tried one expedient after another to restrain output and divide the market. Differentials were often discussed but never settled. Agreements unraveled and were replaced by others: informal deals over 1975–1977, a secret agreement in 1978, a proposed agreement in 1979, a gentlemen's agreement in 1980, the loose and tight quotas of the 1980s, the nonquota quotas of 1990–1991, and the renewed quotas of 1992 and after.

Before 1986, it was OPEC's painful task to divide up a shrinking sales total. Between 1986 and 1993, exports soared from 12.5 million barrels daily (mbd) to 21.5 mbd. Yet repeated increases in demand were almost as hard to handle as decreases. Any change meant a fresh contention over sharing the gain or the loss.

Under competition, market sharing is automatic. Each operator produces all it can, up to the point where the cost of additional output would exceed the price. But a cartel exists for only one reason: to keep the price above that marginal cost. Thereby the cartel shuts off the automatic market sharing mechanism. A joint decision of the members must replace it. But market sharing is a zero-sum game: any member's gain is another member's loss. With much or little excess capacity, there was no respite from meetings and quota deals. One industry journal wrote, "OPEC's perennial dogfights over quotas and prices ... have reached dramatic peaks every two years. Reminded of the squabbling that preceded price collapses in 1986 and 1988 and helped to spark the Gulf war of 1990, traders are nervously mulling the possibility of another major price plunge." [PIW 12-14-92:6] When collaboration breaks down, the trade nervously waits for it to be patched up again.

It was expected that all would be well some day, when excess capacity would be worked off, but it never can be. For bargaining power, a member needs to make credible threats, which require some excess capacity. If a small, local excess leads to a small, local discount, it may soon become a discount everywhere. Those with substantial excess, like Saudi Arabia, demand that others share the burden by cutting output. When the others cannot or will not, the Saudis and others must sometimes prove they are not bluffing. In 1990, with no agreement, the threat of Iraqi force was used to limit output, with some unintended results.

Error and Retreat in the 1980s

OPEC's long-run aim was to raise the price to the cost of synthetic liquid hydrocarbons from coal, oil shale, or tar sands. This is a monopoly goal. It can be reached only when oil-on-oil competition is suppressed. But the necessary condition was not sufficient. Parity with synthetics was far out of reach. The true monopoly ceiling was and is much lower, because consumers reacted more strongly to price increases than had been expected.

Their reaction was slow in coming, as usual, and price controls in consuming countries made it even slower. The lagging demand response misled and entrapped OPEC. Yet the trap was avoidable. The cartel members could have discovered their mistake by slow, cautious price increases, testing the waters each time. A special committee of oil ministers advised this course, but the nations did not follow their own good doctrine in 1978–1979. Their scheduled increase, already too large, was forgotten when the Iranian revolution made feasible an even bigger one.

I have already suggested why: their time horizons were shorter than those of private owners. They were intent on quick gains and heedless of the chance of long-run loss. It was not inevitable, but neither was it strange: they raised the price too much too soon. They were punished for their mistake when revenues fell sharply.

In price fixing, as in singing and mountain climbing, it is easier to go up than to come down. The OPEC nations managed a descent, with a nasty fall on the way. They did not lose their power, they regained their wits, and by the end of 1986 they set a price nearer the monopoly optimum.

They learned their limits and their mission: to trade off some of their market share in return for a higher price and higher revenues. Today they hope for a higher market share, to allow another such trade, some time in the future.

Consumers' Quest for Influence

So-called strategic trade theory is old hat in oil. Consuming countries have invoked economies of scale and learning curves to promote domestic energy sources. They "pick winners" who turn out to be losers and keep subsidizing them to maintain jobs and cover up mistakes. It is an incentive to support higher oil prices.

The consuming countries have also tried to gain the goodwill of producer nations, to induce them to produce "enough for our needs." Oil was fancied to be in tight supply, with permanent excess demand. Therefore producers could allocate oil by grace and favor. They could use a selective embargo to starve countries they disapproved of.

This was wrong in theory, already disproved by experience, yet believed with terrified intensity. In 1973–1974, a U.S. secretary of state bounded all over the Middle East for months, and some of his colleagues hinted darkly at using force to end an "embargo" that in fact never existed. The quest for goodwill and influence with the Middle East-producing nations, including Iraq, was voiced early in 1990 by the U.S. government. An illusory "special relationship" of the United States with Saudi Arabia has been cherished for many years.

In fact, the Saudis have acted as what they are: the leading firm in the world oil cartel. To defend their market share against the nibbling of their fellows, they must sometimes make good a threat to maintain output. The price weakens, until they join to restore it. In 1973, they took the lead in cutting output, which raised the price from $2 to $7. In 1974, they took the lead in raising it to $11. In January and April 1979, their output cuts set off the second price explosion, whose peak price they set at $34 in October 1981. Later they nearly ruined themselves in cutting back output to maintain the price, until they gave up in 1985. They encouraged Iraq in 1990 to enforce lower output, until the enforcer turned robber. They continue to lead in restricting output to maintain prices, while the United States wags its tongue. Jawboning stirs the air, not the listener.

The Saudis and their fellows knew that whatever the price of oil, the United States would protect them from invasion and aggression. They rightly gave it nothing for what they would have anyway. American intervention and "groveling," as an official correctly described it, simply reassured the cartel.

Governments: The Barrier to Expansion

Socialism died in the 1980s as a fighting faith and is scorned as a bad idea whose day has gone. Yet most of the world's oil is still produced by flabby national dinosaurs, OPEC and non-OPEC. The problem is much older. Ibn Khaldun stated it 600 years ago: "Commercial activity by a ruler is harmful to his subjects and ruinous to the tax revenue." [Ibn Khaldun 1377 (1967), p. 232]

Even state corporations that are not corrupt and wasteful cannot follow a rational investment plan. The money they handle is not their own. Their revenues are bespoke for "national needs"—that is, special interests.

OPEC governments invest a tiny proportion of their revenues in oil production. From 1976 to 1987, Middle East and African members spent less than 2 percent. They could raise all the funds needed for oil production investment by cuts in spending on subsidies, consumption, and weapons, yet they find it hard to make those cuts.

The OPEC nations have recognized for years that they would gain by the reentry of foreign companies, with the funds and expertise to do some of the production. The Ayatollah Khomeini felt peace with Iraq was worse than drinking poison, yet he swallowed peace. But the OPEC nations cannot—so far—bring themselves to let private companies assume the high risks and possibly earn high rewards. Those who move first will gain most.

Outside OPEC, the current "low" price is well above the long-run cost of creating reserves in most areas. Moreover, this cost has probably diminished in the past ten years.

More than ever before, the main barriers to discovery and development are state oil companies and high regressive production taxes. Because small fields are overtaxed, large and small fields alike go undiscovered and undeveloped. The former Soviet republics have perhaps the greatest potential and are the greatest underachievers.

But in the non-OPEC areas, state companies are being privatized, let loose to cultivate what provides a return on investment and to weed out what does not. Production taxes are being slowly redesigned to skim rents and increase production and revenues. The faster their progress is, the greater is their output and the greater is the strain on the cartel.

2 Economics of Oil Supply

"Exhaustible Resources": The Nonproblem

Mineral scarcity appears self-evident; there is only so much of the resource. Every unit used today means one less for the future. As the stock shrinks, its value rises. The basic theme has no end of fascinating variations.

The renowned physicist Max Planck once described "phantom problems." One of them "used to keep many a great physicist busy for many years: the study of the mechanical properties of the luminiferous ether." [Planck 1949, p. 56] In time, physicists decided they could not find the luminiferous ether; they also decided they did not need it and had best forget it. (As chemists had forgotten phlogiston.)

Particularly after 1973, the theory and application of "an exhaustible natural resource ... a fixed stock of oil to divide between two or more periods" [Stiglitz 1976, p. 655]; and references at [Adelman 1993, 219] and the resulting "basic upward tilt" to the price of oil kept some fine economists "busy for many years." But the fixed stock, like the luminiferous ether, does not exist. Its optimal allocation over time between us and our posterity is a phantom problem.

No mineral, including oil, will ever be exhausted. If and when the cost of finding and extraction goes above the price consumers are willing to pay, the industry will begin to disappear. How much oil is still in the ground when extraction stops, and how much was there before extraction began, are unknown and unknowable. The amount extracted from first to last depends on cost and price.

Curves like those in figure 2.1 please the eye and sum up the history of many industries. The horizontal axis represents time, the vertical axis production. The lower curve shows production for each period, and the

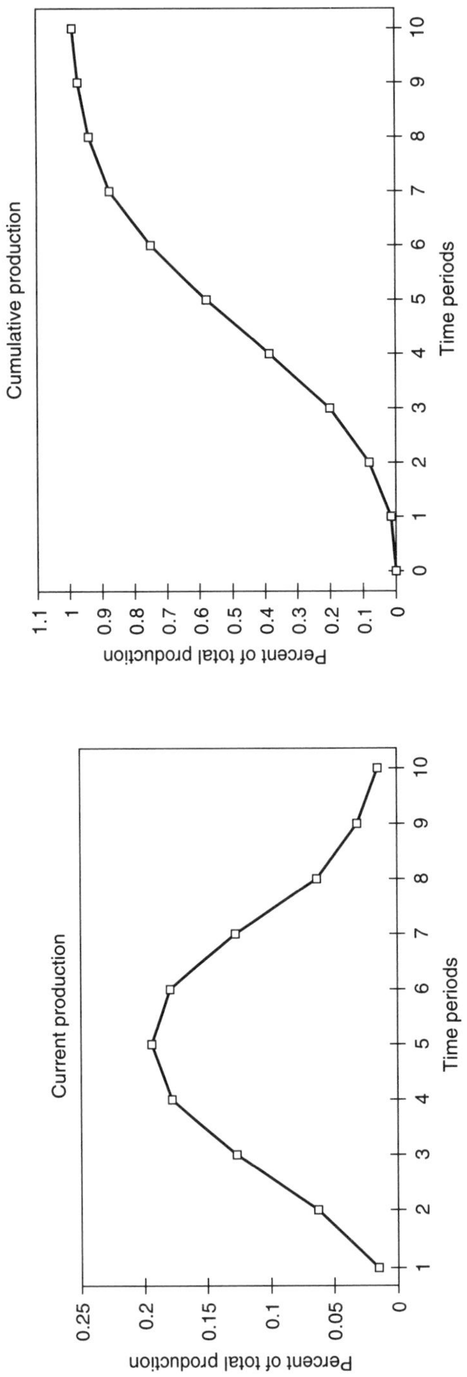

Figure 2.1
Logistical production curves: annual and cumulative.

upper cumulates the total up to the end of the period. Notice that the cumulative curve first grows at an increasing rate, but then it flattens.

For the period 1950–1990, the graph is a fairly good picture of $33\frac{1}{3}$-rpm phonograph record production and for 1950–2000, of mainframe computers, which IBM will cease to produce in 2000 A.D. [WSJ 9-12-94:134]. Many other manufacturing industries have followed similar curves. Nobody suggests that the upper limit is somehow fixed in advance—that the production of records or mainframes (or vacuum tubes, or typewriters, or buggy whips) had to stop because there was "nothing left to produce." It was lack of demand that limited these industries. [Burns 1934] Mineral production in any given place must fear the same fate.

Cost of Inventory Renewal: The Real Problem

Mineral production is a flow from an unknown physical resource, first through exploration from "basins" to "plays," then into identified "fields" and "reservoirs," then through development into current inventories or "proved reserves," to be extracted and sold. Reserves are renewable and constantly renewed, if—and only if—there is enough inducement to invest in creating them. The inducement depends on price and cost. The illusion of a fixed resource, forever running down, hides the real problem: All else being equal, the larger, more accessible deposits are found first, even by chance. Once found, the better deposits (those of lower cost or higher quality) are developed first. As humanity went forever from good ore to bad and from bad to worse, the costs of renewing mineral reserves should keep rising, and prices with them.[1]

In fact, over the long term, more minerals' prices have fallen in real terms than have risen. [Adelman 1993, pp. 220–221] Humanity has wielded increasing knowledge against diminishing returns and has won big so far. This uncertain tug of war for any mineral in any given place at any time is registered in its *cost at the margin*, that is, where it equals the price. The price and the marginal cost of renewing inventory constantly gravitate toward each other in many places; they are all part of a market network. Figure 2.2 shows the prices of six metals: three show statistically significant decreases (aluminum, lead, iron ore), one shows a significant increase (tin), and two are borderline cases (copper, zinc). It does not mean "three downs, one up, two undecided, the downs have it." The figure shows the uncertainty of the struggle at every moment. The price fluctuations make it risky to hold inventory, or to expand it by investment. But

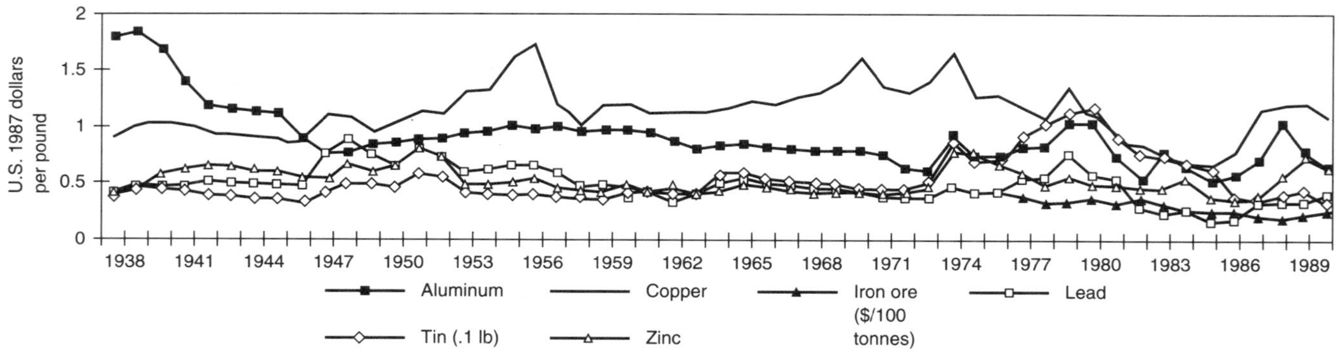

Figure 2.2
Prices of six metals, 1938–1990. Source: U.S. Department of the Interior, Bureau of Mines. Metals prices in the United States through 1991 (Washington, 1993).

the reward of risk is profit, which drives investment. The investment process, and nothing else, explains mineral cost.

The Inventory of Oil Reserves

Petroleum engineers estimate the cost of drilling and connecting new wells into a new or existing reservoir. A well can produce an initial daily amount, which will decline over time because of pressure loss, water encroachment, and other factors, although additional investment in "enhanced recovery" may bounce the output back up. Because operating expenses per well are fairly constant, the cost per barrel must rise as output declines. When cost just equals the market value of the output, production stops at this "economic limit." The estimated aggregate output of the new wells over time is known as the "proved reserves added" or "reserves booked."[2]

In the United States, annual reserve estimates are accurate enough to permit estimates of the annual net and gross additions to reserves. There are also reliable data on annual investment expenditures. They do not always closely match the reserve data, but one can estimate cost per additional barrel added, year by year, over a long period, within tolerable error limits.

But except for the United States and a very few other countries, published reserves are not well defined, and estimation methods are not revealed. Year-to-year changes usually do not mean much of anything. Investment data are even scarcer than reserve data, and dwindling.

Reserve Growth: Some Examples

The reserve in a field grows with time, and sometimes dramatically. In California, the Kern River field was discovered in 1899. In 1942, after forty-three years of depletion, "remaining reserves" were 54 million barrels. But in the next forty-four years, it produced not 54 but 736 million barrels, and it had another 970 million barrels "remaining" in 1986. The field had not changed, but knowledge had: science, technology, and, not least, the detailed local geology learned by development.

England

As the onshore Wytch field was developed in England, it was perceived to extend under the sea. A 1991 development plan for drilling the undersea section from an artificial island was rejected because it was in a scenic area. Two years later, the undersea reservoir was reached by drilling horizontally from the onshore, to a record length. The investment was actually 56 percent less than drilling from the island had been. [OGJ 1-3-94:2] Wytch reserves will be increased accordingly.

These two examples are unusual, yet they help us understand how most reserve growth is in old fields. Reserves eventually booked are many times the initial estimate.

Persian Gulf

A special expert mission estimated Persian Gulf reserves in 1944 at 16 billion barrels proved and 5 billion probable. By 1975, those same fields, excluding later discoveries, had already produced 42 billion barrels and had 74 billion "remaining." Both numbers are much larger today, though not published. Indeed, since 1981, we have had no data from the gulf region, not even production by fields. We cannot tell when fields grow together into one, as in 1948–1957 several grew into the Saudi giant Ghawar. [OGJ 12-30-63:109] Nevertheless, the discovery effort in gulf states has been small and probably most output still comes from those pre-1944 fields, as it did in 1975. Cumulative gulf production between 1945 and 1993 was 188 billion barrels, nine times the 1944 estimate. At the end of 1993, gulf "remaining reserves" were 663 billion, an additional thirty-two times as much.

During this time, predictions of "undiscovered" or "ultimate" reserves have repeatedly been made and then surpassed, sometimes with embarrassing speed. At the end of 1984, it was estimated that there was a 5 percent probability of another 199 billion barrels remaining to be added at the gulf, ever. Within five years, it had already happened.

These estimates are not of the in-ground stock; rather, they are economic forecasts: how much it will be profitable to produce given current costs and current knowledge. The estimator of "ultimate reserves" is doing economics without knowing it. An inspired guess may be right,[3] but there is no way to tell. As knowledge grows, so do the "ultimates."

United States

In the United States, crude oil discovery peaked in 1930, when proved reserves were 13 billion barrels. In the next sixty years, the United States, without Alaska, produced 130 billion barrels. The inventory turned over ten times and is today about 17 billion barrels (with another 6 billion in Alaska). With much effort, many small fields were discovered; more important was the continuing expansion of old fields. Over 1966 through

1977, the only years for which comparison is possible, 19 billion barrels were added to reserves; 17 billion of them were in fields discovered before 1966.

These huge new reserves in old fields were no gift of nature. They were a growth of knowledge, paid for by heavy investment. This history explains why, in 1994, in various parts of the world, there is interest in letting foreign companies develop so-called marginal fields. Much oil can be added in these fields, an additional return on the knowledge gained by operators elsewhere, especially in the United States.

The Sensing-Selection Instrument

At any given moment, reserves are being added everywhere. The industry is a great sensing-selection instrument, scanning all deposits, old and new, to develop the cheapest increment or tranche into a reserve. The reserve increments of any given period are overwhelmingly in existing fields. Nobody "finds" a reserve, just as nobody finds a factory. Oilmen find fields, out of which they may develop reserves over a long time. They find new reservoirs in old fields, and new strata or pools in old reservoirs.

The constant search for least-cost prospects takes the industry to the fringes of known reservoirs and beyond. The search process is driven by cost comparison. We therefore look at the structure of petroleum costs: operating, development, and beyond.

Petroleum Operating and Development Costs

Current operating costs reflect development costs; the conventional allowance is 5 percent of development investment per well per year. This constant annual amount per well becomes a rising cost per unit as output declines.

Development investment expenditures are made to drill and complete wells, install equipment, and connect to a pipeline or tanker terminal. We deal with marginal development investment—per barrel newly booked into reserve inventory or per barrel of newly installed capacity. (Note 2 shows the conversion between reserve additions and capacity additions.)

The harder we squeeze a sponge, the less the additional liquid from squeezing still harder. Similarly, the more intensive the development of a reservoir is, measured by the ratio of production to reserves, the higher is the marginal cost per unit. Development expands reserves and capacity so long as the cost, which is mostly the return on marginal investment in the pool, is below the market price.

But the price, allowing for location and quality, is the same for all pools. Therefore, over any area where capital can flow freely, marginal cost in every project is in competition with marginal cost in every other project. Operators keep expanding the better projects most, driving marginal costs up toward equality everywhere.[4] If the process continued indefinitely, the lower cost would expand most; the marginal cost would rise and would become equal everywhere. This result is forever put off as new choices appear.

But the average cost—the total of all expenditures made from the start, divided by the total of all reserve barrels added from the start—varies enormously among pools. The rent per barrel produced, which is the difference between marginal and average cost, will vary even more, and there is no reason ever to expect equality.

The rent goes partly to the operator, partly to the subsurface landlord, usually a government. Before exploration and production, rent can only be estimated with wide margins for error even in proved areas, and can only be guessed at in new areas. Hence the wide acceptance of a royalty, or percentage of output, is a concession to ignorance. Bargaining over rent, and the difficulty of measuring it and defining it in a law or contract to accommodate the unexpected, will last as long as the industry.

A rich discovery means a large rent and a dissatisfied landlord who wants more of it. He knows that the operator can afford to pay more than promised, perhaps many times more.

A landlord subject to law can only grumble and observe the contract. A sovereign owner can revise the contract, or tear it up. The action will not improve the owner's credit rating, but that may not matter right away.

The Discount Rate (Return on Investment)

Return on investment drives the process of reserve addition. The operator's discount rate or the cost of capital is the minimum acceptable rate on new investment. It equals what could be obtained by other investments with the same degree of risk. For a decision, one must calculate the discount rate that would make the present value of project revenues just barely equal to the up-front investment. If that calculated discount rate equals or exceeds the operator's own discount rate or cost of capital, the project is viable.

Discounting and Development Intensity

More intensive development means a higher ratio *Q/R*, production to reserves, and it raises the required investment per barrel. As well, it speeds the inflow of revenues and raises present value. In this trade-off, a higher discount rate penalizes slower depletion, but it also raises the operator's cost of investing more to deplete faster. Thus, it makes quicker depletion more desirable but less accessible.

Macbeth's porter said of strong drink: "Lechery, sir, it provokes and it unprovokes. It provokes the desire, but it takes away the performance." So too a change in the interest rate affects development both ways, speeding it up *and* slowing it down, though the net effect is probably small. (We will see later that this change may be important to a government.)

Unit Cost and Rate of Return

To calculate unit costs, we assume the present value of the revenues is just equal to the up-front capital investment. Using the rate of return that is barely acceptable to the operator, we can calculate the expected revenues and the break-even price. This is the unit development cost. Or we can take the market price as given, and find the break-even discount rate. These calculations are set out in appendix 2.1. (This is a general method and will be applied later to high-cost fields.)

For the Persian Gulf, a price of $1.21, the 1970 all-time low, development yielded a return of over 430 percent per year. The Arabian American Oil Company (Aramco) made the investment, paid 88 cents excise tax, and kept 33 cents. Its return was over 100 percent per year. No wonder that Aramco in 1970 was planning to expand output from the current 7 mbd to over 20 billion by the end of the 1970s.

Development Cost, In-Ground Value, Finding Cost, and Scarcity Rent

Substitution among Development, Purchase, and Discovery

Operators invest in a wide gamut of projects: improved recovery, more wells into the same pool, wells into adjacent strata or adjacent pools, prospects that are completely known or less completely known, and so on, to the deliberate search for new reservoirs and even new fields or *plays*, which are new areas expected to contain an array of fields. Development thus shades into exploration or in French *recherche* (research).

All of these methods of reserve addition are partial substitutes for each other, and all are in competition for investment funds. If development is becoming more expensive, it pays to look harder for new pools and fields to freshen the mix and moderate the increase in development cost. Conversely, if the newer fields are getting smaller, deeper, or more heterogeneous and faulted, then development cost per unit of reserves booked into those new fields will be higher. This situation pushes operators into drilling more wells into and around the older pools and draining the older pools faster. Thus higher finding cost is registered in higher development cost.

But we cannot calculate annual finding cost per unit as we do development cost. Annual exploration expenditures have long been tabulated in the United States, but we do not know how much is discovered in a given year. Operators calculate the odds on finding a new pool of a given size and development cost in a given place. There is no known way of aggregating those estimates, even if we knew them. (See appendix 2.2 on why one should avoid so-called finding cost per barrel of oil equivalent.)

But we can often calculate a proxy for exploration investment per unit found. An alternative to adding reserves by any combination of developing and finding is to buy them. Reserves of oil and gas are frequently bought and sold, as are the companies that own them. Hence the market value of developed reserves is comparable to all other methods of reserve addition. Because all are substitutes, changes in the cost of any are an indicator of changes in the cost of all the others. Increasing oil scarcity means increasing values and costs across the board.

A higher cost of finding and developing raises the value of a barrel already developed. Conversely, a higher value of a barrel in the ground is a greater incentive to invest more to create more. This drives up the cost. Thus in-ground value and finding plus developing cost always gravitate toward each other.[5]

Structure of Prices, Costs, and Values

Table 2.1 shows the layers of price, cost, and value in the United States in two recent years. Although some of these statistics are subject to wide error and comparison of any two years is chancy, by looking at actual numbers, we can put some flesh on the bones of economic theory. Then we can look at long-term changes.

Table 2.1
Price, cost, and in-ground value in the United States
(dollars per barrel)

	1984	1992
1. Gross wellhead price	25.88	15.99
2. Net price (excluding operating costs, royalties, taxes)	16.67	10.68
3. Reserve in ground, market value	6.94	4.71
4. Development cost	3.84	2.87
5. Discovery value, or "user cost" (line 3 less line 4)	3.10	1.84

Sources: MER; Adelman 1993, pp. 248–250; Scotia Systems Ltd., "pure oil" market transactions, with no gas reserves. (From a research project with G. C. Watkins)
Note: The operating margin includes 15 percent of the price as royalty. This is not a cost but rather a share of the profit. Another 5 percent corresponds to excise taxes, which are in part a charge for services (police and fire protection, etc.) and in part a taking of profit. Therefore the true social current cost is not a third but less than 20 percent of the price. However, the in-ground value of the reserve depends on the net to the owner, not the net to society. The development cost is reduced by 11 percent to allow for the tax allowance. Thus the entries in lines 2 and 4 are private values, comparable with line 3, and permit the subtraction of line 4 from line 3 to arrive at line 5.

Factors affecting the cost of holding the asset oil-in-the-ground, to get from line 3 to line 2:

	1984	1992
Production/reserves	0.108	0.101
Decline rate	0.096	0.091
Holding time (half-life) of asset, in years	6.131	6.574
Annual appreciation in value	0.154	0.133
Riskless rate	0.122	0.062

Production and reserves data from DOE/EIA, decline rate computed by formula in appendix 2.1. Holding time computed from $T' = ln\ (1 - (.5Ra/Q))/-a$. Riskless rate, from Economic Report of the President, interpolating between three- and ten-year Treasury notes.

The net price has been stable for many years around two-thirds of the gross. A traditional industry rule of thumb is that the market value of the in-ground reserve fluctuates around one-third of the gross price, or one-half the net. This has held fairly well on the whole. In the United States, a reserve barrel is held in the ground for production on average in six to seven years. The increase in value from line 3 to line 2 in table 2.1 is compensation for the investment in holding the barrel.

The difference between in-ground value of a developed barrel and its development cost is the value of an undeveloped barrel, called *user cost*. It is sacrificed, over and above development cost, by the decision to develop. When this discovery value equals or exceeds expected finding cost, it is the signal for an investment inflow into exploration. If cost exceeds value, there will be no investment.[6] Thus, user cost is a proxy for finding cost.

Putting aside possible errors in the data of table 2.1, there is a sober comparison for the American oil industry. Development cost per unit of added reserve declined from 1984 to 1992. But reserves added in 1984 were 3.8 billion barrels and in 1992 only 1.5 billion. The lower marginal costs resulted from discarding the poorer prospects.[7] It would appear that the whole supply curve had swung to the left, while the industry has moved down the curve. Discovery value had dropped more than development cost and was probably below finding cost.

Price, Cost, and Reserve Values in the United States: A Test of Depletion Theory

Table 2.1 presented four measures of oil scarcity, in the short and long run, in a large oil-producing area. Figure 2.3 shows them over a long period. The year 1948 marked the end of the repressed wartime inflation and industry distortion. A long-run increase in real prices and reserve values because of the fixed stock of "nonrenewable resources" would cumulate over twenty-four years, even at 2 percent per year, to a 61 percent rise. (Since the price level doubled from 1948 to 1972, the nominal increase should have been by a factor of 3.2.) Instead real prices and values declined.

Additions to reserves were fairly stable before 1972—between 2.5 and 4 billion barrels per year. Incremental development cost declined after 1960, but this was a one-time gain from a gradual easing of wasteful regulation. The stable price, over and above the remaining regulatory waste, was enough to induce an inflow of reserves that was slightly greater than the current outflow. United States oil reserves were in a steady state; production even grew slowly.

Conclusion on Oil Scarcity and Scarcity Rents

To explain the price of oil, we must discard all assumptions of a fixed stock and an inevitable long-run price rise and rule out nothing a priori. Whether scarcity has been or is increasing is a question of fact. Development cost and reserve values are both measures of long-run scarcity. So is reserve value, which is driven by future revenues. They move in the same direction.

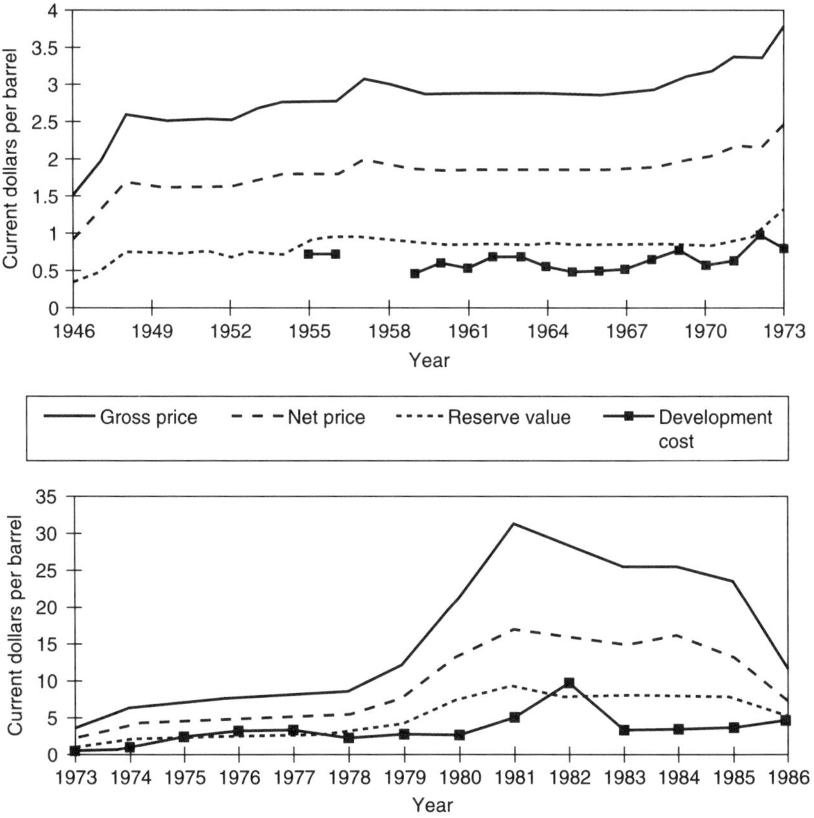

Figure 2.3
Crude oil: Price, in-ground value, and development cost in the United States (*a*) 1946–1973, (*b*) 1973–1986. Source: Adelman [1993, chap. 10].

Because expectations are uncertain, reserves are risky assets. Reserve values (as in table 2.1) are forecasts made by qualified observers with a strong interest in guessing right. The sale of a producing lease or a security concentrates the minds of scientists, engineers, bankers, and oilmen. At any time, they may all be wrong. But the only basis for disregarding them is to assume that private markets cannot value mineral assets properly.

If one assumes a fixed mineral stock, the value of in-ground deposits cannot be determined by any market discount rates, which all relate to the supply and demand for investable funds. But a nonrenewable mineral resource precedes investment. Its value is born not made. (A high discount rate means too-rapid depletion, threatening disaster.) But once we stop assuming a nonrenewable mineral stock, and admit that reserves are a flow created entirely by investment, the usual discounting rules apply. We no longer need to decide "how much of the remaining stock of non-renewable resources to use up." [Solow 1992, p. 9]

Applying the Theory outside the United States

For the rest of the world, we have no such array of data. But having tested the concepts, we can apply them to what we do have: behavior and numbers.

Suppose that reserve owners/producers expect a higher price in the future than they previously expected. Then each would produce less oil and hold back more while not urging others to hold back. On the contrary: the more others produce, the less they will have left, the higher will future prices be, and the greater the gain to those who hold back their reserves for future depletion. Therefore, when reserve owners urge production restraint upon each other, that is evidence that they are trying to talk up the current price, not that they expect higher future prices.

Another test is that the lower the owner's development cost is, the less is the incentive for him to postpone development/production and hold oil for higher prices. For a producer whose cost is near zero, postponement is a dead-weight loss, unless the price is expected to rise at least as much as the discount rate. (If the cost of capital is 10 percent and the price is expected to rise 5 percent per year, the producer loses 5 percent by waiting one year.) At the other extreme, suppose that the price just covers cost. Then even a small price increase is a large increase in the project value. A delay has a big payoff.

Thus if higher prices are expected, production in the higher-cost reserves will be cut back the most. This is only a special case of the general rule of rational conduct: produce the lower-cost mineral first. That was the worldwide pattern for many years, as the growth of the low-cost areas far outstripped the others.

Therefore, if production from low-cost reserves is cut back and production from high-cost reserves is expanded, this change cannot be explained by user cost or price expectations. Sellers' appearing to act against their economic interests is odd or suspect behavior, needing a closer look.

Scarcity in the World Market, 1955–1985

World reserves (table 2.2), in 1944, were 51 billion barrels. By the end of 1993, the world had produced and consumed 690 billion barrels and had

Table 2.2
World production and reserve additions, 1960–1990
(Billions of barrels)

	1944	1945–1960	1961–1970	1971–1980	1981–1993	1944–1993
OPEC						
Cumulative production	—	26	55	103	100	284
Gross reserve additions	—	219	251	128	434	1,032
Reserves at end	22	215	412	436	770	770
Non-OPEC						
Cumulative production	—	51	64	102	190	407
Gross reserve additions	—	98	187	114	207	607
Reserves at end	29	76	200	212	229	229
Total world						
Cumulative production	—	77	119	205	289	690
Gross reserve additions	—	318	439	242	640	1,639
Reserves at end	51	291	611	648	999	999

Sources: *History of the Petroleum Administration for War* (Washington, 1947), appendix 12, table 1; OGJ:WWO; DeGolyer and MacNaughton 1993.

999 billion barrels left. Thus producers have been "emptying out their reserves." This need not continue forever, but cost trends show it is a good bet to continue for years. The production/reserves ratio, imprecise though it is, is half of what it was in 1944.

The principle is that changes in development cost are an indication of change in value, finding cost, and oil scarcity in general. Figure 2.4 is an array, from lowest to highest, of development (and much exploration) investment per unit of new capacity, for each country outside North America and Western Europe in 1955, 1965, 1975, and 1985. We cannot use investment per newly added reserve barrel because the estimates of year-to-year reserve changes are too unreliable.

The estimated production investment is based on 1985 factors in the United States. This controls for inflation but also sets technical progress to zero; hence, it understates the shift to the right and the increased supply. There is also an offsetting down bias: no allowance for higher costs outside the United States, which has a dense infrastructure of supply and service industries. (This is estimated in chapter 7.) Another source of error is that estimated new capacity is based on the assumption that average output per new well is the same as average output per existing well. This probably understates the capacity addition. The error varies by years and

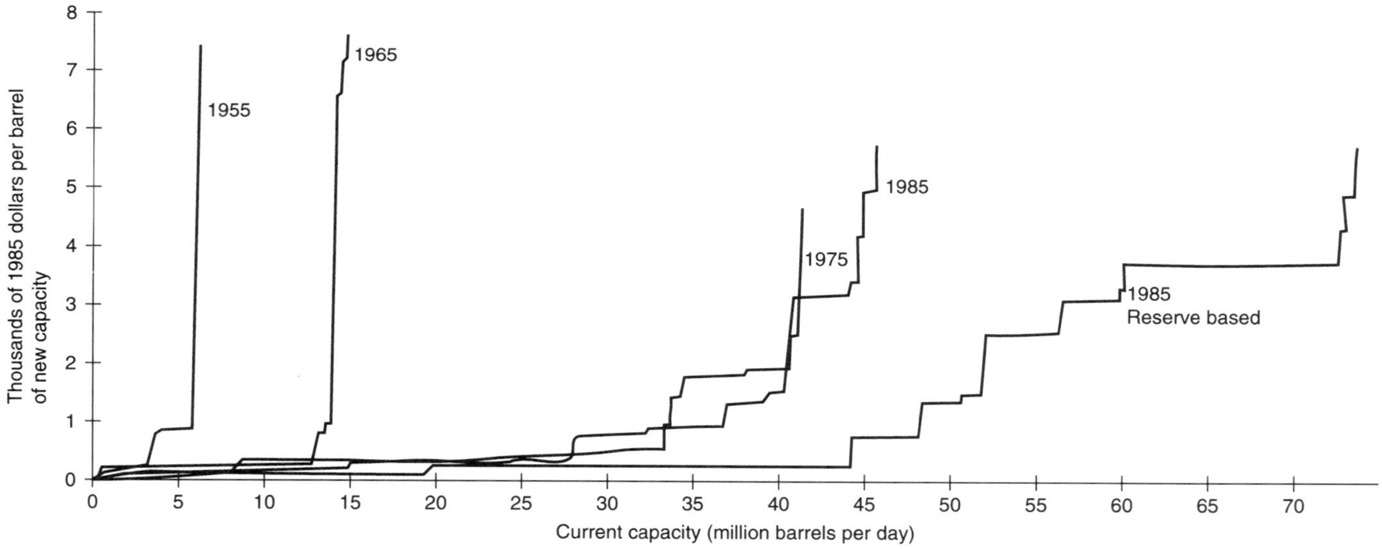

Figure 2.4
Noncommunist world "supply curves," 1955–1985, excluding North America and Western Europe. Source: Adelman and Shahi 1989.

among countries.[8] All in all, the absolute numbers should not be used without making the adjustments applied in later chapters.

Nevertheless, the direction of change is clear for 1955–1985. The expansion of output put no strain on resources; indeed, quite the contrary occurred. The incremental cost of adding capacity did not rise but declined. The values of oil in the ground, if we had such numbers, would also have declined, as they did in the United States, for the alternative to buying a reserve is to create one. From 1975 to 1985, there was little change; we see in later chapters why not.

The line at the far right of figure 2.3 adapts the 1985 line by assuming that all countries depleting at less than 5 percent of reserves rise to that level. This increases their investment per unit, perhaps by several times.[9] But it also increases the output. Taking both effects together, the potential line is far to the right of the actual one and gives us a hint of what unconstrained output would be. For each year before 1985, a similar calculation would show the same picture, because the percentage of reserves depleted each year was extremely low in the Persian Gulf areas. Since 1993 reserves are much greater than in 1985, a more up-to-date estimate would ride far off the page.

Conclusion on Scarcity

Increasing global oil scarcity is an illusion based on the unexamined assumption of a fixed stock and of some value unrelated to investment.

Before 1970 there was at worst stability in the United States and growing plenty in the world outside, where the price was far above any competitive level. Worldwide oil production was over 5 percent of reserves in 1944, 2.3 percent in 1950, 2.7 percent in 1970, and 2.2 percent at the end of 1993. Although strong conclusions should not be drawn from weak numbers, they do not suggest even gradually increased scarcity or shortage.

Investment Cycles and Excess Capacity

One paradigm is that a too-low price leads to underinvestment, hence insufficient capacity. It is assumed that production falls short of demand, raising the price and inducing new investment. But it takes so long to install the new capacity that production lags, raising the price still higher. As the new capacity comes onstream, it forces the price down. But the sorcerer's apprentice cannot turn off the flood; new capacity, initiated years before, keeps arriving.

The price becomes unsustainably low and will not provide an acceptable return on new capacity. Little or none is built. In time, growing demand exceeds capacity. The price rises and there is new building, but the long lead times preclude new production until the price has soared to unsustainably high levels. And the cycle continues.

Supposing the theory to be correct, none of it fits the oil industry. The "long lead times" are only for exploration in new areas, much like research and development in manufacturing. There were no wild upward investment swings before or after World War II. New capacity was added continuously.

In the world market for twenty-five years, the price fell, and capacity expanded sevenfold. In the United States, we saw, the price mildly declined, and reserves were added steadily. Only regulation made for surplus capacity. Many useless wells were drilled because the more wells an operator drilled, the more production he was allowed. After about 1960, better regulation and a price that could not be raised (for threat of imports) shrank the burden.

A variant of the cycle theory holds that since current operating cost is only a small part of total cost (which table 2.1 supports), the price can fall to ruinously low levels under competition. But this confuses operating cost with marginal cost, which rises steeply at high levels of capacity utilization. The theory assumes excess capacity, hence cannot explain it.

Oil even has a built-in stabilizer. The natural decline rate must be offset by fresh investment, or production falls. The rule of thumb of reserves as fifteen times production implies that in four years output declines by nearly one-fourth. (If $Q/R = 1/15$, $a = .067 - .067^2 = .0622$, and $(1 - a)^4 = .77$.) This helps cure any excess supply. Knowledge of the cure speeds it up. Production is held back today for the higher future price, which raises the price today.

Thus, oil is basically a stable industry, and the expansion with declining price between 1945 and 1970 was not an interlude but the normal state of affairs.

Monopoly and Competition

Competition and Its Absence

The competitive process in oil production is a constant movement by each operator toward equating marginal cost to price, which maximizes his profit. Producers can only adapt; those who adapt by innovating may make high profits. The system is kept in motion by the constant intrusion of new pools and new strata in old and new fields and an equally or more important set of shocks from new technology.

If sellers act in concert, they can offer less, charge more, and thereby increase both their profits and the value of their future revenues. Whether sellers are "few enough" to act in concert is an old problem. Fewness is usually measured by either the concentration ratio (the percentage of total sales made by the largest four sellers) or more precisely by the Herfindahl index.[10]

If sellers are few, concerted action is easier, and hence plausible. But this impression of concert must be checked against the sellers' conduct. Perhaps numbers are small yet actions appear independent; possibly the few sellers fear an invasion into their "contestable" domain if they raise prices. Or perhaps numbers are large but actions are interdependent; that is, firms act for the good of the group against their apparent individual interests. Then either the data or the theory or both are wrong; it is back to the drawing board.[11]

In the United States, the number of oil producers is too large and concentration too low for any coordination. In fact, the largest oil-producing states had to do for oil producers what they could not do for themselves: restrain output. After World War II in the Persian Gulf states and in Venezuela there was enormous expansion and a steep price decline. Yet there were only a few sellers. Such a mixed picture needs a fuller explanation, to which we turn in the next chapter.

The Tasks of Monopoly: Optimal Price and Market Division

A group of sellers who restrict output and raise the price have two tasks. The first is to find and reach the optimal price-output combination. Nobody can know this in advance, and they may try successive approximations. An old and good (but not infallible) rule for a monopoly is to find and barely undercut the price of the nearest alternative offered by those outside the group.

Any error here need not be dangerous to the cartel. A single or group monopoly can be at ease over a wide range of prices and outputs; it need only refrain from going too high.

Market sharing among sellers is a different problem; it is unavoidable, difficult, and often dangerous. Competition has an automatic market share mechanism. Every seller produces all he can, up to the point where marginal cost approaches price. A higher output rate would be penalized.

Thus the competitive process generates market shares according to comparative advantage.

The only purpose of group action is to restrict output in order to raise the price above the competitive level. The automatic market sharing mechanism is shut off. It must be replaced by deliberate collective action. The cartel exists to create a price-marginal cost gap, but the gap itself is subversive of cartel order and discipline. Each cartel member would profit by producing more. Day in day out, he must renounce his individual gain for the sake of the collective good.

Maintaining the restricted output total is a zero-sum game. Any gain won by cheating on prices and sales is at the expense of some other member. Collusion brings economic war of all against all. It can be repressed by obedience to quotas, but never removed.

The Leader's Curse

One source of tension is in the predominance of the largest. The small seller can chisel and cheat on price to sell more and get away with it. The larger ones are slow to retaliate, for if they too discount, the whole price level is in jeopardy. It is better to lose some sales. But as they try to offset some losses, they shift the burden increasingly to the largest firm or core group, which has the most to lose.

The buck stops with the largest. If it retaliates, the price-output agreement may collapse, to the injury of everyone. But if it does not retaliate, it may be nibbled far down. It must threaten the others: any further, and you will suffer. They may or may not believe this. They would profit if the largest could be pushed a little more. This game of bluff and counterbluff has been of great importance in OPEC history.

The Instability of Monopoly

A noncompetitive market is like a dammed-up river. If the dike is well built and well maintained, the river may be held for centuries above its "natural" level, but the pressure is always there. As the silt accumulates and forces the dikes to be built ever higher, it is ever more dangerous. A leak not repaired will bring in a flood.

A system is unstable when a small cause may have a large effect. Sellers may anxiously watch for price cheating even on very small amounts because if unchecked it will lead to a spiral of large price cuts. There is constant effort to screen out the random price or market share changes from systematic cheating. The sensitivity to small changes is again a symptom of collusion.

Another source of instability—movements that become self-reinforcing instead of self-limiting—is discussed below.

Governments as Owners of Oil Resources

National governments play a large role in the world oil industry. To appreciate the difference they make, we must first see where they make little or none.

Nonmaximizing Theories

Some believe that governments, unlike private owners, do not try to maximize wealth, hence do not act like monopolists. One influential version is that governments aim only to meet "revenue requirements" because they would rather keep the oil in the ground than maximize wealth.

"Revenue requirements" has the appeal of a horror movie. If governments restrict output to their current "revenue requirements," the price will rise, and with it their revenues. Then the target production level must be reduced further, raising prices still more, causing more production cutbacks.

A nonmaximizing theory confuses ends with means, getting with spending. A state seeks first to survive, then to cultivate its gardens, or spread the true faith, or bash its neighbors, or anything else. But whatever the objectives, the more wealth, the better. Hence each government seeks maximum value from oil production.

We look in vain for an example of a government that deliberately avoids a higher income. The self-serving declaration of an interested party is not evidence.[12] We would rather watch what they do—for example, to see whether the OPEC governments persistently accumulate assets because their revenues exceed their requirements. In fact, they tend to run deficits.

The Unbound Cartel

The group of sovereign states is a far closer fit to the economic model of a cartel than any private group could be. No private companies residing in a modern industrial state would dare to raise the price of a product as important as oil by a factor of 14. They would harm interest groups far more numerous and politically powerful than themselves. In the United States, they would go to jail. Other countries would use other methods.

But in a small, exporting less-developed country (LDC), there is no conflict between its oil sector and the rest of society. All of the benefits of a higher price go to the local economy; the burden is borne entirely by foreigners. The government's duty to its constituents—citizens, royal family, ruling party—is to charge all that the traffic will bear.

Moreover, unlike private companies, sovereign governments are not answerable to anyone. Thus, a cartel of sovereigns faces no barriers between themselves and the optimal monopoly price. A small country that is part of the industrial world, such as Norway, has economic ties that cause its economy to flourish or stagnate in harmony with oil consuming countries. But a small LDC has few ties to the rest of the world except the oil it exports and the goods and services it imports. Even a worldwide recession following a price explosion is partly a benefit to the oil exporter state because it weakens the prices of imports. A loss of oil exports to be factored into calculations of the wealth-maximizing price.

But one government's optimum is not necessarily another's. The choice also depends on how one weights gain and losses, short term against long term.

Governments' Short Horizons and High Time Preference

The discount rate on an expected income stream depends on its risk, and risk depends, first, on the expected volatility. The income of cartel members will fluctuate more than oil incomes generally. They are the residual producers. When consumption drops, it is up to them to reduce output, to maintain the price and then increase output when demand revives. In addition, as cartel members jockey for position, the uncertainty of the sales receipts of any particular member rises. Each must avoid being a residual of the residual.

The effect of oil income volatility on risk depends also on the rest of the receiver's income. If the expected oil income is only a small part of total income and if oil income fluctuations are independent of the other income sources, then even pronounced ups and downs in oil may add relatively little risk. For every up in oil income, there will probably be a down in some other income, and vice versa.

But the economics of LDC oil producer governments are not diversified. Oil exports pay for practically all imports. Indeed, for the largest producers, oil exports pay for practically everything. Nonoil portions of the national income (construction, trade, farming, manufacturing, etc.)

cannot stand on their own. They exist only to serve the oil industry and oil income receivers, or they are subsidized by oil income. If oil income drops, so do they all.

There is a useful contrast. In Texas between 1981 and 1993, the petroleum component of its gross domestic product (GDP) declined greatly, yet the state total GDP grew by 25 percent. Labor and capital shifted out of oil to other self-supporting pursuits. Exports of other Texas commodities—manufactures, farm products, medical services, and others—more than made up for the loss of oil. The oil-producing nations, in contrast, have not been able to create other self-supporting industries.

Furthermore, a private company's political risk is replaced, for an OPEC government, by other kinds of political risk. Nearly all of them are undemocratic. The outs cannot replace the ins except by violence and conspiracy. Moreover, they mostly dwell in violent areas of the world, where attack is always a possibility.

Governments therefore discount oil revenues at a much higher rate than do private companies. They have short time horizons.

High Discount Rates: Little Effect on Investment

In some respects, oil governments' short horizons and high discount rates do not matter. Recall the Persian Gulf example of a company that netted 33 cents, a return of over 100 percent. This was far above its own discount rate; hence there was no constraint on investment. Later, the private company was displaced by the government, which received not 33 cents but 121 cents, for a return of over 400 percent. The constraint was even further away. To this day, in all of OPEC, the price-cost gap is so wide and the return is so high that the discount rate is irrelevant to investment.

High Discount Rates Make Monopoly Pricing More Aggressive

If nations cooperate to fix prices, their time preferences become important. A higher price means higher revenues—immediately. Sales will decrease—later. The higher the discount rate is, the less important is the far-off result, the greater is the reward to raising the price, and hence the greater is the likelihood of its being raised.[13] Thus a monopolist with a high time preference—a short time horizon—will raise prices sooner and more than one with a long horizon.

Furthermore, the short horizon also expresses a preference for high spending and low saving, hence chronic financial problems. A higher price and higher revenues enable members to withstand financial pressure to chisel and cheat on the price. Thus, higher prices promote still-higher prices. Lower prices strengthen the need for cash. Hence they weaken the resistance to price cutting and drive prices still lower. Therefore the system is unstable because a movement in either direction tends to speed up, not slow down.

Summary

This chapter has presented no model of the world oil market; rather, it indicates the principles needed to understand it and especially how not to see what is not there.

Oil is a renewable resource, with no intrinsic value over and above its marginal cost, of which a small fraction is current operating expense. The great bulk of marginal cost is the return on the finding-development investment needed to maintain and enlarge the inventory of reserves. A reliable indicator of scarcity is development investment, both in itself and because it is correlated with finding investment and with discovery value.

Competition drives the price toward the long-run marginal cost. A price many times as high as this cost cannot be explained by a nonexistent intrinsic value; it reflects some block to competition, or control of the market. A greater price-cost gap suggests a higher degree of control. These suggestions must be checked by looking at the number of sellers and the evidence of collusion among them.

Governments are like individuals: similar in getting income, diverse in spending. Each tries to increase revenues as far as it can, to serve its particular ends. Objectives vary greatly from one to another and may even conflict. A collusive group of LDCs will be much less restrained in seeking maximum revenues than a group of private companies. Moreover, LDC governments have short time horizons and will try to take more income faster.

Appendix 2.1: Calculation of Cost and Breakeven Rate of Return, Supply Price

Investment Capacity

R is the new reserve to be developed, in barrels, by investing K dollars. Q is the initial output in barrels per year, and the investment per annual barrel is K/Q (investment per daily barrel, divided by 365). With a decline rate of a percent per year

$$R = Q \int_0^T e^{-aT} dt = Q(1 - e^{-aT})/a.$$

Decline Rate

In theory, the decline rate a, in percent per year

$$a = (Q/R) - (Qe^{-aT}/R)$$

where $Q =$ initial output, $Qe^{-aT} =$ final output, and $R =$ reserves. When decline rates are very low and time to shutdown very long, we can safely neglect the second right-hand-side term, and $a = Q/R$. Otherwise, we approximate:

$$e^{-aT} = Q_f/Q \cong Q/R$$

where $Q_f =$ final output. The theory is that the more intensive the development is, the higher are the fixed annual outlays, hence the sooner the cutoff. Then the formula becomes

$$a = Q/R - (Q/R)^2.$$

As a check, *Oil and Gas Journal,* in the January 31, 1994 issue, gives the Prudhoe Bay field 1993 output as 10.29 percent of reserves. Our formula yields $a = 9.23$ percent, confirmed by [DOE 1992 p. 29] giving Prudhoe Bay an "underlying decline rate" of 9 percent.

Operating Cost

A conventional allowance for annual operating expense per well is 5 percent of the original investment. As output declines, the expense per unit of output will rise. But the more distant higher costs are less important than the nearer ones. We need to balance these effects to find a present value or "levelized" operating cost c per barrel as the basis for planning. Let c' be the percentage of original capital cost to be spent each year as the fixed operating cost; then the levelized percentage of investment per barrel of initial output is:

$$c = c'((a + m)/m)((1 - e^{mT})/(1 - e^{-(a+m)T})).$$

Here m is a market discount rate, appropriate for a project of this degree of risk. Using some conventional rates, if $c' = .05$, $T = 25$, $a = .062$, and $m = .12$, then $c = 1.5c' = 7.5$ percent. (At the limit, as T goes to infinity, $C = 1.56c'$.)

To break even, present value V of revenues must equal investment K:

$$V = PQ/(a + i + c) = K.$$

The alternative break-even equations calculate P as the unit cost (break-even price), or i as the project rate of return:

Cost $$P = (K/Q)(a + c + i) \tag{2.1}$$

or

Return $$i = P/(K/Q) - c \tag{2.2}$$

When the investing company is a private firm subject to royalty or excise tax, the proportion of price taken as royalty $= t$. Then the equations become:

Company "cost" $$P(1 - t) = (K/Q)(a + c + i) \tag{2.1a}$$

or

Company "return" $$i = P(1 - t)/(K/Q) - c - a \tag{2.2a}$$

Persian Gulf in 1970

The price was \$1.21. The needed investment was less than \$100.00 per daily barrel, or less than 27.4 cents per annual barrel. [Adelman 1972, chap. 2] (An allowance for inflation would be about 3.5 times.) The annual decline rate a in Saudi Arabia was 2.9 percent (see chapter 8). Hence the rate of return on new investment was:

$$i = \$1.21/.274 - .075 - .029 = 4.31$$

or 431 percent per year. Alternatively, and assuming a higher risk, the marginal cost of development operation would be:

$$P = (K/Q)(a + c + i) = (.274)(.029 + .075 + .20) = 8.3 \text{ cents.}$$

By 1970, the Saudi and neighboring governments were taking 88 cents per barrel as royalty. Hence $t = \$.88/\$1.21 = .727$. For Aramco:

$$i = P(1 - t)/(K/Q) - c - a$$

$$i = .33/.274 - .104 = 1.06$$

or 106 percent per year.

Appendix 2.2: "Finding Cost per Barrel of Oil Equivalent"

The financial press often cites annual "finding costs" or "replacement costs" per barrel of "oil equivalent," which consists of (1) exploration plus development expenditures, divided by (2) oil reserves added plus the "oil equivalent" of gas reserves added. These estimates are useless.

The addition in the numerator (1) is illogical. Exploration adds knowledge, and development adds reserves. These are different activities, for returns over very different time periods. Moreover, exploration outlays on oil are mingled with those on gas.

This brings us to the denominator (2). Here, the addition is wrong because there is no oil or gas equivalence. Oil and natural gas are not in a stable relation to each other with respect to costs, prices, or reserve values. They can and do move in opposite directions.

Moreover, even if "finding cost per barrel of oil equivalent" meant something for any one year, it would not be comparable with that for any other year. Changes in the exploration-development mix, or in the oil-gas mix, or both together, make comparison invalid. We are told not to add apples to oranges; this is fruit salad.

Notes

This chapter is based on the papers in Adelman [1993, particularly chapters 11–13, and also on two papers not republished in that collection: "Finding and Developing Costs in the United States 1945–1986," in *Energy, Growth and the Environment: Advancement in the Economics of Energy and Resources*, ed. John R. Moroney (Greenwich, Conn.: JAI Press, 1992), pp. 11–58, and "OPEC at Thirty Years: What Have We Learned," in *Annual Reviews of Energy* (Annual Reviews, 1990)].

The data on Texas gross product, 1970–1993, including oil and gas income, were furnished by Gary Preuss of the Office of John Sharp, Texas Comptroller of Public Accounts.

1. One well-known study of the value of U.S. government mineral holdings [cited by Adelman 1993, p. 219] assumed that oil prices had to rise from 1981 levels by 3 percent real per year, or 43 percent from 1981 to 1993. Since the general price level rose 55 percent, the 1993 price of oil should have been $70.42 (31.77 × 1.55 × 1.43). It was in fact about $15.00. If we use a conventional 10 percent discount rate instead of the 2 percent assumed in the study, the total overstatement is by a factor of 23.5 (70.42/15) × (.10/.02).

2. Suppose the estimate for a well is an initial 1,000 barrels daily, or 365,000 barrels per year. If the decline rate is 10 percent per year, production after twenty-five years is only 82 barrels daily, or 30,000 barrels per year. If at current prices lower output will not pay operating expenses, this is the cutoff. The reserve will be booked as 335,000 barrels, its cumulative expected output. A higher price or lower cost will extend the "economic life." In algebra,

$$R = Q \int_0^T e^{-at} dt = Q\ (1 - e^{-aT})/a$$

where R = proved reserves, Q = initial output, a = decline rate in percent per year, and T = time. If T is indefinitely large, this simplifies to $R = Q/a$, or $a = Q/R$, which is usually but not always a good enough approximation. For our example,

$R = 365\ (1 - e^{-(25 \times 0.1)})/.1 = 3350.$

3. "In the calculable future we shall live in an *embarrass de richesse* of both foodstuffs and raw materials.... This applies to mineral resources as well." [Schumpeter 1943, p. 116]

4. Assume the price of oil is $10 per barrel. One oil well produces 10 barrels daily, the other 10,000. The average cost in the big well is only a small fraction of cost in the small well. But under competitive conditions, the marginal cost in both wells is $10. In each well, production is pushed to the limit, where producing one more barrel daily would raise costs on the whole operation by more than $10. Profit is maximized (or loss minimized) in both wells.

5. In theory, the contribution of discovery to in-ground value in any given place ought to stay between a minimum of zero, when available reserves are unlimited, and a maximum of equality with development cost. [Adelman 1993, pp. 243–244] In the United States, discovery value has long fluctuated around 60 percent of development cost. Moreover, exploration outlays (omitting bids for leases, which are not a cost but a sharing of profits) have been around that proportion of development outlays.

6. I estimated in 1986 [Adelman 1993, pp. 155–156] that U.S. industry would shrink because expected finding cost exceeded value. This has in fact happened, but there has been such turbulence that one cannot be sure that the conclusion was borne out.

7. At any given time, capital expenditures have a nonlinear relation to reserve additions. One plausible relation is exponential. Then $K = e^{bR-1}$, where K = expenditures in billions of dollars, R = reserve additions in billions of barrels, and b = a coeffieicnt of greater or lesser cost. Disregarding tax benefits, $K(1984) = 16.2$, $R(1984) = 3.8$, and $b(1984) = .72$. But $K(1992) = 4.9$ billion, $R(1992) = 1.5$ billion, and $b(1992) = .91$, an increase of 26 percent. The precision of these numbers is deceptive. Other mathematical forms would give other results, but all would show a strong increase for oil. Over this period, the coefficient decreased for nonassociated gas reserve additions.

8. The estimates of incremental capacity for the United States are unreliable because of the enormous dispersion of well sizes.

9. It follows from our basic theory [Adelman 1993, p. 223] that investment increases as the square of depletion intensity. If depletion increases from (say) 1 percent to 5 percent, investment increases by a factor of 25. Then the investment per unit of additional output rises by 25/5, or five times.

10. The index is the sum of the squares of the percentages of all sellers. Suppose there are four companies with market shares of 50, 30, 15, and 5 percent. The squares add to: $H = .25 + .09 + .0225 + .0025 = .365$. (The percentage numbers are sometimes squared directly: thus, $2{,}500 + 900 + 225 + 25 = 3{,}650$.) The reciprocal of the index is $1/.365 = 2.74$, a numbers equivalent. That is, the index is the same as would be the result of 2.74 equal-sized sellers. A glance will show that if the smallest 5 percent were somehow missed, the difference would be trifling. Even if the final 20 percent were lost, it would not matter much: H would equal .34, its reciprocal 2.94. The assumption is that the small firms do not count for much, provided they remain small; hence the inaccuracies of counting them do not matter much either.

11. Or the two lines of proof may converge. In 1960, whatever the number of U.S. natural gas producers, the price of gas delivered to Gulf Coast electric power companies was clearly below their best alternative, the delivered price of coal. A gas monopoly would have charged what the traffic would bear, a little below the coal price, enough to ensure the sale. Only competition could explain the price of natural gas. See Adelman [1962], pp. 39–40.

12. "The Saudis have long made it clear that they regard oil production much in excess of what is required to cover current revenue needs to be a concession on their part to the Western countries, the United States in particular." [CIA 1979, p. 5] (Also see [DOE 1989, q. OGJ 1-9-89:22, 11-27-89:32] This nonsense was important because it was believed.

13. Suppose that the long-run percentage of sales loss equals the percentage of price increase. Over the long run, annual revenue is unchanged. The half-life of this adjustment is assumed at seven years. With a low discount rate (5 percent), a 50 percent price rise increases the present value of future revenues by 12 percent. With a high discount rate (25 percent), a 50 percent price increase raises the present value by 30 percent. For a full explanation and the derivation of the valuation formula, see Adelman [1993b, pp. 438–439].

3 The World Oil Market to 1970

In 1945, the world outside the communist blocs had an inventory of 44 billion barrels of oil reserves; by 1970 it had consumed 150 billion barrels and ended with reserves of 521 billion. Production rose from 7 mbd to 38 mbd. The inflation-adjusted price fell by three-fourths.

The predominant view has long been that during these years the multinational companies kept world oil prices untenably low for the long run; consumers lived in a fool's paradise. "Limited resources" were bound to make oil more scarce and ultimately force up prices after 1970. The energy shortage or crisis was unavoidable. Governments could at most anticipate or exaggerate it.

None of this is true. Oil companies wanted higher, not lower, prices. Lower prices meant lower profits and political grief in the United States and Western Europe, where they were accused of destroying coal miners' jobs, and domestic oil companies. [PPS 1968a; Gordon 1970]

"Limited resources" raise prices by forcing a persistent increase in marginal cost: the amount that must be newly invested to develop an additional barrel of reserves and capacity. We have seen that investment needed for an additional daily barrel of capacity fell, even with excessive allowance for the value of resources used up, over and above actual investment. But even had cost been rising, it would have had a long way to go before it could equal the price and force it up. In 1970, when oil reached its all-time low price, new investment would pay out several hundred percent per year. The erosion of the price-cost margin alarmed the new owners of the oil fields: the host governments. In later chapters, we will see what they did to ward off competition.

The World Industry Structure before 1945

The U.S. Core and the World Fringe

As late as 1940, and omitting the Soviet enclave, the United States produced 70 percent of world oil and was a large exporter (table 3.1). The U.S. price was the world price.

U.S. oil production (unlike refining) was always competitive.[1] The law of capture made it more costly than it needed to be. [Adelman 1993, chap. 2; McDonald 1971] Subsoil mineral rights in the United States, unlike all other countries, were not reserved to the ruler (hence the payment called "royalty") but belonged to the landowner. Fluid minerals like oil and gas migrated from beneath one property to another and belonged to anyone who could capture them. A landowner could drill a well and legally draw oil from under a neighbor's land. This led to much overdrilling and then to even more wasteful regulation and state-imposed cartelization, which raised the domestic cost and price to above-competitive levels.[2]

The United States did for the producing industry what it could not do for itself: restrain output to keep up the price. The political power was supplied not by the large companies but by the far more numerous domestic producers, royalty owners, and local oil service and other support industries. Then a federal law did for the states what they could not do for themselves: it prohibited interstate shipments of oil produced in excess of a state allowable.

Outside the United States, the industry was very different. There were only a handful of producers. In the beginning, this was required by economies of scale.

Economies of Scale

Economics of scale exist in all industries at all times. On a graph, if size is on the horizontal and unit cost on the vertical, the relation between size and cost resembles a bent **L** shape, _. At some point, early or late, the cost decline stops, and more size does not bring more efficiency. The downsloping segment varies greatly, and industry structures vary with it.

The crucial relation is between economies of scale and size of the market. At an isolated crossroads, the market is too small to hold more than one food store, which is a minuscule monopoly. At the other extreme, world oil today is large enough for many huge companies. But it was not always so. History plus inertia explain the world oil market structure.

Table 3.1
Leading oil producers to 1970
(thousand barrels daily)

Year	United States	Mexico	Venezuela	Russia	Indonesia	Persian Gulf	World	United States (%)
1910	575	10	0	170	30	0	898	64
1920	1,214	430	1	70	48	34	1,887	64
1930	2,460	108	374	344	114	126	3,868	64
1940	3,707	121	508	599	170	280	5,890	63
1950	5,407	198	1,498	729	133	1,755	10,419	52
1960	7,055	271	2,854	2,957	419	5,255	20,967	34
1970	9,637	487	3,708	6,985	854	13,957	45,688	21

Sources: American Petroleum Institute, *Petroleum Facts and Figures*, Centennial Edition (1959); and DeGolyer and MacNaughton 1991.
Note: Peak Mexico production was 530 mbd in 1921.

Economies of Concentration and Integration

Only a few firms were capable of the risky search for oil in remote often harsh places. In each consuming country, refining and marketing was a small industry, protected by distance and government, making entry difficult and unprofitable. Production was too risky without an assured outlet, known as "finding a home for the crude." Refining was too risky without an assured supply of crude. Hence in each country the few sellers were confronted by few buyers, and neither side wished to be at the mercy of the other. The obvious solution was vertical integration. Refining-marketing joined with production by merger or by branching upstream into production or downstream into refining. Vertical integration saved the transaction costs of incessant search and negotiation. Buyer-seller contracts do it today, and could have done it then, if there had been any appreciable number of buyers and sellers. But the need to enter both production and refining-marketing together made entry more difficult. Thus inertia preserved a concentrated integrated structure even as the original cause—economies of scale in small markets—slowly disappeared.

There was important interwar entry. To the original four (Shell, British Petroleum, Exxon, and Mobil—their most recent names) were added Chevron, Texaco, Gulf, and the Compagnie Française des Pétroles (CFP, now Total).

World Structure of Trade and Prices

Through 1947, the United States was an exporter of crude oil and products, and the worldwide price was "Gulf of Mexico plus" transportation or "Texas Gulf plus." With the unimportant exception of Costanza (Rumania), the U.S. gulf was the only place in the world where tanker loads of crude and products were sold at an arms-length price, sometimes publicly reported. Elsewhere crude oil was shipped only within integrated channels, with no arms-length price. But the 1932 price collapse in Texas was viewed as a threat everywhere. [FTC 1952, p. 249]

The Texas price and the transport cost were beyond the companies' control. Local delivered prices were higher outside than inside the United States, and costs were lower. Hence there was a very wide price-cost gap in the world market. It was at first a location rent—the reward to risk or luck—but it persisted for decades.

The companies in the world market did not need to cudgel their brains over the optimal monopoly price, as would OPEC afterward. Texas-plus was the best they could do, and it was very good. Naturally they wished to keep the margin from competitive erosion, and it suited them that the non-U.S. world remained a net importer. Investment was restrained, and outsiders continued to find it difficult to enter.

Insiders needed to abstain from rivalry on price, a constant temptation because the margin over cost was so wide. Since they were vertically integrated, there were no markets in crude oil and few in products. The companies competed mostly in selling finished products to consumers in many local markets. Sellers at each point of sale were few. The value of a purchase (mostly gasoline at the pump) was very small in relation to the buyers' income, so buyers had no incentives to hunt for bargains. Motor gasoline outside the United States was a luxury product and was heavily taxed as such, even before World War I. Hence a cut in the pretax price was not fully reflected in the posttax price. Reactions to price changes were limited and slow, and they provided little incentive to discount prices. Rivalry was decorous, even sluggish.

No sales area was of crucial importance. A seller was better off yielding some sales and making it up elsewhere instead of cutting prices to maintain share.[3] For a long time, the companies relied on a general understanding to restrain rivalry and to keep it away from price. It gave the system a flexibility they rightly wished to keep.

By the late 1920s, the tacit understanding was becoming harder to maintain. The 1928 Achnacarry meeting of Shell, Exxon, and BP became famous because any meeting at all was unusual. The preface to the "As-Is" agreement made there sounds very up-to-date: "Each large unit has tried to take care of its own overproduction and tried to increase its sales at the expense of someone else. The effect has been destructive rather than constructive competition." The solution was to maintain the existing country-by-country market shares (outside the United States). This brings us down to earth. "As is" was not a triumph but a confession: spontaneous coordination was not enough. When a price-fixing group is unable to achieve a market sharing agreement, their second-best solution is to freeze or roll over current shares.

The agreement to agree took them from tacit collusion to a loose cartel. The antitrust laws, as then enforced, were no barrier. After 1928, there were numerous meetings. In May 1932, the American Petroleum Institute called a three-week meeting, described by the European press as preliminary to a "world oil conservation conference." The Soviet Union, a small exporter that had cooperated with the cartel in various countries for years [FTC 1952, p. 278] sent a negotiating team. It was expected that the international majors would buy and distribute Russian oil products, thus taking them off the world market, but the negotiations broke down over the Russian insistence on periodic increases in their quota. [FTC 1952, pp. 239–240] (This sounds like the attempts to bring the non-OPEC nations into cooperation with OPEC.) The companies went from one expedient to another to maintain as is, until World War II put everything on hold.

Producing Areas

To the eve of Achnacarry, Persian Gulf oil was simply Persian. In Iraq, we know of a natural gas flare from the Book of Daniel. Later, Alexander the Great was shown a bitumen lake near Kirkuk. During the three-week orgy of welcome for his troops in Babylon, naphtha seeps were used as fireworks. But in the twentieth century, the good prospects actually slowed development. Rival groups counterlobbied and stultified each other at the Ottoman court, during the Young Turk government, then under the Iraq monarchy, after prolonged negotiations among themselves and their governments. As late as 1922, although oil discovery was expected in Iraq, "there [was] apparently no expectation ... of securing a large initial production from any particular well, as is the case in South Persia and Mexico." [FTC-Church p. 521] Then the first Iraq discovery, in 1927, was made; the initial production per well was spectacular.

As for the flat plains to the west in Kuwait and Saudi Arabia, Exxon and Mobil "were all sure that there could be *no oil at all*" there. [FTC 1952, p. 114, emphasis in original] As late as 1940, the whole Middle East produced only half as much as Venezuela, and in 1944, an expert mission estimated total Persian Gulf proved reserves as 15 billion barrels, where the United States had 20. [OGJ 3-23-44:45; API 1950]

Cycle of Mutual Mistrust

Mexico's collapse from 23 percent of the world total to less than 3 (see table 3.1) resulted partly from the primitive state of reservoir engineering, which permitted the northeast Golden Lane fields to water out. It is also an early example of a vicious cycle of mutual distrust. The government, aggrieved by too little oil revenue, demands more. The company feels threatened and decreases investment and production. The government feels more aggrieved, threatens more, and so on, until the government expels the company.

In particular, Mexican "attacks on the industry tapped powerful anti-American sentiments." [Krasner 1978, pp. 155–172; Meyer 1977, pp. 9, 241; Mancke 1979, pp. 17–58] During the turmoil of 1910–1920, taxes were raised; then the oil companies' ownership of reserves was questioned and then revoked in 1918. Exploration and development stopped in 1915, resumed, and stopped again in 1927. By 1938, when oil was nationalized, there was not much oil left to quarrel about.

Similarly, Iraq and neighbor governments thought Iraq Petroleum Company (IPC) was developing too slowly and producing too little. In 1938, IPC's general manager reported conversations with various rulers: "The future leaves them cold; they want money now." [8 Church 531]

Oil and the Pacific War, 1941–1945

In July 1941, the United States and Britain forbade oil shipments to Japan, which went to war to break the blockade.[4] In Europe and Japan, Anglo-American oil companies were feared and distrusted as potential instruments of their governments. It was not understood that these companies had no power. The 1941 boycott was effective only because British and American armed forces controlled the oil-producing regions. After the war, they no longer did.

The Postwar World Market

U.S. Imports Restricted 1948–1970

In 1948, the United States became an oil importer, but its imports were limited from the start in an escalating struggle. First, there were congressional hearings to expose and intimidate the multinational companies, which were importing some of their foreign production. Next, the Texas Railroad Commission required detailed import reports. The importer companies were also large domestic producers and knew that "excessive" imports would breed lower "allowables" (production quotas) in Texas and other states with production controls.

But this limit failed when some nonproducing companies began to buy and import oil. In 1954, the U.S. government began to fix and administer "voluntary" import quotas. Like As Is, they were based on past imports. A company with no previous import record had no quota. But more companies began to import, quotas were flouted, and the noncheaters threatened to retaliate. In 1959, import controls became mandatory.

It is often said that the 1959 import limits provoked excess capacity abroad, which forced down world prices. This is wrong, first because prices declined before and after 1959. Second, imports had been limited for a decade. Mandatory quotas merely formalized and enforced previous controls. Furthermore, there was no excess of producing capacity outside the United States. On the contrary, the rapid expansion during the 1950s and 1960s shows a constant need for more capacity.

The Changing Price Structure

Nevertheless, the price decline after 1947 needs to be explained because there was good reason to expect a strong price rise. The world industry was much more concentrated. With the United States no longer an exporter, the eight majors—five Americans plus Shell, BP, and CFP—were now practically the only suppliers.

The world price after 1947 was no longer tied to Texas-plus. It could have gone higher, toward an optimal monopoly level, but the road actually taken was downward (table 3.2). Thus a huge gap grew between the potential (monopoly) price, which was much higher than the old Texas-plus, and the actual price, which went much lower.

The market in fact moved in the competitive direction. The Persian Gulf price fell to where it just paid to ship to the U.S. East Coast and Western Europe. From being U.S.-plus, the price became U.S.-minus. To be sure, continued competitive erosion could have brought the whole structure even lower. The floor was still a long way down—the price that would yield an acceptable return, taking account of risk, on marginal in-

Table 3.2
Oil price FOB Persian Gulf and United States, 1947–1970
(current and constant 1990 dollars per barrel)

	Persian Gulf price		U.S. price	
Year	Current dollars	Constant 1990 dollars	Current dollars	Constant 1990 dollars
1939[a]	2.21	22.87	1.02	10.56
1947	2.22	13.21	1.93	11.48
1949	1.75	9.79	2.54	14.21
1960	1.63	6.94	2.88	12.25
1970	1.21	3.78	3.18	9.95

Sources: Adelman 1972, pp. 134–135; PIW.
[a] A rough guess. The 1939 real freight cost is taken as about twice the 1949 level, or in 1939 dollars $1.19. This is added to the U.S. price.

vestment. [Adelman 1972, p. 133] We need first to explain why the price fell when market conditions favored a rise.

The Marshall Plan Drastically Lowers Prices

Texas-plus had been somewhat eroded during World War II. Then starting in 1947, under the Marshall Plan, the U.S. government financed a large volume of European oil purchases. The plan was carried out by a separate temporary office, with highly competent economic advice. The United States did not question the price level set at the U.S. Gulf, but it would accept only the price structure that would exist under competition, with substantial exports to the United States: the Texas Gulf price minus freight from the Persian Gulf. In a quiet revolution, the Persian Gulf price was lowered drastically, and henceforth it was the effective world price.

Downdrift after the Marshall Plan

After 1950, as the Marshall Plan wound down, the U.S. government ceased to finance oil purchases. Its influence was gone, yet prices were not restored; they even continued to decrease slowly. Only competition can explain this effect. The companies' old concert of action was much weaker. Since oil was now imported into the United States, any action affecting its price was subject to the U.S. antitrust laws. In the postwar political climate, the American majors did well to stay out of the courts if possible. Whatever the verdict, no big antitrust case could be won; at best, one could limit the damage.

The major companies still had great defensive market power and used the simplest possible price strategy: do nothing. In a small group, when each individual is resolved to hold the price line until somebody else moves first, it may be quite a while before anybody moves at all.

To do little was, of course, to do much. The companies' orderly retreat was possible because they were few and had overlapping membership in several joint ventures. The production plans of the Iraq Petroleum Company (Exxon, Mobil, Shell, BP, CFP) were known to two Aramco partners (Exxon, Mobil), hence to the other two, Chevron and Texaco. They were known to one of the Kuwait partners (BP, Gulf) and to the companies extracting 95 percent of the oil in Iran (BP, CFP, Exxon, Mobil, Shell, Gulf, Texaco, Chevron). The great bulk of Venezuelan production was by Exxon, Shell, and Gulf. Thus, each company and group could restrain capacity expansion knowing that others would do the same. But they could only slow the price erosion, not stop or reverse it.

The entry of new firms increased pressure on them. The degree of concentration (Herfindahl index) declined from 0.204 in 1950 to 0.104 in 1969, and the number of "firm equivalents" increased from 5 to 10.[5] [Adelman 1972 p. 81] The Soviet reentry into world oil drew far more attention than its size warranted. A symptom of an above-competitive price is sellers' hypersensitivity to small amounts of uncontrolled offerings.

Thus, the competitive process explains a slowly declining price through mid-1970. But even then, as will be seen, the world price was still far above the competitive level. Even the defensive strength of the companies was not as important as the tax system as it evolved up to 1960.

To understand it, we need to look at the relations of companies and host governments after World War II. Before 1960, taxes were a share of profits, with no effect on price. From 1960, taxes changed from income taxes to excise taxes and became the floor to prices.

The 1950s: The Income Tax Decade

The oil concessions (or contracts) originally provided for a payment per unit produced. But in 1948, Venezuela legislated fifty-fifty, an equal sharing of profits between company and government. Saudi Arabia soon decreed it, despite the contract violation; "there was little Aramco could do to resist." [8 Church 341–378, esp. 346–347][6] Kuwait and Iraq followed suit. In Iran, it was feared that the breakdown of negotiations "could result in collapse of the country or confiscatory action." [8 Church 345] That soon arrived. Thus, within a short time, the tax system had been completely changed: by a South American *caudillo* (dictator), a fundamentalist Muslim monarchy, two relatively secular Muslim monarchies, and others. Political or religious sympathies were irrelevant. Each regime used a power that it had not previously had.

Income Tax Credits

The U.S. State Department suggested that the income tax payments to the host governments be deducted from U.S. income tax. If applied, say, to Aramco, there would be "in effect ... a subsidy of Aramco's position in Saudi Arabia by US taxpayers." The National Security Council agreed, and the Treasury duly changed the regulations. [4 Church 87]

It was not done by backstairs oil company influence. In 1967, the elder statesman John J. McCloy, then an oil company lawyer, reminded Secretary of State Dean Rusk, without contradiction, that the income tax credit was recommended ... [by] the Department of State and the Treasury Department [which] recognized that it was in the national interest of the United States to keep such nations stable and friendly to the United States and thereby ensure American access to the vast oil reserves there located. If the oil companies did not provide the necessary revenues by paying substantial taxes to producing countries large amounts of foreign aid might well be involved.... In 1950 ... the Department of State took the position, when consulted by Aramco, that a 50–50 income tax arrangement ... appeared to be to the advantage of all concerned. [9 Church 115–116]

The stated objective was to provide foreign aid and to bypass Congress. Therefore, there was no process whereby anyone asked how much subsidy was needed. In 1950, Saudi oil revenues were $113 million; in 1970, $1300 million.

A supporting Internal Revenue Service (IRS) memorandum distinguished income taxes from other types. Taxes on mineral production were not deductible "where the tax attached even if the miner made no profits." [8 Church 354] In a few years the tax was, by the Treasury's criteria, no longer an income tax and should no longer have been deductible. But the tax credit was not withdrawn. The foreign policy had not changed, to support the producing governments. Events in Iran strengthened it.

In Iran, the feared expulsion of BP (then Anglo-Iranian) took place early in 1951, and the shah fled. But Iran was unable to produce much oil, the companies blocked its sale, the much-feared Soviet support never materialized, and in 1954 the Iranian government was overthrown and the shah restored.[7] The new arrangement gave Iran the same half share of profit as the other governments. BP kept only 40 percent of Iran production rights, and 55 percent was transferred to the other majors (Shell, Exxon, Chevron, Texaco, Mobil, Gulf, and CFP). Five percent went to the Iricon group of smaller American firms. (Their role is explored later.)

Producers' Needs Grow with Revenues

The larger the oil revenues were, the more dependent on them the governments became and the more sensitive to their fluctuations (figure 3.1). Higher revenues were anticipated and spent in advance. The producing countries were unable to develop any autonomous export industries. Nonoil activities were supported by the oil sector. Oil paid for increasing imports of food, manufactures, and services done by foreigners, from the most menial to the most highly skilled work.

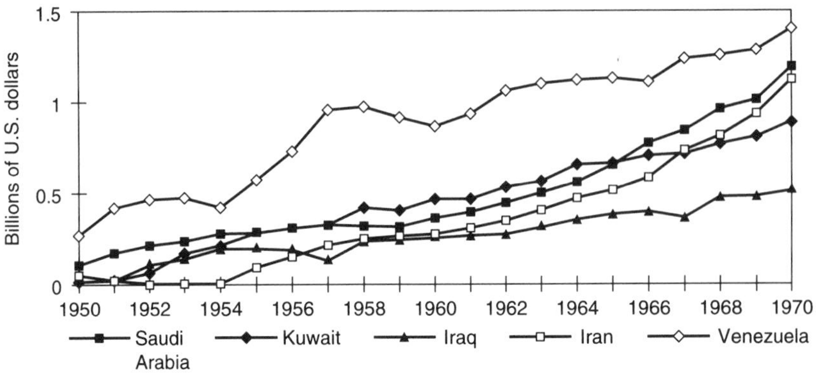

Figure 3.1
Oil revenues of the five big producers, 1950–1970. Source IMF:IPS.

The host governments, unlike private companies or companies' shareholders, could not draw on other incomes to stabilize their spending; hence they could not wait, and their time horizons were short. The 1938 Iraqi complaint became louder: "The future leaves them cold; they want money now." (See figure 3.2.)

Saudi Arabia was forced to borrow over 1956–1958. Its current account surplus then went positive but declined steadily and went negative over 1968–1969. Iran's imports exceeded exports in eighteen of twenty-one years under review. Venezuela overspent in the late 1950s as soon as its income growth slowed and again in the late 1960s. Iraq managed better throughout, both under the monarchy and after the 1958 revolution. In general, however, it overspent current income in anticipation of future income. Lower revenues meant a divisive internal struggle over who bore the burden of spending cutbacks.

The governments could not save oil revenues to build a fund from which to smooth out the flow of expenditures, nor could they invest to create a flow of nonoil income. Hence the higher oil revenues generated a felt need for still higher revenues. Price fluctuations aggravated the nations' conflict with the companies and with each other.

The Ratchet: 1953 Succeeds, 1957 Fails

In 1953 and again in 1957, following the temporary closure of the Suez Canal, there appeared an important phenomenon: the price ratchet. A temporary shortage or pretext raises the price. The sellers keep quoting

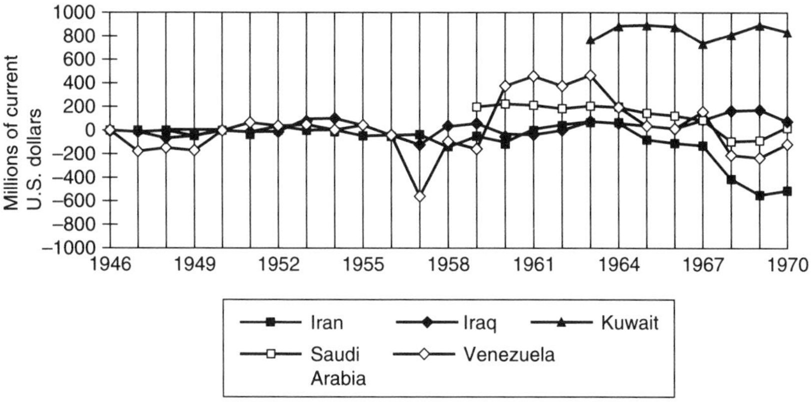

Figure 3.2
Current account balances of the five big producers, 1946–1970. Source IMF:IPS.

the higher price, cutting production to equal the amount demanded, as necessary to hold the price line. The ratchet held both times in the United States, because the States had the sovereign power to cut back production volumes. The average crude price in nominal dollars went from $2.53 (1952) to $2.78 (1954), and from $2.79 (1956) to $3.01 (1958). [DOE: AER 1992:157]

Outside the United States, the companies followed the U.S. increase. The ratchet held in 1953 but not in 1957. They lacked sovereign power and could not maintain the higher prices. The Venezuela-realized price, the only consistent series for this period, was $2.11 in 1951 and $2.30 (±.01) between 1953 and 1956. In 1957 it was $2.59 but then fell to $2.08 in 1960. [PODE 1970] The price drop brought on a crisis in government-company relations.

The Crisis over Posted Price Reduction in 1959

In order to reckon income tax in a producing nation after 1950, it was necessary to post a price at which the crude oil was deemed to be sold. Taxable income per barrel was defined as the posted price less cost. The bulk of the oil was transferred to refining affiliates, and small amounts were sold to refiners as crude oil. Large amounts were sold under long-term contracts from one major to another, at low prices.

Until late in the 1950s, the prices realized on the small amounts of oil sold at arms length to third parties and the prices of refined products sold to distributors (less transport-refining-marketing costs) seem to have been

Table 3.3
Taxes and posted prices, 1958 and 1960
(dollars per barrel)

	1958		1960	
	Posted	Actual	Posted	Actual
1. Price	2.08	1.80	1.80	1.63
2. Less cost	0.10	0.10	0.10	0.10
3. Profit pre-Tax	1.98	1.70	1.70	1.53
4. Tax at 50 percent nominal	0.99	0.99	0.85	0.85
5. Company profit	0.99	0.71	0.85	0.68

close to posted. But then the major companies were unable to follow the ratchet outside the United States. Soon one heard that "only fools and affiliates pay posted prices." Arms-length crude prices went and stayed lower.

Table 3.3 approximates the tax effect. By 1958, the actual price realized for Saudi crude oil was probably below $1.80 and the company profit below 71 cents. By 1960, posted price had been reduced to $1.80, but actual price had kept falling.

When a company cut its posted price, it thereby cut its income tax. This was resented as a unilateral act. The government, which had counted on revenues rising at the same rate as production and was probably already committed to higher spending, found itself receiving less than expected.

The companies were obviously afraid to reduce posted prices in the face of government anger. Exxon did so in 1959 and again in August 1960. BP publicly stated its disagreement with Exxon but matched the cut. In September 1960, the governments formed OPEC.[8] Whether in consequence or not, the immediate and important event of late 1960 was an unspoken agreement: posted prices could not be reduced, whatever happened to real prices.

Tax was now levied on a fictional income: the fixed posted price (less book costs) multiplied by barrels transferred or sold. Since tax was no longer a percentage of price or profit, the nominal income tax was no longer a percentage of profit. It was an excise tax, in cents per barrel.

Had the U.S. Treasury followed its own rule, taxes payable to the host governments would have been considered excise or royalty payments, as they were in fact. They would have become deductible from income but not from income tax. But there was no such change. In 1960 as in 1950, the U.S. government aimed to subsidize the oil producer governments, in amounts now many times as large as in 1950.

Before turning to the economic effects of the return from income tax to excise tax, a word of warning is due. It cannot be stated too often: *after 1959, no posted price number has any meaning as a price.* It can serve only as an interim figure for calculating the per barrel tax. Many writings on oil markets are spoiled by reliance on meaningless posted prices.

The 1960s: Excise Tax as Price Floor

The excise tax put a floor under prices. An income tax is not a price floor because it declines even faster than the price. At the limit of zero profit, there is zero tax. But a tax per unit is a cost per unit to the producing company, just like the outlay for labor and supplies. The floor to price became tax plus cost, or as the industry called it, tax paid cost.[9] Moreover, the floor was effective at a distance, because no prudent company would consider signing a term sales contract at less than tax plus cost plus some contingency allowance.

The price continued to decline toward the floor throughout the 1960s. In 1967, following the Six-Day War, an attempted Arab boycott of the United States failed and was quickly ended. Customer swapping and evasion were too easy. The OPEC secretary-general drew the moral: "There just can't be an effective selective embargo."[10] [WSJ 9-8-68:9]

It was later asserted that the embargo was "lifted through the efforts of Saudi Arabia." [Akins 1973, p. 468] This was not true. The Saudis joined with the Libyans to urge that the companies raise Mediterranean posted prices, hence taxes. The Suez Canal blockage had raised tanker rates, which had raised Mediterranean prices. [OGJ 8-28-67:46] But no such action was taken.

After 1960, arms-length crude sales became more visible. Product prices in Western Europe (denoted "Rotterdam" for the largest receiving port) began to be published regularly, to the annoyance of the integrated oil companies. One could then reckon a composite product price per barrel by subtracting the refining costs and a market-determined transport rate from producing country to refinery, to compute an FOB crude price equivalent, commonly called a *netback price.* My first published estimates appeared in 1963. Since netbacks and arms-length crude prices should approach one another, I could use them as a mutual check. [Adelman 1963, 1972, chaps. 4–6] With an approximate arms-length price, one can

Table 3.4
Persian Gulf "Big Four"*: Cost profit, and tax
(current dollars per barrel)

	1960	1970
Price	1.63	1.21
Cost		
Capital[a]	0.10	0.06
Operating	0.08	0.05
Total	0.18	0.11
Profit	1.45	1.10
Government payment		
Amount	0.765	0.881
Percent of profit	54.3	80.1
Tax-plus-cost		
Amount	0.945	0.991
Percent of price	0.58	0.819

Sources: PIW 4-12-82:11; Adelman 1972, p. 63; PE 1966, 1975.
*Iran, Iraq, Kuwait, Saudi Arabia.
[a] Refers to 1959–1960 and to 1968–1969, taking the necessary rate of return at 20 percent, other capital charges 2 percent, decline 2 percent.

calculate the approximate producing profit per barrel and its division between company and government. Such approximations appear in table 3.4.

By 1970, the host governments were getting around 80 percent of the profits, over and above a 20 percent return on the companies' investment—and appropriately. Not company power but their excise tax system was the floor under prices, far above the competitive level.

Governments' Struggle for Market Share

In the 1950s and 1960s there was a constant struggle among governments for higher market share and higher revenues. Figure 3.1 shows that Venezuela started and ended as the largest, but most Persian Gulf nations increased much more. The exception, Iraq, fell far behind, for reasons to be seen.

Allocation in a Noncompetitive Market

Under competition, allocation among sellers is automatic. Each producer expands output to the point where incremental cost equals the price. But in the world market, the incremental cost was far below the price everywhere, and expansion was highly profitable. Hence, companies had to allocate production among the various countries in which they operated. Each government always demanded more output so that it might receive more taxes. The sum of their demands exceeded what the market would take at current prices.

The myth that the OPEC nations wanted to produce less than did the companies belongs with the myth that the companies deliberately held down the price before 1973. [Askari 1990, p. 28; El Serafy, in Ahmad et al., 1989, p. 12; Heal and Chichilnisky 1991, p. 57] That would have meant less profit and more political danger, in both the host countries and home countries, from local coal and oil producers.

The 1950s' experience in unilateral contract revision and the 1960 understanding that taxes were excise taxes showed that each government had unlimited power within its borders. But higher taxes or threats of expropriation were a disincentive to invest. A company might slow expansion or actually cut back. Mexico was an example, as were Iran and Iraq.

Iran

The shah of Iran used the lever provided by the 1954 settlement that created the Iran Consortium. As noted earlier, it gave a 5 percent share to the Iricon group of newcomer firms. Their behavior (and the record in Libya) is a window through which we can see the threat from newcomers' competition.

The Iricon firms had negligibly small shares of the world market. Hence, lower prices would lose them little revenue on existing sales. Had they been free to lift as much Iranian output as they wished, production would have greatly expanded. Even at lower prices, there would still be a wide margin over tax-plus-cost on fresh investment. But the major firms in the consortium would have been forced to sell all their previous output at the lower prices, considerably lowering their profits. The consortium decisions were made by and for the major firms, which produced 95 percent. Their first concern was protecting the price level.

The shah tried to pressure the U.S. government to pressure the major companies in the consortium to make them produce more or to allow Iricon to produce more.

A valuable glimpse of the 1960s is in the testimony of an impressive oilman, Howard Page of Exxon. His company did not want a share of the

Iran Consortium: "As a business deal ... it was for the birds.... We had to spend money for capacity and reserves that we already had." [7 Church 289] But the alternative was to leave the oil to those who would have increased output, thereby bringing down prices and greatly reducing the majors' profits.

Page explained how every additional barrel lifted through the consortium in Iran was a barrel less for Saudi Arabia, Kuwait, or Iraq. Cost was not a factor, since it was the same in Iran and Saudi Arabia. [7 Church 282] Excessive total output would bring down the price of oil. But offending a host government by too little output might bring expulsion. As Page summed up: "The Iranian Government was, as every other government, always trying to get more and more ... but ... this is like a balloon, push it in one place, it comes out in another, and so if we acceded to all those demands, all of us, we would get it in the neck." [Page, at 7 Church 288]

Later, of course, the shah would declaim that he preferred to keep the "noble" material oil in the ground for posterity. Others would say that Saudi Arabia had been producing more than it wished, for the good of the world economy.[11] Such statements are refuted by all available evidence. Others alleged that the companies depleted the oil too quickly. In fact, Persian Gulf extraction rates were the lowest in the world.

Continuity and Change in Iraq

Another Cycle of Mutual Mistrust

After the 1927 Kirkuk discovery, tension between the Iraqi government and the resident Iraq Petroleum Company (Exxon, Mobil, Shell, BP, CFP) followed a cycle of mutual mistrust, similar to the one in Mexico.

On the eve of World War II, production in Iraq was less than half that in Iran, which doubled again by 1945, while Iraq had recovered only to prewar levels, having slumped when a pro-Axis regime took over briefly. After 1945, growth had to be shared not only with Iran but with the new countries, chiefly Kuwait and Saudi Arabia. Each of the IPC companies compared expansion in Iraq with expansion elsewhere. Even Shell, with no other Persian Gulf production, had long-term contracts at very low prices with BP and Exxon for non-Iraq oil. By the late 1950s, in part because of the hiatus in Iran, Iraq had caught up. But the Iraq monarchy demanded that IPC produce more and that it relinquish some of its concession, which extended throughout the country. Then other companies could enter and would produce more because they had no divided loyalties and were not pledged to produce in other countries. These demands were pending when the monarchy was overthrown in 1958. The new regime pursued them.

In 1961, "Law 80" stripped IPC of its concession area, leaving it only the existing operating fields, plus the newly found North Rumaila field.[12] The government tried to use North Rumaila development as an inducement for IPC to expand production more rapidly but failed, and it nationalized North Rumaila in 1962. The dictator Kassim was overthrown and killed in 1963 by the Baath (Arab Socialist) party, headed by Saddam Hussein, although until 1979 his title was only vice president.

Iraq tried to get outside companies to explore in the IPC concession, and there were discussions with at least half a dozen, including several American firms. The U.S. government discouraged both American and other oil companies from seeking to drill and produce in Iraq. (Many relevant documents were later declassified and printed in [8 Church 529–590].) The American concern was partly Iraqi cooperation with the Soviet Union after 1958 and partly reconciling the conflicting demands of the Persian Gulf countries, each constantly demanding that its resident company produce more. Additional output from Iraq would make a hard task even harder. There is no evidence that the State Department was trying to protect oil company profits. There is abundant evidence of its concern that the more was produced in Iraq, the fewer outlets there were for Iran and Saudi Arabia and the greater was the threat to their existing good relationship.

Iraq itself was partly or mostly to blame for no entry. It demanded more than companies were willing to give. In 1967, the French were invited into the confiscated area, ostensibly because of their friendly attitude toward the Arabs during the Six-Day War. A concession to the French national company Elf-ERAP "was ratified with great pomp in Baghdad [and hailed throughout] the Middle East as a great victory over Anglo-American imperialism."[13] [*Le Monde*, weekly ed. 11-23-67] But nothing came of this overture. Elf-ERAP found three fields but received no development rights on any, or on North Rumaila. Iraq was too grasping for "even the anxious French" and found no comers. [PIW 12-18-67:3] Negotiations with Italy also failed because Iraq asked too much. [8 Church 554] The new Iraq National Oil Company needed Soviet help to get started.

After the 1961 Law 80 seizure, IPC investment dwindled. By 1968 exploration and drilling activities had "almost ceased." Only one oil well was completed in 1968 and none in 1969. [WO 8-15-70:170, 188] At the end of 1971, no rigs at all were operating in the IPC area. [AAPG *Bulletin* 9-72:1805] Output increased but not nearly as fast as elsewhere in the gulf region. By 1971, ten years after Law 80, Iraq oil production was up from 354,000 to 624,000 barrels per day, but its percentage of the Middle East was down, from 18 to 10 percent. In 1972, as we will see in chapter 5, Iraq raised its excise tax; IPC cut output and was expelled.

The U.S. Government Attempt at Reconciliation

As the 1960s drew to a close, the industry and the State Department increasingly worried about reconciling the governments' conflicting demands for higher output. They wished to rebut the impression that American companies were to blame for not expanding fast enough in Iran and Iraq because they preferred to produce in Saudi Arabia.

Early in 1968, the premier of Iran denounced the consortium plan to expand by only 6 to 7 percent. "We cannot permit our wealth to remain below ground. [The consortium plan] would jeopardize the whole five-year development plan." [NYT 3-9-68:10] A year later, Iran threatened to seize "a substantial share of oil production within her borders unless the [consortium] ... voluntarily [*sic*] expand their production" by at least 10 percent per year. [NYT 4-16-69:7] (We will see that the pressure paid off in early 1972.)

A memorandum by Secretary of State Rusk in January 1968 cited the opinion of Mobil, Exxon, Chevron, Texaco, Gulf and BP: "Increases in Libyan production will make it exceedingly difficult [to] meet [the] estimate of 9% increase [in] Consortium off-take this year." In the next month, he drew attention to the fact "that greatest percentage increase among Gulf producers in 1968 will be not Aramco but Abu Dhabi offshore, where British Petroleum and CFP" were the only producers. Therefore Aramco, and the US government, should not be blamed for Iranian frustration "when there is [no blame] to shift." [8 Church 590]

The U.S. government sought to maintain the status quo and reconcile the Persian Gulf governments to the division of markets, to prevent their taking action against the companies that were trying to prevent more output than could be sold at current prices. This was consistent with the U.S. government's support for commodity agreements. [NYT 2-6-68:55]

In 1969, President Nixon, after a private meeting with the shah, asked the consortium companies to produce more in Iran. They refused. Later that year, the shah made another such request, but this time the administration refused. "An increase in Iran's quota was possible only at the expense of Saudi Arabia and other friendly Persian Gulf countries." [Kissinger 1982, 857]

The 1970 Oil Market: All Signs Down

The cost of expanding Persian Gulf and Venezuela reserves and capacity was so far below the 1970 price that continued competition could mean only continued price decline.

Low costs are not disputed but are often ignored or waved off as somehow not applying to "a limited resource." As I showed in chapter 2, there is no such thing.

"Proved reserves" are inventory. The industry rule of thumb for efficient operation is a fifteen-year reserve-to-production ratio. In the Persian Gulf area it was over sixty-five years. But companies without reserves wanted to build their own inventories: "New sources of crude are sought ... vigorously, because it still costs less to produce crude than to buy it. The margin between the cost of production and ... even discounted selling prices is still relatively large in ... the Middle East and North Africa." [PPS 5-70:162]

To see what would happen if Persian Gulf reserves were put under great strain, I did a crude simulation in 1966. [PPS 1966] Assuming a stronger than previous increase in demand and zero new discoveries, investment cost would rise little between 1966 and 1985.[14] As seen in the preceding chapter, the value of a developed reserve in the ground depends on expected prices, less extraction costs. Reserve values in the United States (the only place where reserves were freely traded) stayed level in nominal terms (falling in real terms) between 1946 and 1970.

Another leading indicator is the level of prices on long-term contracts as compared with current prices. The longer-term offers or deals tended to be slightly below the current spot price. [Adelman 1972] In 1967, no one wanted "to sign up long-term contracts except at far lower prices." [PIW 9-18-67:4, 8-21-67:1]

In the United States, the standard contract for heavy (residual) fuel oil delivered to electric utility plants gave the buyer the right to a lower price if oil was offered by another seller; it gave no corresponding right to the seller to receive a higher price.

In 1969, the current spot price was $1.27 per barrel. The shah offered to sell the United States 1 million barrels daily over ten years at $1.00 per barrel. [Kissinger 1982, p. 857] This 21 percent discount equated roughly to prices declining at nearly 5 percent per year—faster than the just over

3 percent decline from 1960 to 1969. An examination of the trade press shows that no higher prices were expected [PPS 9-70:318], a conclusion confirmed by interviews with oil executives.[15] [Helfat 1988, p. 82] Thus every economic measure and indicator pointed to stable or decreasing prices.

Some internal Chevron and Exxon documents reprinted by the Church committee [8 Church pp. 592–604] confirm the burden of plenty, current and expected, far past 1970. The Chevron group warned in December 1968 that "there would not be a market for the production increases desired by the governments of Saudi Arabia and Iran without displacing production expected by the governments of other producing countries." [8 Church 760] In February 1970, "growth in production at the rates desired by Middle East governments for 1970 would be difficult to achieve. A considerable margin of surplus productive capacity would continue to overhang the Free Foreign [non-US, non-Communist] crude market." [8 Church 756, 759] They did not mention excess capacity in North America. (See appendix 3B.) In September 1970, cutbacks in Libya were factored into the detailed estimates, but there was no change in outlook. [8 Church 755–756]

Later in 1970, the near-term situation had eased but not basically changed. No barriers to expansion were seen. Aramco capacity, then at 8 mbd, was expected to reach 21.1 mbd (including 1.1 mbd natural gas liquids) by the early 1980s.[16] [7 Church 452–453, 519–520, 560, 587]

Exxon's internal supply-demand assessments over the 1960s parallel Chevron's and confirm Page's vision of the continuing pressure of excess supply. The 1968 document forecast continued loss of market share for the "international majors." They would lose slightly in the Middle East and Caribbean, but Libyan output was slated to rise "dramatically.... In contrast with historical ownership patterns in North Africa and the Middle East the bulk of the new increment of production will be produced by companies considered 'newcomers' to the international oil trade without established captive outlets and without a significant stake in the Middle East. [Accordingly, Persian Gulf production] is expected to fall sharply." [7 Church 602–603] The 1970 assessment continued to expect "an inordinately low growth rate available for the Middle East as a whole that must be further divided among the new Middle East concessions as well as the established producers. *No known method of allocating the available growth is likely to simultaneously satisfy each of the four major established concessions, i.e. Iraq, Iran, Kuwait and Saudi Arabia.*" [7 Church 606, emphasis added] Their only comfort was the hope of decreasing supply nearly twenty years later, by the late 1980s.

The Producing Nations: Frustrated and Threatened

The market was working. Rivalry among the old majors and the newcomers had worn away most of the old monopoly margins. Current prices were still far above long-run marginal cost, ample to attract large new investment to increase production in new areas, displacing less efficient fuels, and helping world economic growth. But from the viewpoint of the producing governments, the market was working much too well.

"Dependence" and Long-Run Frustration

The disappearance of U.S. exports meant that the world price was no longer limited by Texas-plus. A monopoly could have charged much more, even in 1948. Since then, the world had grown much more dependent on oil from Venezuela and the Persian Gulf. But the dependence was not being exploited; quite the contrary. Far from rising, the price had dropped. The gap between potential monopoly price and actual market price was enormous, and growing.

Threat to the Excise Tax Structure

By changing the income tax to an excise tax in 1960, the OPEC nations had insulated themselves from the price decline. They had even raised per barrel taxes. But the buffer between price and tax-plus-cost was the companies' margin, which had been greatly eroded from both sides.

A glance at table 3.4 shows the problem. The price was $1.21 and the excise tax 88 cents per barrel. The 11 cent investment-operating cost was truly a cost, and as such not reducible. That left only some 22 cents margin between tax-plus-cost and price. The price decline since 1960 had been about 4 cents per barrel per year. At that rate, a crisis was due in about five years. If the governments acted as individuals, they had to reduce taxes. If they acted in concert, they might increase prices, directly or by controlling output.

The prospects appeared poor. OPEC had established an output allocation plan once, in 1965, but the member countries had not paid much attention to it. The resolutions of June 1968 stated that members would not be bound by past contracts. They claimed "participation" or at the limit expropriation. Saudi Arabia announced it would seek 50 percent ownership in Aramco, and Kuwait soon followed this example, although no dates were set. [NYT 10-8-68:15]

Expropriation presented some obvious dangers. Ahmed Zaki Yamani, the Saudi oil minister, was not the least of those who saw the pitfalls of sellers' facing customers without the intermediation of the multinational companies. Wherever the governments had had direct responsibility for selling oil, they had been too ready to discount. [Adelman 1963; Yamani, in PIW 6-16-69; Mikdashi Cleland and Seymour 1969, pp. 211–233; Schurr and Homan 1969, pp. 140–151]

In June 1970, OPEC passed resolution XX.122, calling for "rational increases in production from the OPEC area to meet estimated increases in world demand during the period 1971–75." The plan was not even mentioned in Chevron and Exxon memoranda, and one trade journal did not take it very seriously. [PPS 9-70:318] It was no wonder, since those same governments were pushing their companies and the U.S. government for higher output and for relinquishment of areas to let out to newcomers, so they could produce even more.

In February 1971 Chevron sounded a new note: "The oil industry is currently faced with a possible cutback or embargo on production and exports by the OPEC nations," although a "cutback for any extended period of time" was viewed as unlikely. [7 Church 751] Exxon's 1972 assessment was that after 1980 "governments will limit production as a means of avoiding any future sharp declines in oil revenues." [7 Church 606] Neither Exxon nor Chevron expected actual price increases, but the reversal in outlook is startling. In the next chapter, we see how it happened.

Appendix 3A: The Special Case of Venezuela

Venezuela is a mild, or even doubtful, case of the cycle of mistrust because its output was limited by an outside force: U.S. import policy. Unlike the Persian Gulf countries, it had long had several producing companies, but Exxon, Shell, and Gulf accounted for 80 percent of production. [PPS 2-70:62]

Venezuela had a freight advantage on shipments to the United States but a disadvantage on shipments to the Eastern Hemisphere. The government came gradually to frown on shipments to the latter since they required a lower price, which might be generalized to all buyers. Accordingly, Venezuelan output was limited largely to what it could sell in North America and subject to the informal, then "voluntary," then mandatory quotas.

The long-time minister of hydrocarbons, Perez Alfonso, later the father of OPEC, was in favor of restraining output. So was his successor, Perez Guerrero, who thought more revenues meant more wasteful spending. But theirs was a minority view. Successive governments, dictatorial and democratic, tried to have more oil admitted into the United States. Throughout the period, Venezuela maintained a considerable presence in Washington and tried in vain to obtain better treatment under the quota system, in particular, equality with Canada and Mexico. They were supported by Exxon and by some of the operating companies. [Shultz et al. 1970, Appendix N]

In 1956, Venezuela granted new producing concessions. Some new companies made good discoveries, but the climate for exploration and development was poor. First, the market was limited. It had been expected that U.S. production would decline, hence imports would rise even more rapidly than consumption. But U.S. domestic production kept rising through 1970. The second damper on foreign investment was that from 1958 there were no more new concessions. By 1970 nationalization under existing law was approaching. Between 1975 and 1985, 79.2 percent of concession acreage was to be turned back. Another 13 percent was to be surrendered by 1989 and the remaining 8 percent by the end of the century. [Perez 1969, unnumbered table, "Expiration of Concessions"] As in the Persian Gulf region, "In the government view, existing concessions had not been adequately explored, while the companies contended that they had." [PPS 2-69:55]

In 1960, the government proposed a system of service contracts, but the terms were not actually set out until 1968. Venezuela would pay the company a fee per barrel or a percentage of the offtake for services rendered. The company would no longer decide output volumes. [NYT 3-26-68:59] It took two more years for the Venezuelan congress to pass the authorizing law. [PPS 9-70:344] No such contracts were ever proposed. The discussion stopped in 1960, to resume in 1991.

With little prospect of expansion, drilling fell off rapidly, from 1,200 wells completed in 1958 to 300 in 1967, and exploratory wells declined even faster than development. The reserve-to-production ratio fell from 17:1 to 12:1. [PODE, various years] This "proved" that Venezuela contained little oil. The decline in drilling and reserve additions lasted until after the 1976 nationalization.

Appendix 3B: U.S. Excess Producing Capacity

After World War II, there was large-scale excess capacity in the United States, due mostly to the system of market-demand prorationing, which induced much unneeded investment. But this excess gradually shrank, precisely because the real price was gradually declining.

I believe (but cannot prove) that there was an understanding in Congress and the executive that the oil-producing states would no longer raise the price; in return, imports would continue to be restricted. Whatever the reason, the decline in real prices checked the endless upward spiral that such schemes impose. Higher costs "justify" the authorities' raising prices, which in turn induce more wasteful investment, which raises costs, and so on. Because of the price cap, there was less wasteful drilling and also a sloughing off of the most uneconomic projects. The one-time decrease in cost per new barrel of reserves developed was partly a movement along the curve, partly a shift. [Above, p. 22]

The disappearance of U.S. excess capacity was not important. The Chevron and Exxon assessments never mentioned it. One reason may have been that the capacity was unreliable. During the Suez crisis in autumn 1956, the Texas Railroad Commission had refused to allow higher output. While additional oil could be shipped out from some areas readily, it would have accumulated in other areas, which lacked the gulf-bound pipeline capacity. To allow all areas to expand equally would aggravate the surplus; to allow some areas to increase production more than others was "unfair." Justice prevailed. Higher production was not permitted until February 1957, when West Texas inventories were drawn down, and prices had risen. By that time, the crisis was over. [Adelman 1972, pp. 157–158] By 1970 there had in effect been no available surplus capacity for nearly fifteen years.

Notes

1. In its early years, U.S. oil refining was monopolized by the Rockefeller group, which became the Standard Oil Company. It was broken up by the U.S. Supreme Court in 1911 and the entry of newcomers.

2. Under "market demand prorationing," the state limited total production and assigned each well its share or "allowable." This was an incentive to drill unnecessary wells for additional allowables, hence to install excess capacity, which raised costs. Where tracts were large and reservoir subdivisions few, as on much of the federal lands and particularly offshore, the problem was small. It was minor but nonnegligible in Canada and nonexistent elsewhere.

3. This fact is often interpreted to mean that a price war could easily develop in any given locale because a seller "could afford" it. But sellers do not cut a price because they can afford to, but because it pays.

4. The Japanese strike on Pearl Harbor, triggered by an oil embargo, should have aimed primarily at the oil storage tanks: large, immobile, and impossible to miss. Their destruction would have pushed the U.S. Pacific Fleet 3,000 miles eastward and, according to Admiral Nimitz, prolonged the war another two years. This outcome should have been apparent to Admiral Yamamoto, who planned the attack, and had once commanded the Yokosuka naval air station. [Agawa 1979; Wilmott 1982] Like many other decision makers before and since, he and his advisers missed what was right in front of them.

5. As explained in chapter 2, the reciprocal of the Herfindahl index is the number of equal-sized firms that would generate it.

6. The Church hearings are cited by volume number and page; for example, "2 Church 3–4" means volume 2, pp. 3–4.

7. The belief, true or false, that the restoration was mainly due to American action, "contaminated America's relations with the Islamic Republic of Iran following the revolution of 1978–79." [Bill 1988, p. 86] But this cannot explain why the United States is viewed as the Great Satan. It "had come to bear the brunt of an increasing clash between Western and Islamic cultures.... The radical Islamic clergy had come to view the United States itself as a greater danger to Islam than any other external power had ever been—and certainly more than the Soviet Union could ever be." [Fuller 1991, p. 252]

8. There had been contacts between Iran and Venezuela in 1947 and an Iraqi-Saudi agreement in 1953 whose terms were not revealed. [Mikdashi 1982, p.23] The members at the end of 1992 were Algeria, Ecuador (since resigned), Gabon, Indonesia, Iran, Iraq, Kuwait, Libya, Nigeria, Qatar, Saudi Arabia, United Arab Emirates (Abu Dhabi, Dubai, Sharjah), and Venezuela.

9. I pointed out that the excise tax plus cost was a floor to price. [Adelman 1963, 1964] Making overallowance for rising incremental cost with continued heavy depletion and a guess at continued rent retention, my opinion was that the price (in 1963 dollars) would approach $1 per barrel. It actually went lower by 1969. I did not expect the OPEC nations to pass from the defense of price to a successful offensive.

10. Unlike an embargo, a cutback in output might be serious; I suggested a strategic reserve, including coal stocks for electric power plants. The suggestion was not taken up anywhere. The paper was published in French, Italian, and Japanese; but not in English. For an excellent summary, see PPS [1968a].

11. "During this period [1960–1971] there were also pressures [on Aramco] from the King of Saudi Arabia to increase liftings of Saudi Arabian crude oil to increase the revenues of the Kingdom." [Irving 17:30]

12. The new regime also claimed Kuwait and mobilized on its borders. The British flew in a brigade, about 5,000 troops, and the threat ceased.

13. The French, *Le Monde* continued, had not flinched when mighty IPC "showed their teeth." The Anglo-Saxons had been outmaneuvered; they could not block a French company "without provoking a grave political crisis." It served them right because "on the morrow of the last war they would not let France into the game in this region." This rodomontade, from a sober anti-Gaullist journal, tells us much about French opinion.

14. The estimates were briefly summarized [PPS 1966], which drew on extensive industry contacts, particularly in British Petroleum. No one raised any objections to them.

15. "This study assumes that, during the pre-embargo period, petroleum companies expected the growth rates of real prices and costs to equal zero." Nothing in the published literature suggests that anyone expected real oil price increases. "Although executives may have anticipated the sharp price increases, they denied this in interviews with the author." [Helfat 1988, chap. 4]

16. Later allegations that Saudi fields had been overproduced were inconsistent and unsupported. [Adelman and Ward 1980]

4 Price Reversal, 1970–1971

After a twenty-five-year decline, the world price of oil began to rise in 1971 (fig. 4.1). No strain of demand on supply explains the reversal. New reserves were being created, at costs and prices that provided handsome to lush returns on investment. Long-term contract prices tended to be lower than short-term ones. Reserve values, reflecting future prices, were stable to declining.

The prospect of continued price decline was a reason for action against it. The ball was in the governments' court. Their excise tax was the principal price support and was about 73 percent of it. Another 12 percent was cost. The companies' remaining margin would be gone in a few years if the price kept ebbing at the 1960–1970 rate. Yet the governments' constant struggle for larger market shares had helped push down the price (fig. 4.1).

The mutation of 1970–1971 was the producer governments' turn from increasing their share of revenues to the much bigger and more rewarding task of raising the floor to the whole price level. They found, early on, that the U.S. government would support them.

Until 1971, the producer governments and the U.S. government were worried over future excess supply. So were the oil companies, at the center of the market. But from February 1971, the Chevron and Exxon memoranda cited in chapter 3 point to a new willingness or ability of producer governments to restrict output. We need to explain it.

Continuity in Libyan Oil Policy

The first break was in Libya, where exploration had begun in the 1950s. In sharp contrast to the Persian Gulf, the monarchy let out concessions to five majors (Exxon, Mobil, BP, Texaco, Chevron, and later Shell) and several independents, most notably the Oasis group (Continental,

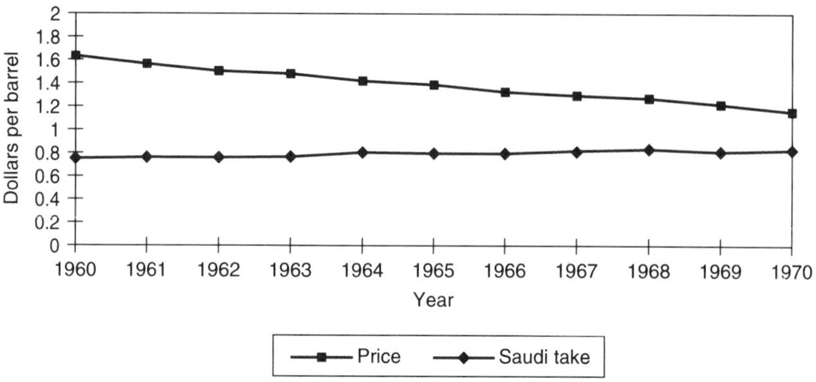

Figure 4.1
Arab Light, market prices, and government take, 1960–1970. Sources: Arab Light price, PIW 4-12-82:11; Saudi take from [PPS(PE)], various issues, 1961–1971.

Marathon, Amerada Hess), and Occidental shortly after. In Libya, unlike the gulf, there was no limit on any company, and investment and development went ahead with unprecedented speed. By 1965, output exceeded 1 mbd; by 1970 it was over 3 mbd.

In November 1965, the monarchy unilaterally changed the tax code to conform with the OPEC system. Taxes were based on posted prices, not on actual market prices. A tax based on fictional income became in effect an excise tax, in cents per barrel.

The majors were not displeased, since the newcomers with low market shares had been the leaders in price erosion. When the independents protested that the monarchy was violating their contracts, they "were rolled in carpets and all their bones broken." They gave in "when it became obvious that it did not matter whether they were going to or not." [*Economist* 1-14-66:232] The 1966 "agreements" did, however, stipulate that no further tax increases were to be made without the companies' approval.

As noted in chapter 3, the Libyans and the Saudis in 1967 wanted to raise taxes—by raising posted prices—in the Mediterranean, to take advantage of the tanker shortage and higher spot prices there, but no action was taken. [OGJ 8-28-67:46]

The demand was made soon thereafter by Libya alone, despite the 1966 agreements. This was a breakthrough, for it went beyond the OPEC formula. By raising posted prices and eliminating some discounts, it would raise the excise tax. [PIW 8-7-67:6, 8-14-67:6, PPS 3-68:89] While this was being considered, the companies agreed to accelerate tax and royalty payments. In return, the government retroactively approved prices that had existed since 1961 but that it had called unfairly low. [PPS 11-68:423] This commitment lasted nine months. In a press interview in August 1969, the Libyan oil minister threatened that unless postings and taxes were increased, the government would claim payments for underpricing since 1961. [PPS 9-69:344]

The next month, the regime was overthrown by a group of young army officers who established what they called "Arab socialism" similar to Egypt and Iraq: a one-party dictatorship with a controlled economy, a large public sector, and a militant foreign policy, cooperating with the Soviet Union.[1]

The U.S. State Department in 1970 viewed Gadhafi as "a fanatic anti-Communist." [Akins 1973, p. 470] When opponents tried an early coup, the U.S. embassy in Tripoli warned him in time. The dominant view in the U.S. government was that "the real danger of radicalization resided in our *opposition* to Gadhafi." An interdepartmental report said, "Libyan oil ... is literally the only 'irreplaceable' oil in the world." [Kissinger 1982, p. 860] This was obviously false and was soon proved as such.

The advent of the Gadhafi regime has often been called a watershed or break with the past, but in fact it continued the policy of the monarchy: first hit the more vulnerable independents, then make the others match the terms, then disregard the contract, and start again. One need not be "radical" to be rapacious.

The new regime nevertheless probably had more oil know-how. The new prime minister was a lawyer who had previously worked for Exxon, and the new petroleum minister had an engineering degree and had worked for Oasis. [PPS 10-69:384] Moreover, at this time, Algeria was engaged in similar negotiations to raise the posted prices of the French companies operating there "and sent an oil delegation led by a senior government technical adviser to ... co-ordinate the two countries' efforts to secure higher revenues." [PPS 3-70:102] The Libyans could tap into eight years' Algerian experience, which is worth review.

The Algerian-French "Special Relationship"

The 1962 agreements establishing Algerian independence granted Algeria such high taxes on oil production that only the high internal French price level would make it profitable for the French and the few foreign operators. *Le Monde* agreed it was expensive but worth something to "shake up" the supposed cartel of the multinational companies. [LM 8-29-62] In

1965, when the treaty was made even more favorable to Algeria [LM 7-30-65], new demands were made. By 1970, most non-French companies had left, and Algeria made a unilateral increase in the per barrel tax from 79 cents to $1.21. *Le Monde* called this a clear violation of the accord and regretted the breakdown of what had seemed a promising new area of producer-consumer cooperation. The foreign editor wrote a long review of Franco-Algerian oil relations in the framework of world politics (which is worth reading, decades later).

In 1962, de Gaulle had reckoned he would keep Algeria inside the *mouvance française*. It would show the world a new type of relation between metropolis and former colony and would be an implied rebuke to the United States, which had just gone through the Cuban missile crisis. It gained him great credit:*

> From one end of the third world to the other, they applauded the boldness and generosity of him whom they gladly honored, in the Arab countries, with the title of "sheik." Formerly the butt of UN resolutions, France now had a starring role. [But all the principal elements of the accord were disregarded.] Harassment and plunder of all kinds, disregard of many agreements, make the cost look heavy of the self-imposed French burden of buying Algerian crude oil at prices above market, importing unneeded wine, and giving Algeria considerable economic aid of various kinds. [Fontaine 1971]

They could neither explain the policy nor give it up.

In January 1970, a Libyan-Algerian meeting issued a joint communiqué calling for close cooperation. Libya then demanded a 44 cent increase in posted prices for its oil, about 22 cents more excise tax, to equate to the Algerian demands on France. [OGJ 4-27-70:49] This was 12 cents higher than the last demand by the monarchy. Frequent consultations between the two continued throughout 1970. In July, Algeria confiscated the holdings of Shell and Phillips and announced a unilateral increase in posted prices, which amounted to an increase in the excise tax of approximately 38 cents per barrel. "Because of the U.S.... interest in gas imports, the U.S. government had reportedly been reluctant to make representations on behalf of American companies whose properties were nationalized by Algeria." [6 Church 2–5]

Natural gas price controls in the United States had depressed supply and increased demand. The shortage, to policymakers, "proved the need" for natural gas imports, hence the "need" to please the Algerians, so that they might be willing to sell more gas. The ruling dogma is that a seller

---

*DeGaulle's voice was like no other. One longs to hear him declaim Le Fontaine's fable of the crow and the fox, and the moral: it is costly to listen to flattery.

allocates oil or gas by grace and favor, by his good or ill feeling about a buyer.[2]

Libyan Squeeze

In May 1970, Libya proposed to follow the Algerian example. It would set the posted price, hence the per barrel tax. [PPS 5-70:182] Other producer governments warned that any such price/tax increase "would trigger similar moves elsewhere. Anything Libya gets, the others want too." [OGJ 6-8-70:3]

On May 3, there was a break in the Syrian section of the Trans-Arabian Pipeline (TAP), which carried 875 thousand barrels daily (tbd) from Saudi Arabia to the eastern Mediterranean. Normally it would have been repaired in twenty-four hours, but Syria refused permission to repair, demanding higher transit fees. The Saudis, whose oil was thus interdicted, threatened to cut the subsidy to Syria, as a front-line state against Israel [OGJ 6-1-70:33], but their threats were not taken seriously.

The Trans-Israel Pipeline (TIP) bridged the gap between the gulf and the Mediterranean. Startup capacity was 400 tbd; in June, an increase to 800 tbd was announced, with 1.2 mbd planned for January 1971. [OGJ 6-8-70:NL3, 69] But to placate Arab opinion, use of TIP to relieve the shortage in Europe was ruled out. Spot tanker rates jumped, and the price of Libyan oil increased, but term tanker rates did not change, for as additional shipping was drawn in and the break was repaired, spot rates, and delivered prices with them, were expected to head down quickly. In January 1971, this is exactly what happened. Thus it was foreseeable and foreseen that the extra Mediterranean premium was temporary. Still, the demand was for a higher permanent tax.

On June 12, before negotiations even began, Libya imposed production cuts, amounting to 600 tbd by the end of July, on Occidental, Oasis, and Amoseas (Texaco/Chevron). Occidental was hit most heavily, losing 300 tbd out of 800 tbd in Libya, nearly nine-tenths of its worldwide production. The accusation was that it had been overproducing the field. As in 1966, "Libya has certainly applied [production cuts] where they will hurt most." [PPS 8-70:280] Libya still expected to expand output. In early June, it demanded that operating companies increase drilling "or face legal action." [OGJ 6-8-70:63, 6-15-70:43]

In August and again in September, additional cuts brought the reduction to 660 tbd, nearly 18 percent of the total. Libyan revenue losses were about $241 million annually, compared with actual revenues of $1,320 million.[3] But Occidental's cuts were nearly half of its output. "Final" offers for higher excise taxes (via new higher posted prices) were announced. Everyone was ready for "some dramatic move ... particularly in view of the close coordination ... between Tripoli and Algiers." [PPS 9-70:339]

By early October, all the companies had agreed to higher taxes. The usual figure mentioned for tax increase is 30 cents per barrel. By cutting production and revenues 17.8 percent for about five months (an annual rate of 7.4 percent), Libya gained many times its investment.

This was a milestone. First, the 1960 fictional posted price was not only maintained but actually increased. Second, the percentage of the fictional income (posted price minus costs) was also increased, from 50 to 55 percent, for the first time since 1950. Finally, the higher-priced crudes FOB Libya were in close competition with Saudi and Iraqi crude oils "lifted" (loaded) at the eastern Mediterranean, whose prices would be promptly raised. Hence the Libyan success meant a spread of the rise to the Persian Gulf and a crisis there.

How the Libyan Squeeze Prevailed

Had the U.S. government regarded concerted price raising as a threat, it could easily have convened the oil companies to do what they did six months later (and too late): work out an insurance scheme whereby any single company forced to cut back would receive crude oil from the others at tax-plus-cost from another source. Had that been done in June, losses to the Libyan regime would have been gains to all others. This situation would have severely tested OPEC unity at a time when its members were unprepared for cooperation. The Libyan regime had been in power for barely a year. Any Libyan commander of a division or even a regiment could consider how he might gain a billion or more dollars a year by issuing the right marching orders.

In July 1970, as we now know, Occidental appealed to Exxon to make good its loss in Libya by supplying Persian Gulf oil, at tax-plus-cost, plus "a reasonable profit, such as ten percent." It took two weeks for Exxon to respond with an offer of oil at "market price." [Hammer 1987, p. 345; Sampson 1975, p. 253] On its face, the offer was meaningless, since anybody can buy at market price. But it probably meant an undiscounted "market price," higher than what was being realized in the open markets.

We do not know whether Exxon was afraid to help Occidental, or wished to damage it, or had some other reason. Nevertheless, Occidental's request proves that Persian Gulf oil was an acceptable substitute for Libyan oil. Of course Persian Gulf would have imposed higher refining costs, but the increase would have been small in relation to the price. Oil at Persian Gulf "market prices" was not as good a buy as Libyan oil even at the new high prices, especially since the lack of resistance meant the Persian Gulf "market prices" would shortly rise.

In addition, the companies might have tried to block or diminish the sale of their Libyan oil in Europe. We do not know what success they would have had. Confident State Department predictions to the contrary notwithstanding [5 Church 6–7], nobody knew. But the heads of Shell and BP said that "without Libyan oil the companies should be able to supply Europe with 85 to 90 percent of its needs for at least six months, probably without need of rationing; by that time there would probably be either a settlement, or further alternative sources." [Sampson 1975, p. 214] There was, in short, enough additional producing capacity already in existence or available on short notice.

In mid-1970 there was no sign of the OPEC countries' attaining a solidarity that had entirely eluded them so far. Therefore the policy question was, since resistance would probably succeed, should we resist?

In fact, the U.S. government continued its support of the Libyans. James E. Akins, head of the State Department Office of Fuels and Energy, testified that he considered the Libyan demands "reasonable." [5 Church 6] It is his one-word summary of what he said at "frequent meetings with the Libyans" during 1970. [Akins 1973, p. 472] We do not know what he actually said; no papers have been made public.

Allowing for quality and transport cost, Libyan oil was priced below Persian Gulf oil. If it was "underpriced," by the same token Persian Gulf oil was overpriced. In fact, the long-continued price decline and the chronic excess of offerings over demands meant that both oils were obviously overpriced, although Libyan oil less so. To call the Libyan demands reasonable implies a policy premise; oil prices must be raised. The next chapter will show in detail that this was indeed U.S. policy. Akins's argument that supply could be assured by a redress of Arab grievances against Israel echoed Gadhafi himself. The Libyans would have been dull people if after "frequent" meetings they had not known that the U.S. State Department favored their demands.

The September 20 Meeting

Occidental agreed to the higher taxes on September 4. [5 Church 11] Three days later, some major oil companies had a meeting with the secretary and undersecretary of state, and Akins, who did not remember who was there or what was said. [5 Church 12] Akins failed to mention a meeting at State on September 20, attended by representatives of the seven major companies. "For the first hour, the oilmen were lectured [by Akins] about the problem of Jordan and the Palestinians, with the implication that *a Middle East settlement would also settle the oil problem.*" [personal interviews, in Sampson 1975, pp. 214ff, emphasis added]

Sir David Barran, managing director of Shell, argued that appeasement was even more dangerous than resistance. He predicted "an avalanche of escalating demands from the Producer Governments and we should at least try to stem the avalanche.... The U.S. Government officials were not at all convinced ... whereas of the oil company people present, some were less than others and some not at all." [Barran to Church, August 16, 1974, in 8 Church 771–773] Akins, speaking for State, opposed resistance. "Akins, according to one of the British oilmen 'was hypnotized by the Saudi Arabians. He said that there was no question of Saudi Arabia following Libya. I said you must be joking and nearly walked out.'" [Sampson 1975, p. 215]

In the end, nothing was done. The discussion was vague and inconclusive. As Barran wrote later: "I really cannot express any authoritative or informed view on why other companies, and others such as Mr. Akins, should have formed a different opinion from ours on this issue." [Barran to Church letter]

The Libyan oil demands came when the State bureaucracy was trying to change U.S. policy on Arab-Israeli relations (see appendix 4A). State believed that its "1957 solution" would solve the oil problem in Libya and elsewhere—above all, in Saudi Arabia. Officials did not explain why anything done by or to Israel would lead the Arabs (and non-Arabs) to produce any more oil or charge any less, but the idea remained fixed.

From Tripoli to Tehran

The new Libyan taxes were immediately matched for Persian Gulf oils lifted at the eastern Mediterranean; the OPEC nations demanded the new terms everywhere. At the beginning of December, the Venezuelan Congress raised taxes by setting a higher scale of posted prices and a 60 percent rate on the fictional income tax. [PPS 1-71:4] The OPEC countries held a special meeting in Caracas December 9 through 12, calling for a general 55 percent tax rate. They demanded negotiations within thirty-one days with the Persian Gulf producer nations and resolved on "concrete and simultaneous action." [MEES 12-18-70]

But even before negotiations began, Libya indicated that it would demand still higher prices and taxes. [MEES 1-1-71] If the gulf countries secured 55 percent of the fictional income, that would match what Libya had received in September. They had to have something more. [Cf. 5 Church 80] This was called "leapfrogging." Still, there was no threat of cutoff of supply by anyone in OPEC even as late as January 13. [NYT 1-14-71:2, interview with Iran finance minister; WSJ 1-14-71:2] The date is significant.

Locking the stable door in the hope that some horses were still inside, the Libyan producers now entered into a safety net agreement providing that if one company was forced to cut production because of its resistance to further demands, others would replace it with production in Libya or the Persian Gulf, at or near cost. [Hammer 1987, p. 349; 5 Church 81; 6 Church 224–230] There seemed to be the elements of a coherent defensive strategy: avoid leapfrogging by insisting on global negotiations, and support the independents. The safety net was, in the view of one independent, "like any insurance policy.... [W]e thought we would never have to collect on the insurance because we were going to have solidarity of approach." [5 Church 82]

Even this late in the day, a show of solidarity had an effect. The Libyans had threatened "appropriate action" against Occidental and Hunt-BP if they failed to meet a demand for higher payments by January 24. But when the deadline passed, nothing happened. [OGJ 2-1-71:38]

On January 7, a joint company–State Department meeting resolved on a plan for joint negotiations. "Our best hope ... would lie in the companies refusing to be picked off one by one in any country and by declining to deal with the producers except on a total, global basis." [Barran, in 8 Church 772]

On January 16, the Persian Gulf–producing companies submitted their proposals for higher taxes. [NYT 1-17-71:1, 1-19-71:2] The next day, Undersecretary of State John Irwin II, in office only a short time, arrived in Tehran. His briefing was mostly by Akins. [5 Church 145–174] On January 17, Irwin met with the shah and his finance minister. A personal letter from President Nixon to the shah was not delivered because it was feared the shah might not like it.

Irwin set forth the plight of Europe and Japan if oil supplies were cut off. [PONS 3-1-71, but referring to January 17] Surely he did not mean to incite to riot, but he did just that. There is hardly a stronger inducement to act against your opponent than knowing that it will damage him severely.[4] Before this reassurance there had been no threats to cut off oil;

afterward there were plenty. After talking with Irwin, the shah rejected any idea of OPEC-wide bargaining. Insistence on it was a sign of "bad faith." But he promised that if the companies bargained with the gulf producers only, the latter would abide by their agreement, no matter what others did. [5 Church 147–149] Irwin did not even wait for the arrival of the two chief company negotiators two days later but immediately accepted the shah's assurances. He cabled Secretary of State William Rogers that the companies should bargain separately with the gulf countries. Rogers immediately endorsed this. [5 Church 264]

On January 19, the company negotiators arrived to learn of the U.S. government's about-face. Douglas MacArthur, the U.S. ambassador to Iran, endorsed the argument of the shah and of Jamshid Amuzegar, the Iranian finance minister, that OPEC-wide negotiations would be a mistake because the gulf "moderates" could not restrain the "extremists," Venezuela and Libya. He therefore "urge[d] in the strongest terms [to] settle with the Gulf." He did not share the companies' doubts that the gulf producers would abide by any agreement. In fact, additional Libyan demands had already been endorsed by OPEC. [6 Church 64] Irwin, MacArthur, and Secretary Rogers ignored this.

The companies now tried for an orderly retreat by dividing the team into two parts. But since much gulf production was shipped by pipeline to be lifted at the eastern Mediterranean, in direct competition with Libyan output, it was impossible to decide gulf prices without reference to the Mediterranean, and vice versa. Now the gulf nations demanded immediate negotiations and refused to discuss their eastern Mediterranean postings. The companies accepted this too. They expected no help and were not "too impressed ... by the attitude of the U.S. Government." [5 Church 265]

While the companies retreated, the U.S. government convened a meeting of the nations of the Organization for Economic Cooperation and Development (OECD) on January 20. The record has never been made public, but there is no doubt that the American representatives assured the other governments that they could count on five years' supply at stable, albeit higher, prices.

The OECD meeting could have kept silent, to keep the OPEC nations guessing and hence moderating their demands, for fear of counteraction. Instead, their statement praised the companies' offer and declined to estimate its cost. Moreover: "[C]ontingency arrangements for coping with an oil shortage ... were not discussed." [WSJ 1-21-71:26; NYT 1-21-71:11; OGJ 1-25-71:82; PONS 1-21-71]

This was not selling the pass, but giving it away. Previously an open threat by the OPEC nations would have carried little credibility in view of the failure of even mild attempts at production control. They had privately discussed an embargo, which was supported by the allegedly "moderate" king of Saudi Arabia and shah of Iran. [6 Church 71] But only after the capitulation by Irwin, Rogers, and MacArthur—and then the OECD—were embargo threats credible, and then they were made often. The culmination was an official OPEC resolution February 7 providing for an embargo after two weeks.[5] [NYT 2-8-71:9; WSJ 2-8-71:18]

While the threats were being made, the State Department praised itself: "Mr. Irwin's mission persuaded the oil producing nations of the need to enter real negotiations." [NYT 1-26-71:5] Minister Amuzegar was more accurate: "There is no question of negotiations or resuming negotiations. It's just the acceptance of our terms." [WSJ 2-9-71:4, 2-12-71:11]

The companies wanted assurances from the Gulf producers that whatever they agreed to would hold for five years. [WSJ 2-9-71:4; see also WSJ 2-12-71:11; NYT 2-13-71:7] It is hard to believe that they took these assurances seriously. George Piercy of Exxon had asked the Saudi oil minister Ahmed Zaki Yamani about them: "I did probe very bluntly as whether they would stick to that. Q. What was the response? A. As I recall, I got silence." [5 Church 227]

One must respect Yamani's refusal to lie. Soon after, there were also company agreements with the African producers. In late January, Venezuela followed by legislating new tax rates. [OGJ 1-25-71:82] Table 5.1, in the next chapter, summarizes the changes for Saudi light crudes. Readers can use it to distinguish "posted price"—an artifact, not a price—from the per barrel tax, which, added to the small item of cost, was the floor to price.

The next day President Nixon said he was pleased. [WSJ 2-16-71:4] The State Department had previously stated the objective as "stability, orderliness, and durability.... We don't want anybody ... badly hurt." [PONS 1-19-71; 1-28-71]

At a special conference and later too, State emphasized Undersecretary Irwin's role and hailed the agreement for ensuring "stability" and "durability." They "expected the previously turbulent international oil situation to calm down following the new agreement." [NYT 2-17-71:3, PONS 2-17-71, 2-18-71] Later State claimed credit for Secretary of State Rogers. [WSJ 3-5-71:1]

These predictions and self-congratulations imply the following. First, higher prices and producing-country revenues, gained by collective action, are favorable to the United States and other OECD countries. It was not explained why. Second, the producing countries will keep their word. But anyone who could believe this and disregard "persistent commercial bad faith"[6] was incompetent. "If such agreements were worth anything, the present crisis wouldn't exist." [PIW 2-8-71:1] It was foreseeable and foreseen that the agreements would be scraps of paper, "a solemn promise that must hold a world record in the scale and speed of its violation." [Kissinger 1982, p. 865] And third, without the State Department, there would have been no price reversal. This is plausible. Had the United States sponsored a Libyan safety net, Gadhafi (in power less than a year) might have been discredited and overthrown. Had the U.S. government not destroyed the new-found solidarity of the companies and not won the OECD blank check, the Persian Gulf producers might have been frustrated. Even as late as January 25, the shah said, "If the oil producing countries suffer even the slightest defeat, it would be the death-knell for OPEC, and from then on the countries would no longer have the courage to get together." [Shahanshah 1971, p. 9] This could be mistaken, of course; the OPEC nations might later have tried again. The historian can only record that the U.S. government helped the cartel in its hour of greatest need.

Tripoli-Tehran: The Great Reversal

Unilateral tax increases were not new. Before Tripoli and Tehran, the OPEC nations had exerted defensive market power. Their excise taxes had put a floor of tax-plus-cost under the price. But raising taxes in concert, to raise the worldwide price floor, was indeed new.

The increase had no relation to supply and demand of crude oil, and it reflected no scarcity, present or foreseen. The oft-cited growth of "dependence" on OPEC oil simply states the monopoly rationale. In the previous chapter, we saw how the end of U.S. exports erased the price ceiling and made the monopoly price much higher. But to move toward that price took effective collusion. That was the new element. As the *Middle East Economic Survey* put it:

> Even as late as a few months ago, the very idea of the producing countries of the Gulf achieving an across-the-board price increase ... [as] agreed upon in Tehran on 14 February would have seemed almost inconceivable.... This victory—and victory it certainly was—was mainly due to two factors: firstly the *unprecedented degree of unity* shown by the OPEC member countries; and secondly the great skill and nerve of the three-man ministerial committee which negotiated the deal on behalf of the Gulf states. [MEES 2-19-71:4, emphasis added]

The Price Ratchet

Like the price increase following the 1956 Suez crisis, the 1970–1971 rise was an example of the price ratchet. A temporary scarcity raises the short-term price level; then the sellers act together to hold the price even after the scarcity has passed.

The 1970 starting point was a temporary high price in the Mediterranean, caused by the temporary tanker shortage, itself a result of the Syrian pipeline break. In late January 1971, when Syria received the additional revenue it had demanded, the pipeline was repaired almost overnight. [WSJ 2-1-71:7] As a result, during the first six months of 1971, the drop in tanker rates made Libyan oil overpriced. Its output fell by 16 percent, while Iran and Saudi Arabia increased by 20 and 28 percent. [WSJ 8-16-71:2] But the higher price level remained. The 1970s should be called the "ratchet decade."

Foreign Policy Background

The Libyans were encouraged by the State Department, intent on making friends with Gadhafi and on imposing a "1957-type solution" to the Arab-Israeli dispute (See appendix 4A). State also aimed to keep the gulf out of the Soviet orbit and to smooth company-government relations, particularly in Iran, the principal U.S. ally in the area.

The shah purchased "all the most expensive equipment." [8 Church 571–574] Higher oil prices were an easier way to support him than voting funds in the Congress. [Church Report 28, 33–44] The mind-set of that time is suggested by a simulation of a third world war, where the number of index references to Iran greatly exceeds the references to Italy, a fair reflection of the relative military importance of the countries as then perceived. [Hackett 1978]

Appendix 4A outlines the tension over Arab-Israeli relations between the president (and national security adviser Kissinger) on one side and the State Department on the other, but there was no dissension over Tehran.

Increasing Market Control: A Cartel Learning Experience

The host governments now knew what they could only have hoped for: the consuming nations would not resist, and the United States would actually help. The tendency was self-accelerating. The more wealth they had, the easier it was to forgo some income temporarily by cutting production to force the price still higher, and incomes and wealth much higher.[7]

Before February had ended, Libya had made fresh demands. This time the oil ministers of Libya, Algeria, Iraq, and Saudi Arabia threatened to cut off Mediterranean oil together if there was no acceptance. [NYT 3-20-71:5] The next day the threat was repeated. [NYT 3-21-71:7]

A few days later, Algeria seized the French oil companies, but the French defense minister, Michel Debré "insisted that 'Mediterranean solidarity' be held paramount. Otherwise 'insecurity and foreign powers will prevail.' M. Debré did not specify whether he meant the Soviet Union or the United States or both." [NYT 2-26-71:3]

Once the agreements were in place, the oil companies saw them as a gain. Higher prices worldwide meant higher values for crude oil produced everywhere. In Great Britain, the objective was to cover the tax increase "and leave some over"; the February increase was matched by a product price increase perhaps half again as great. [MEES 2-19-71; *Economist* 2-22-71; PONS 2-22-71] A financial analyst called the agreements "truly an unexpected boon for the world-wide industry." [OGJ 5-10-71:46] The secretary-general of OPEC said there was no basic conflict between companies and producing nations. [NYT 2-11-71:14] Sir David Barran of Shell, who knew there was a time to resist and a time to cooperate, spoke of a "marriage" of the two interests. [Barran 1971] Most precise of all was the chairman of BP, who called the companies "a tax collecting agency" for both producer and consumer governments. [BP 1970] Some would see a difference between serving a government to collect revenues from its own citizens and serving a government to tax citizens of other countries.

Prices and oil company profits in 1971 were the highest since 1963, although there was a profit slide later in the year as competition in products (not crude) reasserted itself. [FNC Energy News October 1972]

Appendix 4A: U.S. Middle East Policy Issues

Immediately after World War II, experts believed that ... Middle East oil would be vital to American security, and that therefore the United States should do nothing to jeopardize friendly relations with the Arab states. Above all, the United States should try to keep its distance from Israel."

[Quandt 1977, p. 12]

In 1950, the State Department persuaded a New York State commission to exempt Aramco's New York office from a state law against discrimination, because, it said, "a state of war exists between Saudi Arabia and the state of Israel." [NYT 1-9-62] In an official statement to the commission, "The Department had defended the company's refusal to employ Jews as a matter of 'national interest.'" [NYT 7-14-63] The courts eventually enforced the law.

We do not know what was said to Saudi Arabia off the record. "Wall maps in the U.S. Embassy in Saudi Arabia—including the Ambassador's office—did not show any country known as Israel." [WO 10-87:4] Saudi Arabs who saw this hint that the United States would approve or tolerate the dissolution of Israel would have some reason to complain, in later years, of American duplicity.

After the 1956 Suez crisis, there was discussion of another pipeline from the Persian Gulf to the Mediterranean, to bypass the canal. The State Department opposed a Trans-Israel Pipeline TIP; its choice was Iraq. [WSJ 2-20-57] In 1968, after TIP was built and operating, State discouraged its use. [PONS 4-22-68] During the tanker shortage of 1970, utilization of TIP was higher but still below capacity. [PPS 5-71:176] Its use would have alleviated the tanker shortage and lowered delivered prices in Europe, thereby making the Libyan demands appear less "reasonable."

When the Israelis joined in the 1956 Anglo-French attack on Suez and occupied the Sinai Desert, the Eisenhower administration forced them to evacuate it. American pledges, of canal transit for ships going to or from Israel and of free passage through the Straits of Tiran, could not be kept. "The methods employed in gaining Israeli withdrawal had profound implications for the future.... The Arabs were convinced that the proper amount of threat and enticement would persuade the Americans to repeat their 1957 performance.... Within the American bureaucracy a fascination with a 1957-type solution remained." [Spiegel 1985, pp. 80–81]

Through several administrations, "the foreign policy bureaucracy [tried] to get the president to deal with the Middle East in terms of 'on the ground' realities" as they perceived them. [Taylor 1991, p. 44] "Middle East experts ... were very critical of ... a policy ill-suited to regional trends and to the maintenance of healthy U.S.-Arab relations." [Dowty 1984, p. 151] (Both Taylor and Dowty are in strong agreement with State.)

In August 1967, after defeat in June but expecting a 1957–type solution and subsidies from the oil-producing states, the Arab nations resolved at Khartoum on "no peace, no recognition, and no negotiation" with Israel. [Quandt 1977, p. 65] The State Department reported that month that Arab attitudes were good, but "the Israeli position is hardening. They have stated publicly and privately that they must have a peace treaty with the Arabs and they are prepared to wait it out." [8 Church 547]

The Johnson administration did not repeat the 1957 solution and intended "rather that the territories should be exchanged for a genuine peace agreement." [Quandt 1977, pp. 63–65] The policy change was repugnant to State. Some (see appendix 4B) "promised" the Arab governments to expect another 1957. At the beginning of 1969,

> Ire at Israel reaches a new peak in the State Department.... Some State Department men offer a sinister analysis: Israel is deliberately plotting to destroy the moderate Arab regimes in Jordan, Lebanon, Saudi Arabia, Kuwait, to shove those nations into the Soviet orbit; Israel would then be the only U.S. client in the Mideast. One official says the upset of Lebanon's moderate government has "enormously strengthened" this view. [WSJ 1-10-69:1]

The Nixon administration moved toward a 1957 solution (the Rogers plan of December 1969), which was rejected by Arabs, Israelis, and the Soviet Union. [Quandt 1977, pp. 70, 77, 92; Spiegel 1985, p. 188]

Early in 1970 there was fighting along the Suez Canal. State arranged a cease-fire August 7.

> Within days, however, the provisions of the cease-fire were being violated and a new crisis was in the offing.... By the time [it] was over, the views of ... those in State who had urged "even-handedness" were virtually banished from center stage.... [State was discredited because it refused to admit the evidence of Egyptian movement of missiles, which] at a minimum, showed bad faith on Nasser's part. [Quandt 1977, pp. 104, 107–108]

On September 3 came the decision to sell Israel additional jet fighters, to offset French aircraft sales to Libya. Between September 6 and 9, there were three successful and one failed airline hijacking by the PLO, with nearly 500 hostages taken. An invasion of Jordan from Iraq and Syria was feared. But Jordan, "to the horror of the State Department— ... twice [indicated] that if matters reached the boiling point they would acquiesce in Israeli intervention." [Spiegel 1985, p. 196] Jordanian aircraft heavily damaged invading Syrian tanks, which quickly withdrew. With the danger passed [Quandt 1977, pp. 110–118] Nixon and Kissinger were pleased, but "State saw the crisis as a diversion from the Arab-Israeli settlement that would transform the U.S. role in the area." [Spiegel 1985, p. 201] The president and

> Kissinger had been prepared to allow the State Department to pursue the logic of an 'even-handed' approach to the Arab-Israeli conflict, and this had led to the Rogers Plan and the August 1970 cease-fire. But ... the Soviet Union had failed to

cooperate ... and the Arabs displayed little gratitude for United States restraint in providing arms to Israel....

Over the next three years, the White House refrained from day-to-day involvement in Middle East diplomacy, but a careful eye was kept on the State Department to ensure against excessive activism." [Quandt 1977, pp. 120–122, 128]

Thus the Libyan oil crisis was an incident in the continuing effort of the State Department to reverse national policy, from the 1967 model to the 1957 solution. There was no explanation why they thought this had anything to do with the supply and price of oil. The connection was assumed, never analyzed.

Appendix 4B: James Akins, Window into U.S. Policy

The ideas and assumptions governing US policies toward the oil markets are almost completely undocumented. [Goodwin et al. 1981, p. 446] A window on those ideas is found in the career and statements of James E. Akins, head of the Office of Fuels and Energy over a long span of time. [5 Church 127] He drafted the energy portion of the president's State of the Union Address in 1973, became U.S. ambassador to Saudi Arabia in September 1973 (but was dismissed by Kissinger in August 1975), and became an adviser to President Carter, serving as his special envoy to the Persian Gulf nations [PONS 11-24-76:1], although he broke with the president by opposing the Camp David agreements between Egypt and Israel.

Akins is a highly valuable resource to the historian; he said what the State bureaucracy thought and assumed. Opposing the 1967 change in policy that linked Israeli withdrawal from Arab lands to a peace treaty, Akins and others continued to "promise" Arab governments a 1957 solution. He wrote, "The United States could continue to promise to take political action desired by the Saudis, i.e., put pressure on Israel to withdraw to its '67 borders, if Saudi Arabia would keep up oil production. *The United States has made such promises to Saudi Arabia consistently for the last fifteen years.*" [Akins 1983, p. 33, emphasis added] Like the unfulfilled hint of the wall map, these unkept promises must have seemed like American duplicity.

Akins believed that "only with [Arab-Israel] peace can we have petroleum." [*Oil Daily* 10-28-76:1] He predicted in 1982: "Should there be no movement toward peace [defined as Israeli "withdrawal from the West Bank, Jerusalem, Golan, and Gaza"] ... producers in the Arabian peninsula will eventually find it expedient to cut down, if not to cut off oil production. If they do not, they could face revolution or sabotage." [Akins 1982b, p. 6]

He repeatedly alleged, with no proof, that Israel was planning to invade and occupy the Arab oil fields. [*Arab Oil and Gas*, 10-16-81:19–28, 10-14-83:28–37; OGJ 12-19-83:59] As noted in chapter 3, he stated without proof that the 1967 boycott of certain Western nations was ended by the efforts of Saudi Arabia. In fact, the Saudis pressed at the time for higher Mediterranean taxes and prices. I noted as well his statement about threats, also completely contrary to the public record, and his belief that Gadhafi was "a fanatic anti-Communist." [Akins 1973, p. 470] In fact, Gadhafi proved more Communist than Gorbachev.

Akins's opinions on Middle East politics, right or wrong, gave him no economic insight. He had access to economic information but did not know what to ask for, hence was often poorly informed. Ignoring economic principles, he knew with certainty many things that were not so.

On oil prices, Akins did not realize that for Libya to charge a permanent excise tax margin on its crude oil greater than any permanent delivery cost premium would mean an overpriced crude and loss of sales when the temporary advantage disappeared. This happened within four months. His lack of knowledge is also evident in other statements before and after his 1972–1975 career.

He did not understand production cost as an economic fact with consequences. For him, it was an idle debating point: "To say that the cost [price] of oil should be very low in the Persian Gulf because the cost of production of the oil is very low is very nice as far as consumers would be concerned, but I don't think you are going to get any oil producer to admit that this has any validity." [Akins, at 5 Church 4] He had no idea of the money involved in oil investment: The prospect of lower crude oil prices "frightens me [he said] more than any other.... These [producing] countries have demands for armies and roads and schools, and they would be putting nothing into oil production." [Senate 1972, p. 122] In fact lower crude prices were irrelevant because revenues were so vastly in excess of oil investment. In 1970, one-sixth of Middle East wells were drilled in Saudi Arabia. [WO 8-15-71:51] Therefore approximately one-sixth, or $46 million, of Middle East capital expenditures of $275 million on crude oil production was in Saudi Arabia. [CMB 1970] This was only 3.4 percent of Saudi 1970 revenues of $1,359 million [PPS 9-71: 327]. Later revenues were, of course, much larger.

Other statements reveal the ideas underlying Akins's positions. "Bribery is not necessary to do business in the Middle East." [PONS, 5-6-76:3] "The international oil cartel kept the prices of crude oil low assuring profits in the downstream areas." [Akins 1983, p. 2] He ignored that downstream (refining-marketing) was competitive, and profits were taxed and that cheap imports were politically dangerous.

Government price fixing was a necessity, he thought: "Without price supports or acreage limitations on U.S. grain ... farm production would decline, and the world would face serious food problems." [OGJ 7-6-79:29]

Notes

1. By 1990, nearly three fourths of the Libyan labor force worked for the government. [WSJ 3-9-90:A10] Gaddafi granted the Soviet fleet a base. The demise of the Soviet Union meant that Libya lost a "powerful ally." [NYT 12-12-89:A2, A10] In August 1991 Gadhafi sent congratulations to the leaders of the "magnificent" coup against Gorbachev. [Economist 8-24-91:20]

2. France agreed in 1982 to pay Algeria a price for natural gas far above even the inflated crude oil equivalent, to promote "economic cooperation." After the collapse of crude oil prices and the gradual freeing up of the natural gas market in Europe, these prices were grotesquely high. In August 1988, a French diplomat was appointed as chief executive officer of Gaz de France, with a view to settling this "long-simmering dispute." [WSJ 8-29-88:15] Twenty-six years after their first "agreement" with Algeria speedily proved worthless, the French, like the Americans, could not grasp that these "special relationships" bring nothing but economic burdens and political humiliation.

3. Revenues per barrel were just $1 in 1969, which we multiply by the production numbers. [PPS 9-70:324]

4. In 1986, Vice President George Bush told the Saudis that their excessive output was hurting the world industry. Since the Saudis' purpose was to inflict pain until the industry cooperated, it was reasonable for them to increase output still more, and they did. See chapter 7.

5. Akins said that "the approach we [State] made in the Persian Gulf was primarily because of the threat to cut off oil production.... We informed the countries that we were disturbed by their threats, and these were withdrawn very shortly after our trip." [Senate 1972, p. 123] (See also [Akins 1973, pp. 472–473].) He cited no threats before the visit, and offered no evidence for anything he said. The documentary record contradicts him; it shows no threats before the visit and many afterward. [NYT 1-18-71:4, 1-24-71:18, 1-25-71:1, 1-26-71:5, 2-3-71:1; WSJ 2-2-71:11, 2-4-71:2; see also Shahanshah 1971, pp. 10, 12, 15, 37] There was an additional threat in early March.

6. Producing country claims "are now being pitched so high that they could really damage Europe's long-term economic growth.... Once upon a time, had a group of backward countries, with highly unstable governments and a reputation for persistent commercial bad faith tried to hold the Western economies to ransom as the oil producing governments of the

Middle East are now doing, they would have seen the gunboats steaming up the Gulf.... Gunboat diplomacy is now a thing of the past.... [A]greements struck with Iran and other oil producing countries are unlikely to last more than a few months." [Economist 1-30-71]

7. My view was that prices would rise toward the monopoly optimum, but the gains would be hard to hold. Without the companies, "unless the producing nations can set production quotas and obey them, they will inadvertently chisel and bring prices down by selling incremental amounts at discount prices." [NYT 3-29-71:49]

5 Price Breakout, 1971–1974

From mid-1970 to mid-1971, the price of Saudi Light crude oil went from \$1.20 to about \$1.70. The producing company excise tax paid to the Saudi government went from 88 cents to \$1.26. But by 1975, the government was the seller; it received nearly \$11.00 and allowed the company about 30 cents. This revolution went far beyond the original five-year schedule of tax increases in the February 1971 agreements (table 5.1).

Price and Excise Tax, 1971–1974

The agreements were soon violated, with tax and price exceeding the schedule (figure 5.1). Nevertheless, they stayed close together until the start of 1973, when prices began increasing more rapidly. In October 1973, tax was doubled, closing the gap. For months after October 1973, prices could not—and still cannot—be observed at all. On December 31, the tax jumped to \$7, effective the next day.

Figure 5.1 shows the excise tax only for 1974. There are no relevant price series, only a few observations that I will make where available for comparison with the tax. Published company data are useless because they are retrospective, revised after increases were in effect.

We consider first whether the price rose because of the pressure of demand on insufficient supply. If that was the case, the tax increases would have been a capture of rents. We then consider the opposite possibility: that the governments raised taxes and thereby prices.

No Strain on Capacity

Excess demand means capacity strain, with additional output costing more than price. In that case, competitive pressures force the price up toward the marginal cost. The reverse is true with excess supply; the resulting slack would force prices down.

Table 5.1
Price increases under Tehran agreement: Saudi Light 34° API ($/barrel)

Date	Posted price	Excise tax
11-14-70	$1.800	$0.908
2-14-71	1.800	0.989
5-31-71	2.180	1.261
12-31-72	2.285	1.321
12-31-73	2.392	1.390
12-31-74	2.502	1.456
12-31-75	2.615	1.525

Source: MEES 2-71.

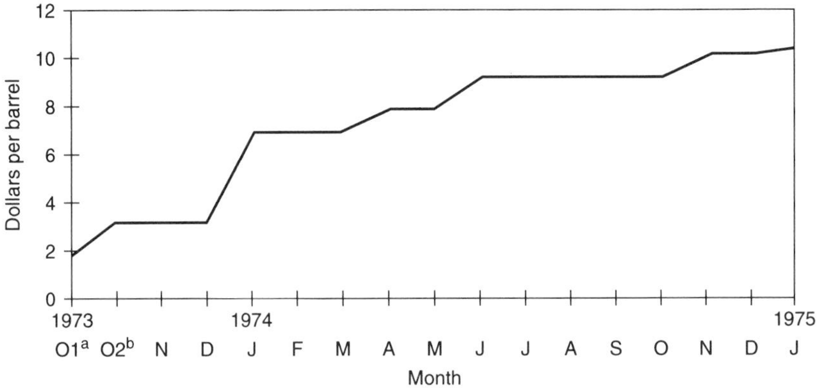

Figure 5.1
Saudi tax on Arab Light, 1973–1974, (*a*) first half of October, (*b*) second half of October. Sources: table 5.5.

Slack and strain may occur in either the short run, with output available only from current equipment, or the long run, given additional investment. The analysis up to now has looked at the long run. With more investment, much more could have been produced, at a cost far below price. In this chapter, we look at the short run for 1971 through 1974 and ready capacity.

Before 1971 excess capacity outside North America was small and fleeting (figure 5.2 and 5.3). In the Persian Gulf area, production had been growing at around 10.5 percent annually. Here even 20 percent excess capacity, say from completion of a big block of additional facilities, would

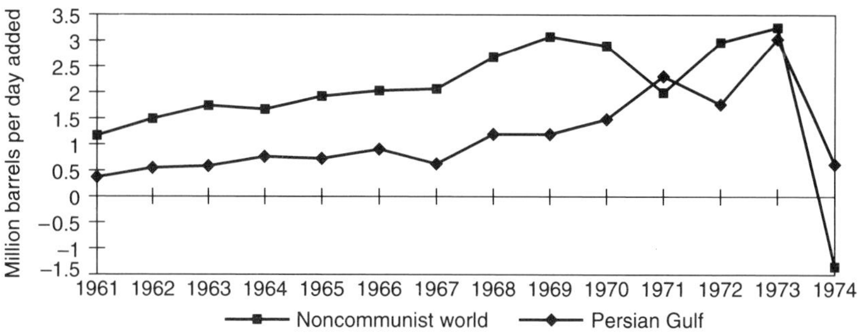

Figure 5.2
Annual absolute growth of noncommunist world consumption and Persian Gulf output 1961–1974. Source: BP.

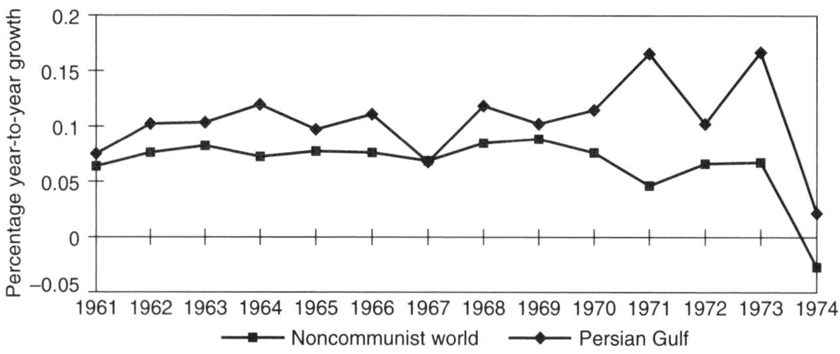

Figure 5.3
Percentage annual increase: noncommunist world consumption and Persian Gulf output, 1961–1974. Source: BP.

be liquidated in twenty months. During the 1967 Arab "embargo," Iranian and Venezuelan output was at a premium. Output increased in those countries by only 5.7 and 5.9 percent respectively, a rough measure of available short-run spare capacity. [Adelman 1972, pp. 163–164]

During the 1960s, the rate of world consumption growth had increased and then turned down. Persian Gulf expansion had increased steadily through 1971, then paused. Figure 5.3 shows how the relative consumption growth rate had been quite steady at just under 8 percent per year; then it declined mildly, never regaining the 1969 peak. Even the higher growth rates of the 1960s had not strained producing capacity, and prices had persistently decreased. There is no indication after 1969 of any strain on the system or rising cost of expansion.

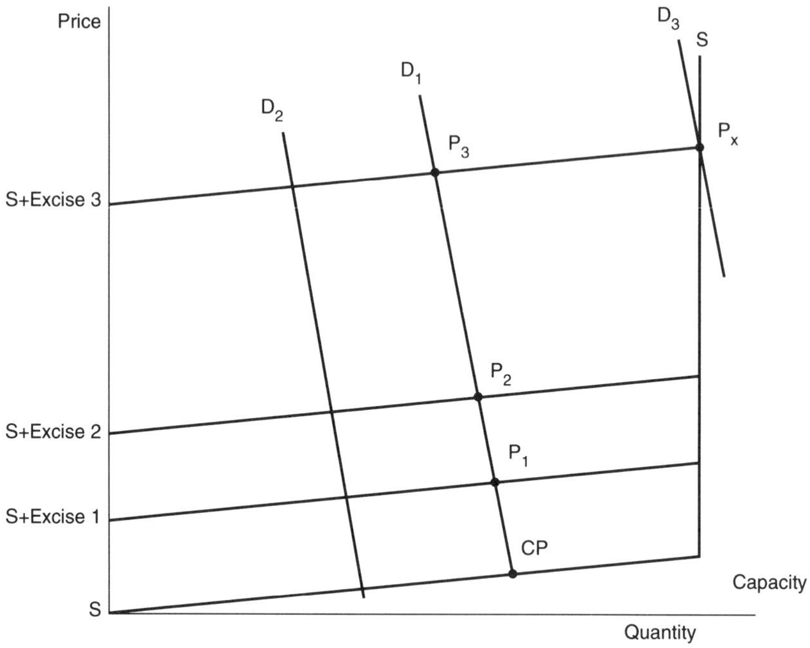

Figure 5.4
The price ratchet: short-run demand, supply, and excise tax.

In summary, growth indicators give no hint of any strain on capacity between 1971 and 1973. But because capacity stayed only a step ahead of production, as it should under competition, the system was vulnerable to short-term stress, accidental or contrived.

Tax Changes Permit Volatility Upward, Not Downward

Since 1960, government take had been an excise tax. To the operator, it was like labor, supplies, or any other variable cost, except that it was more easily calculated.[1] Figure 5.4 shows how it was used as a ratchet, to prevent a downward movement.

The lowest curve is the short-run supply curve, *SS*. Like all other short-run curves, it is very inelastic at capacity. *SS* goes vertical; no more can be produced at any price. D_1 is the initial short-run demand curve, and *CP* is the competitive price. Let the governments now impose Excise 1. For the private companies that produce and sell, their new supply curve is $S +$ Excise 1, and the price rises to P_1. If the governments again raise the excise tax in concert, the price rises by nearly the amount of the tax, to P_2, with no collusion among the companies. The effect of demand on the price is very small compared to the effect of the tax.

So far we have considered only *consumption demand*, assuming that buyers had no interest in building up or drawing down inventories. But now assume a burst of *precautionary demand*, for inventory, generated by fear of dearth. It may be reinforced by *speculative demand*—overbuying to gain from expected higher prices. Together, the temporary demands shift the demand curve far to the right (D_3), where it intersects the vertical section of the supply curve, at price P_x.

As these temporary stimuli disappear, demand reverts back to D_1 or even D_2 to work off excess stocks. The higher price P_x cannot remain long, unless there is another concerted increase in the tax, to Excise 3. Then the companies' supply curve is $S+$Excise 3. From now on, even a drastic shift leftward in the demand curve, or a drastic shift rightward in the supply curve, has only a minor effect on the price. The process is a ratchet: market forces raise the price temporarily to P_x, but the excise tax keeps it permanently at P_3. (In the longer run, demand will be more elastic, as we will see later.)

Excess Supply But Rising "Take" Raises Price, 1971–1972

From early 1971 to nearly the end of 1972, prices increased despite continuing substantial excess supply.[2]

Mediterranean and Caribbean

In March 1971, Libya made new demands, which the companies rebuffed. The industry expected no reduction in supplies available for Europe. [PIW 3-22-71:3] Late in 1971, there were "bulging crude stocks in Europe, low tanker rates, and an oversupply of crude in world markets." [OGJ 11-1-71:40] The Venezuelan ambassador to the United States called for a hemispheric energy policy and preferential access to the U.S. market: "Our normal production is 3.5–3.6 million barrels daily, a level to which we must return by 1972." [OGJ 12-6-71:46]

Venezuela increased taxes and imposed penalties for underproduction. Producing companies were required to deposit money to offset the deterioration of producing properties in anticipation of their reversion to the state in 1984. [PIW 12-27-71:2, 4-12-72:4]

In November 1971, Iran seized two small islands in the Persian Gulf from the local emirate. It was not clear why Libya blamed Britain for this "anti-Arab" measure, but it "retaliated" by seizing BP's share of the big Sarir field. [OGJ 12-13-71:41] The Sarir shutdown "was a dud.... It didn't even cause a ripple in the tanker market—usually so sensitive to any shutdown of 'short-haul' oil." [PIW 12-13-71:1] (Indeed, it was welcome.) "Companies have been anxious to reduce production in Libya but 'no company had the courage to do it and now Libya is making BP do it 100 percent,' one observer said." [PONS 12-9-71:1]

Algeria and Libya were both discounting prices in early 1972. [PPS 2-72:53,64] Persian Gulf output actually declined, especially in Iraq. The premium on Gulf exports through the eastern Mediterranean exceeded the savings on the tanker haul from the gulf, because tanker rates were back to normal. [PPS/PE 2-72:46].

Mediterranean oil was "increasingly priced out of a competitive market by their high tax burdens." [PIW 5-22-72:3] The short-term downswing was the mirror image of the upswing of two years earlier, when the State Department called Libyan demands "reasonable." By that criterion, State should have called for a reduction in Libyan and Iraqi excise tax.

Iraq Expropriation

The drop in Iraq exports was the last chapter in the forty-five-year vicious cycle of government threats and IPC underinvestment. The government demanded not only that output cuts be restored but also that IPC promise an expansion of 10 percent per year. [NYT 5-18-72:67] When IPC refused, Iraq seized about 62 percent of its production on June 1.[3] [PPS 7-72:238] The rest of IPC served as a hostage but for only a short time.

Iraq was not expected to have any production difficulties, since there had been only about two dozen expatriates left in the fields there. [PPS 7-72:238] Iraq offered to sell "at reduced and competitive prices," at spot or by long-term contract. [WSJ 6-6-72:6] Other Arab OPEC nations promptly loaned them enough money to tide Iraq over the period of low production after the exit of IPC. "The offer by Iraq to sell its newly nationalized oil at a cut rate ... would have driven down the revenues received by the other countries for their oil." [NYT 6-10-72:37, 6-12-72:6]

Deficient Demand

The managing director of Shell said in May 1972 that the industry was suffering "from having provided the facilities for an increase in trade which did not materialize." [PPS 6-72:222] The shah unsuccessfully tried to get special access to the American market. [Kissinger 1982, p. 868] In August, Sheik Yamani expressed his concern with falling prices. [IHT 8-22-72]. Later, an oil executive resumed the familiar theme: crude would remain in surplus "for at least the next few years." [OGJ 10-2-72:38] A Shell executive wrote: "The underlying situation of supply and demand remains one of potential surplus. Yet the producing countries manage to reap the rewards of a sellers' market by creating a producers' monopoly." [Chandler 1973]

In November 1972, an internal Aramco memorandum questioned whether the rise in output should be accelerated. The "incremental cost" was the increase in investment plus operating cost:

> [I]t would appear that an incremental cost of about 4 cents per barrel ... would be easily justified provided the crude can be sold.... Not only do we still have substantial surpluses of Arabian Light yet to be sold in 1974, but ... we are experiencing considerable difficulty in selling availabilities of Arabian Heavy and Arabian Mediums. [7 Church 487]

Yet despite excess supply, prices and tax take kept rising and by the end of 1972 exceeded the levels set at Tehran (table 5.2). It is time to see how this was accomplished.

In August 1971, the Persian Gulf OPEC nations demanded additional tax increases though higher posted prices. The ostensible reason was the devaluation of the dollar. But the Tehran and Tripoli agreements had both allowed for worldwide inflation, of which the devaluation was only an incident. [Tehran agreement paragraph 3(c)(2)] Moreover, much of the Persian Gulf revenues were in sterling, not dollars. Thus the five-year Tehran agreement had lasted for about six months.

The president of Exxon charged that the new demands violated the Tehran agreement, but "the industry will solve these problems just as our differences with them were reconciled earlier this year and before." This was precisely correct, both as to substance and as to ritual. The companies made an offer; the governments refused it and broke off the talks; the companies made a better offer; taxes were raised again and crude oil prices with them. [WSJ 10-8-71:7, 11-5-71:16, 11-9-71:2, 11-11-71:6, 1-12-72:7; NYT, 11-3-71:14, 1-14-72:4; PIW 1-31-72:1, 2-7-72:3]

Before that was settled, the governments had already made an additional demand, for 20 percent "participation," or part ownership of production. The companies said they were distressed that the agreements "have not led to the long peace ... that they had anticipated." They

Table 5.2
FOB prices: Arab light crude, 1970–1973 (U.S. $/barrel)

Date		Chevron third-party sales			Saudi government take
		Minimum	Maximum	Average	
1970				1.21	0.88
1971	J				0.99
	F			1.41	0.99
	M				0.99
	A				0.99
	M			1.57	1.26
	J				1.26
	J	1.71	1.79	1.66	1.32
	A	1.60	1.79	1.68	1.32
	S	1.60	1.79	1.68	1.32
	O	1.60	1.79	1.75	1.32
	N	1.71	1.80	1.75	1.32
	D	1.71	1.92	1.72	1.32
1972	J	1.83	1.92		1.44
	F	1.83	1.90		1.44
	M	1.82	1.91		1.44
	A	1.82	1.95	1.84	1.44
	M	1.82	1.91	1.84	1.44
	J	1.82	1.91	1.84	1.44
	J	1.82	1.91	1.87	1.44
	A	1.82	1.95	1.89	1.44
	S	1.78	1.95	1.91	1.44
	O	1.78	1.95	1.94	1.44
	N	1.78	1.95	1.95	1.44
	D	1.83	2.08	1.97	1.44
1973	J	1.96	2.22	2.13	1.53
	F	1.95	2.09	2.18	1.53
	M	2.05	2.28	2.25	1.53
	A	2.05	2.35	2.30	1.63
	M	2.15	2.28		1.63
	J	2.18	2.33		1.63
	J	2.25	2.50		1.63
	A	2.35	2.68		1.63
	S	3.35	3.36		1.77
1973	O1	3.35	3.36		1.77
	O2	3.51	4.18		3.04
	N	3.41	4.12		3.04
	D				3.04
1974	J				7.00

Sources: Irving 17:30; PIW 12-25-72, 6-4-73, 6-11-73, 10-22-73, 12-31-73; MEES 2-15-71; PPS 11-73:403.
Note: Because of rapidly increasing take in 1973, no record can be wholly accurate. An approximate check is provided by Saudi annual average Saudi take per barrel, which was, respectively, 1.26, 1.44, and 1.57 during 1971–1973 [PPS [PE] 11-73:416, 9-76:338]

would resist the demands as a violation of the agreement. Exxon president J. Kenneth Jamieson, who a year earlier had rebuffed Occidental and helped blast open the door, called the new demands contrary to the 1971 agreement, which had explicitly stated it satisfied all previous OPEC claims and resolutions.

The governments responded that "they would take part in a 'combined action' if they didn't receive 'satisfaction,'" and the companies agreed to negotiate. In December, OPEC set January 20, 1972, as the final date. [PIW 12-13-71:5] King Faisal warned that he and other gulf countries would legislate if the companies failed to grant participation. [NYT 2-19-72:41] In March, the Aramco companies conceded it "in principle." [NYT 11-3-71:4; WSJ 12-7-71:19, 1-24-72:5; PIW 9-13-71:4, 3-20-72:4]

At this stage, "participation" was a misnomer. The governments (excepting Iraq) had no plans to produce or sell. "Participation" was merely a means of raising the take.

As of March 1972, 20 percent "participation" would have meant 9 cents more per barrel.[4] Terms were generally agreed by October. [PPS 11-72:398] Compensation was less than $1 billion. [PIW 10-30-72:5] At the beginning of 1973, additional payments amounted to between 9 and 18 cents per barrel, with costs expected to "rise in future years as prices escalate under the Tehran agreements." [PPS 2-73:44] Moreover, there was even some talk, admittedly vague, of the governments' ultimately doing some of their own marketing.

Supply-Demand and Cartel Tax Increases in 1973

By the beginning of 1973, excess supply had disappeared from the Persian Gulf, and worldwide consumption was increasing, although at a slower rate than in earlier periods. Libya and Algeria had priced themselves out of much of the market, thereby increasing the demand for Persian Gulf oil (table 5.3). Chevron and Exxon had expected Libya to produce over 4 mbd [7 Church 332, 339]; production was only 2 mbd in 1973.

But as the picture improved, it became more complex:

Table 5.3
Market economies' oil production and consumption, annual increase

	1965–1970	1970	1971	1972	1973	1974
Consumption	8.2%	8.1%	5.2%	7.2%	7.4%	−2.8%
Production	8.7	9.8	5.1	5.1	9.3	−1.4
Venezuela	1.4	3.6	−3.7	−8.7	4.7	−11.6
Algeria	12.6	8.9	−25.0	37.2	2.3	−5.0
Libya	22.1	6.8	−16.7	−19.0	−2.7	−30.0
Nigeria	31.6	100.9	41.0	19.0	12.9	13.4
Persian Gulf	10.6	12.0	17.1	10.9	17.4	2.9

Source: BP, various years.

Reported sales and negotiations indicate supply and demand alone will add 3 cents to 7 cents a barrel to price rises of about 7 cents for Tehran-Geneva, and probably at least another 7 cents for participation costs.... Exact impact of supply and demand is difficult to pin down, however, since many suppliers have recently been reluctant to sell crude, due either to their own internal needs or to awaiting a clear determination of participation costs. [PIW 12-25-72:3]

This shows the double effect of taxes in short-run price determination. Irrespective of supply and demand, taxes were due to increase, by 14 or more cents; supply-demand would have added another 3 to 7 cents. But some of the higher demand was not for consumption but for inventory building, to avoid the expected higher taxes. Moreover, experience in 1971–1972 proved that even with deficient demand, prices would rise because of higher taxes. It was prudent for buyers to overbuy for inventory before the tax hike took effect.

Henceforth this was to be a familiar pattern. There was no downside risk in ordering too much. Expecting higher prices to come, buyers demanded more oil at the current prices, and sellers offered less. The effect precedes the cause: the market price rises before the expected tax increase takes effect. In 1973, the higher excise taxes drove prices up directly. The expectation of still-higher taxes and prices fueled higher demand and indirectly drove prices up even more.

The 1973 production increase exceeded the recorded consumption increase because of inventory buildup. But consumption itself is overstated because it includes the buildup of downstream stocks of refined products by distributors and consumers. We cannot measure downstream stocks, but storage capacity was substantial. [Burton 1982]

Thus, inventory accumulation increased demand but threatened to weaken it later. It scarcely looked promising for the price level. To understand what was happening, and the anticipations in the market, we must backtrack somewhat to the beginnings of what became known as the energy crisis.

The "Energy Crisis"

Despite long-run and short-run oversupply, there was a mounting preoccupation in 1971–1972 in consuming countries, and especially the United States, with the "energy crisis." In fact, there was nothing unreasonable about expecting an increase or decrease in the price of anything, including fossil fuels. It was a question of fact. But why a "crisis," and whence did it come?

Allegedly the world was running out of minerals, especially fossil fuels, with a growing gap between supply and demand. The director of the U.S. Bureau of Mines said that by the end of the century, "We will be competing for those foreign mineral products with other mineral-hungry nations all over the world. The situation could lead to a global minerals shortage that would make our current energy crisis look like the good old days by comparison." [PIW 9-10-73:12]

Secretary of State Kissinger said that the "crisis" was "the inevitable consequence of the explosive growth of world-wide demand outrunning the incentives for supply." [PIW 9-17-73:3] Unlike Akins, the secretary could see cost not as a debating point but as a fact with profound consequences. If one tones down "crisis" to "price rise," the Kissinger statement is coherent and can be tested with timely data.

Early in 1973, Aramco had announced that it would triple capacity in only five years. Haste makes waste; even so, its investment would be only $147 per daily barrel of capacity. An industry journal pointed out that since Aramco received 30 cents per barrel net of operating cost, "$2.2 billion would be recovered in about 16 months—a payout that can only be considered fantastic." [WO 4-73:15] But Aramco took only one-seventh of the price. For the owner, Saudi Arabia, the payout was in 70 days, *if* it had put up the money. If a 75-day payout or even a "fantastic" 16 months was an insufficient "incentive for supply," what would be sufficient?

The Gap: Natural Gas in the United States

"The gap" was especially plausible in the United States. For a decade, natural gas prices for interstate shipment had been fixed by regulation below the market-clearing level. The result was a true gap of excess demand. The Federal Power Commission (FPC) therefore allocated gas, first shutting off less urgent customers, then juggling regional requirements and availabilities. The obvious result was many unsatisfied demanders, and thereby a general model was fixed in the public consciousness. Higher prices were not only "unfair"; they could not close the gap anyway, since demand was inelastic and gas resources were dwindling. [FPC 1972] (Some economists echoed this and argued that public policy should consider only income distribution.) [Dirlam and Kahn 1958]

Because of the gap, gas imports were badly "needed." Pipeline executives flew in search of Algerian and Soviet gas supplies, and the secretary of commerce made a trip to Moscow to discuss imports of Soviet gas in a "frantic search for supplementary supplies of natural gas to meet the threatened 'energy crisis.'" [OGJ 12-6-71:121; NYT 12-2-71:1] Gas imports were vital because the "energy crunch" was coming in three to eight years, said the director of the Office of Emergency Preparedness. [WSJ 12-10-71:5] The Federal Power Commission (FPC) predicted a gas deficit of 9 trillion cubic feet (tcf) by 1980 and 17 tcf by 1990. [NYT 2-26-72:46] By late 1972, subsidies were being granted for liquefied natural gas (LNG) tankers to bring gas from Algeria, at prices considerably higher than domestic.

The pipelines' resale prices were required to average together high-priced additional supplies with low-priced controlled gas. Thus, buying high-priced gas permitted pipelines to raise their resale prices on low-priced gas. Pipeline companies could make large profits buying gas at prices several times higher than what they would receive for it. Naturally they were ready to buy at very high prices.[5] By mid-1973, U.S. firms began "hard bargaining for Soviet gas supplies," after a protocol signed during the Nixon-Brezhnev summit. The Nixon administration was committed to transport subsidies, but there was still dispute over prices and interest rates. [PIW 6-4-72:6, 6-18-73:6, 7-9-72:6] [Kosik 1975]

Fears of Import Cutoff

In the United States, oil imports had been limited since 1948 in the name of national security. The debate was made sharper by the Tehran agreements, and the many subsequent threats to cut off supply.

Security had been undermined by low prices, and here was the proof. Dire predictions were heard with increasing frequency between 1971 and 1973 of "the recent history of sky-rocketing price hikes and threatened cutoffs of foreign oil. The OPEC cartel has shown it can make this oil both expensive and scarce." [OGJ 11-1-71:19] "The threat of a massive production shutdown used very effectively during last winter's negotiations still hangs over OPEC crude."[OGJ 11-15-71:121]

The State Department

The State Department had claimed credit for the Tehran agreement, following which the previously turbulent oil market would now settle down. It did not apologize for its forecast or admit that the threats came after capitulation in Tehran, not before. In May 1972, Undersecretary Irwin warned an OECD meeting of a worldwide shortage of 20 mbd by 1980. [NYT 5-27-72:35] Iraqi sources referred to this soon afterward in explaining the seizure of IPC. [OGJ 6-12-72:75; NYT 6-16-72:65]

In May 1972, Akins said that "by 1976 our position could be nothing short of desperate." Persian Gulf reserves would decline after 1980. [OGJ 5-15-72:50] But he gave no support for either proposition and also mixed policy advice with nonfacts—for example, "[I]t is not in the Arabs' interests to allow the companies to continue expansion of production at will, and the producing countries, most notably Saudi Arabia, must follow Libya's and Kuwait's leads in imposing production limitations." [OGJ 6-12-72:66] Iran no longer desired higher production. "Venezuela has restricted production.... Saudi Arabia has shown some indications that it may not increase production to the extent that the world will require." [OGJ 9-25-72:70]

Only the statement about Kuwait was true. It had indeed restricted output to 3 mbd because of fears, later seen to be groundless, that reserves were overstated. [PPS 4-72:131, 5-72:180] But no other nation followed Kuwait's lead. Iran, for example, had rejected output restriction [PPS 6-72:215] and announced it expected output to double under the new agreement with the consortium. [PIW 7-3-72:6; PPS 8-72:283]

In the Middle East, 53 percent more wells were drilled in 1972 than in 1971; Saudi well completions nearly tripled [*World Oil* 8-15-73:65] as Saudi Arabia aimed at 20 mbd capacity by 1980. [PIW 10-9-72:3] The Aramco board chairman had congratulated its personnel for meeting the challenge of a huge construction program during 1971. [*Aramco Review*, 1972, p. 1] We have already seen that Iraq and Libyan production was down only because they were priced too high and that Iraq expelled IPC for producing too little. As for Venezuela, it had decreed "stiff" penalties for exporting too little. [PIW 4-15-72:4]

Akins was unaware of all this. He feared a world shortage because "the Arabs would find it difficult or impossible to raise the enormous sums of capital" needed for production. [OGJ 6-12-72:66]. In 1970–1971, total Middle East production capital expenditures (nearly all by companies) were $725 million. [CMB 1971] Middle East government revenues were $10.8 billion. [PPS 11-73:416] In 1972, the ratio was even higher than 15 : 1.

The Worldwide Scare: Preempting Supply

In October 1972 Sheik Yamani visited Washington to propose an agreement that would give Saudi oil a "special place" in the U.S. market, to "practically guarantee its continuous flow to these markets." With no "special place," there was no guarantee. The proposal drew much attention, "coming as it does amidst awakening fear of an energy supply shortage." [PIW 10-9-72:3] The State Department was "enthusiastic." [OGJ 10-9-72:35, 39; see also WSJ 10-16-92:A1]. It was obviously illegal under the General Agreement on Tariffs and Trade (GATT). Later State said it wanted no formal agreement; informal arrangements would suffice. [OGJ 12-11-72:53] The proposal nevertheless scared other nations:

> Fears of Europe and Japan that Middle East oil they were counting on to cover their needs will be diverted by the United States for its own use in a few years weren't dispersed by the recent explanations of [Undersecretary Irwin to an OECD meeting]. If anything they're now even more worried because the U.S. clearly warned them that nothing will stop it from importing the oil it needs. [To counter the "threat"] France may try to do something ... when ... King Faisal makes a state visit next year. [PIW 11-6-72:1][6]

Prince Saud al-Faisal, deputy oil minister and the king's son, said that "just opening the U.S. market for large volumes of Saudi oil would mean a shortage for Europe and Japan." [PIW 11-20-72:7]

Thus the message was driven home that consumers should fear preemption and should bid for a "special relationship" for "access" to adequate supply. The overtures of the French (recently kicked out of Algeria) toward Iraq allegedly "assured" and "guaranteed" a continuing supply of Iraq oil. [NYT 6-14-72:65; WSJ 6-12-72:9, 6-19-72:7] In December, the Japanese government committed $780 million for a 22.5 percent interest in Abu Dhabi offshore, for which BP had declined to pay $200 million. [NYT, WSJ 12-11-72] The Ministry of International Trade and Industry (MITI) wanted an international agency to control the flow of crude oil. [PPS 2-73:74]

Official Truth in 1973

Oil Supply Depends on Foreign Policy

President Nixon asked Kissinger, Secretary of the Treasury George Shultz, and White House adviser John Ehrlichman "to study the relationship between energy policies and foreign and security concerns." [Kissinger 1982, p. 870] Accordingly, Akins was loaned to the White House staff in January. He told an Aramco employee

> something that was not known to the State Department and known to very few in the White House. The President's chief domestic aide, Ehrlichman, was planning a trip to Saudi Arabia around April.... Akins wanted Ameen [of Aramco] to tell Yamani it was very important that he, Yamani, take Ehrlichman under his wing and see to it that Ehrlichman was given the message we Saudis love you people but your American [Middle East] policy is hurting us. [Aramco cable, quoted at 7 Church 423]

Thus, a government official used a private person and a foreign minister to change his own government's foreign policy. He would do it again. Soon Yamani threatened that any attempt by consumer nations to ward off OPEC price increases would mean "war ... their industries and civilization would collapse.... [However, OECD is prepared] to do everything for the sake of the continuation of the flow of oil." [PONS 2-22-73] He repeated the threat of economic collapse in March. [OGJ 3-26-73:42]

One began to read articles headlined like this one: "Oil Diplomacy: Shortages of Energy may Spur US Foreign Policy Shift." [WSJ 1-30-73:18] In the Nixon administration, some saw oil as a bilateral problem with the Saudis. "More and more experts were arguing that the level of Saudi oil production would be heavily influenced by our attitude toward Israel." [Kissinger 1982, p. 871] Akins had said earlier that the sellers' market in oil had arrived in June 1967. [OGJ 5-15-72:50] He was wrong, but the statement showed his political agenda, as did his message to Aramco.

During 1973, repeated Aramco briefings to visitors and representations to the government voiced the same line: there was an "energy crisis," U.S. imports would grow very rapidly, and the only country that could supply them was Saudi Arabia, but it was not in the Saudis' economic interest to expand. Therefore it was necessary to offer them a special inducement, to ensure that their "pro-U.S. policy" of higher output would continue.[7]

The main themes of President Nixon's April message [PPS 5-73:164] were that Saudi Arabia was sacrificing its economic interests by expanding output, hence the United States owed them, and that the United States needed to be more self-sufficient. However, there was no indication of how to translate themes into policies.

Natural Price Is Monopoly Price

In April 1973, Akins gave what he explicitly called the State Department rationale of higher prices: the price "would rise by 1980 to a level equal to the cost of alternate sources of energy" (p. 464). Such a level assumes monopoly control, for only when there was no longer any oil-on-oil competition could its price rise to equal the nearest alternative.[8] But monopoly was a necessary not sufficient condition. We will see how the cartel learned this lesson the hard way. Akins never learned it.

The Potent Myth: Selective Embargo

Akins's most important thesis was that the Arabs might institute a selective embargo (or boycott), starving the United States while supplying friendly nations. [Akins 1973, pp. 467–468] In fact, it was impossible to boycott only one or two countries while keeping the rest supplied. The world market, like the world ocean, is one great pool. Ships could be diverted from their nominal destinations. Crude oil would be charged into a refinery that commingled crude oil from various sources and shipped them to various places. Most important, if the target area was undersupplied, prices there would rise, giving the nonboycotters incentive to ship there, foregoing less profitable sales. The result would be a customer swap.[9]

In remarking on the failure of the 1967 embargo, the secretary-general of OPEC, Francisco Parra, had pointed out that a selective boycott could not work. To explain the failure, Akins had to invent the myth that it was "lifted through the efforts of Saudi Arabia." In fact, Yamani said "that non-Arab producing areas are realizing 'undreamed of profits' in supplying oil to 'those we consider to be our enemies.'" [PIW 7-31-67:8] It is sometimes claimed that Akins's article stated only his personal views. [Skeet 1988, p. 84] In fact, it stated the view of those in authority: it codified what Akins had said many times, without their rebuke or counterstatement; his superior in rank, Undersecretary Irwin, had echoed his views; his being loaned to the White House was a stamp of approval; and Secretary Kissinger believed in the reality of a selective boycott and acted on his belief, to his eventual discomfiture.

An *Economist* supplement said Akins's views "are apparently the model" for oil executives' speeches and reports, and "have heavily influenced public comment." [*Economist* 7-7-73: Survey: 11] It later summed up:

> Many Americans ... think that the OPEC countries should be granted substantial increases. The most forthright spokesman among them has been Mr. James E. Akins. They maintain that only steep increases will spur America and other industrialized countries to start a crash program to develop alternative sources of energy [which] ... are needed, they argue, if the world is going to avoid an energy crisis in the 1980s or 1990s when oil production, even assuming perfect cooperation from the Arabs, may not be able to keep pace with world demand.... Price increases hurt competitors, Europe and Japan, much more than they will hurt America. [*Economist* 11-26-73]

So certain was "an energy crisis in the 1980s or 1990s" that it was vital to take drastic action immediately. An industry journal ventured a mild remonstrance: "In the past year or so, exaggerated talk of an energy crisis—in effect, anticipating problems that may not become pressing for another decade or so—has greatly strengthened the bargaining power of the Arab states." [PPS 11-73:404]

Market Tension in 1973

There was increasing reason to expect higher prices as 1973 wore on, and it had nothing to do with consumer demand or adequacy of supply.[10] OPEC demands accelerated before the year began. On December 14, 1972, Saudi Arabia had demanded an increase in the price Aramco would pay it on "participation" oil sold back to Aramco (above, p. 97, and note 4). "Yamani has made the decision to break the agreement [of June 1972]." [Church Report 136][11] He had already made the threats about "collapse."

In January 1973, the shah of Iran told the consortium he would observe the June 1972 agreement only until 1979 and only if it increased production from the current 4.5 mbd to 8 mbd. He cited "the widespread talk of a world energy crisis, and particularly the shortage predicted for the United States." [PPS 2-73:47] The consortium agreed to be a service company and to give up all producing, refining, and export facilities "in return for a guaranteed long-term supply of crude." [PPS 4-73:145]

In March, Iraq expropriated the rest of the IPC concession, without compensation. [PPS 4-73:124] Then it seized a Kuwaiti border post and claimed two Kuwaiti islands. It had long claimed all of Kuwait but did nothing at this time to enforce either claim.

In May, the Japanese minister of trade and industry (and future prime minister), Yasuhiro Nakasone, after a visit to the Persian Gulf, declared the need for "strong and continuous petroleum diplomacy.... Japan [is] a major consuming nation with clean hands in Mideast oil.... Japan will not take part in any consortium of consuming nations." [PIW 5-14-73:5] U.S. senator William Fulbright warned that U.S. Middle East policy "could well lead to a selective boycott of the United States, coupled with the establishment of exclusive political and business arrangements with Europe and Japan." [PIW 5-28-73:6] The United States, he said, "could virtually assure access to Middle East oil by changing its policy." [OGJ 6-4-73:37]

Controversy on what Middle East policy ought to be solidified the illusion that there was in fact an "energy crisis," or an "access problem," or that there could be a "selective boycott." In mid-April, Akins proposed a world commodity agreement to set oil prices and ensure "availability." [NYT 4-16-73:29]

During the spring and summer, King Faisal sent messages through diplomatic channels and through Aramco warning that oil production might be reduced, or at least not expanded, if the United States did not modify its Middle East policy. [7 Church 504; PIW 7-16-73:5] He also gave a television interview on August 9, as did Yamani on September 2. To be sure, "each time King Faisal made some reference to the use of oil as a weapon against the West ... one of his aides would hasten to inform United States officials that this was meant only for domestic Arab consumption." [Quandt 1977, p. 160n.]

In July, Kuwait repudiated an agreement, signed in January, for 25 percent participation, rising to 51 percent in 1982. "Thus the participation agreements in the four countries [Abu Dhabi, Qatar, Saudi Arabia, Kuwait], only 6 months old, already seem doomed. They were the result of nearly a year's negotiations, led by Saudi oil minister Yamani, and they provided the pattern for a similar agreement in Nigeria." [OGJ 7-16-73:64]

Agreements that took a year to negotiate and only six months to cancel show the rapid acceleration of OPEC power. In fact, the agreements were also repudiated in Iran and Libya. President Nixon proposed a cabinet-level Department of Energy and named a national energy policy coordinator, George C. Love, who argued that no revenues were high enough to induce the Saudis to agree to big production increases; something extra must be done for them. [WSJ 8-15-73]. "There are sound economic reasons for the Saudis to say oil is better in the ground.... [W]e have to take a closer look at what we can do to make it to their advantage to export oil to us." [OGJ 9-10-73:41] This was a self-fulfilling prophecy, for if we believed it and were willing to do something extra for the Saudis, they would be glad to demand it. The buyer is asking to be had who tells a seller, "I know you don't want any more sales, but just to do me a favor, won't you sell me something?" There are few such private buyers because they do not stay in business very long.

During the summer, National Security Council memoranda "called for more distance between the U.S. and Israel." [Spiegel 1985, p. 243] At the end of August, Akins was appointed ambassador to Saudi Arabia and was expected to leave for his post soon. Saudi officials were pleased. [PIW 9-3-73:10]

On September 4, the State Department revealed:

> The Saudis ... have been threatening to slow their expansion of production unless Washington modifies its support for Israel. [State] had made a discreet effort to nudge Israel toward rethinking her position.... A number of department officials feel that this effort should be continued and intensified because of the United States' growing requirements for Saudi oil.... "There is a growing recognition in all circles here that some accommodation has to be reached," one official said. "It will be a settlement less attractive than Israel could have had last month." [NYT 9-5-73:1]

The next day, September 5, the president blamed both Israel and the Arab states for the impasse and stressed U.S. dependence on Middle East oil. [PIW 9-10-73:1]

Prices kept rising through September. Because the concession-holding companies feared they would lose some of their output to "participation" oil and knew prices were rising, it was prudent, and also profitable, to hold the oil instead of selling to third parties, which were thus forced to go directly to governments and bid for the "participation" oil. In May, June, and July, various governments sold to independents at prices ranging from $2.39 to $2.60 per barrel. [PIW 6-25-73:1, 7-30-73:1, 8-6-73: Supp 1] This switch of suppliers concealed an inventory buildup by the concession companies and added to excess demand, which at the start of the year reflected mostly the anticipation of a still-higher tax and price. As fears increased following Saudi threats, so did the demand for hoarding.

We can make an approximate calculation of inventory buildup in 1973. Worldwide noncommunist crude oil inventories increased by 419 million barrels through the end of 1973. [IPA 1976, p. 23] But there was a large drawdown in the last quarter: 133 million barrels for nine OECD countries excluding Italy, through the end of January. [CIA-IESR 1978] The worldwide drawdown must have been greater. Therefore the worldwide inventory buildup for the 273 days from January through September must have been substantially above 552 million barrels, or 2 mbd. Outside the oil industry, there must have been substantial buildup of product inventories by consumers, who had substantial storage capacity. The incremental precautionary and speculative demand must have exceeded 5 percent.

In mid-August, PIW said prices were becoming "unprintable"—that is, so surrounded with special and surreptitious terms as to make even experienced observers unable to compare one with another. [PIW 8-13-73:3]

In a competitive market, volatility up is matched by later volatility down. But the OPEC countries' ratchet would prevent the relapse and make the higher prices permanent. The more they raised taxes and prices, the greater was the fear of more price increases, the more the precautionary demand for oil, and the more the price increases (figure 5.4).

In June, there was yet another "supplemental" agreement to increase the take. An oilman summed up the situation: "Sometimes I wish we would straightforwardly pay the host governments more money outright, and stop trying to find excuses, principles, and rationales to justify increased payout." [PIW 6-11-73:2] This common sense was incorporated at the end of July in the new agreement of Iran with the consortium. It included a "balancing margin": Iran was to receive total payments "no less favorable than those applicable (at present or in the future) to other countries in the Persian Gulf." [PIW 7-23-73:1] Thereby tax changes in Iran became a record, sometimes made public, of tax changes elsewhere. For the historian, it is a bit of luck.

The Tehran agreement was effectively buried on August 4, 1973, when Yamani told Aramco that "the next change [in tax] will be a very large one and not a real negotiation....The companies will have no choice." [cable cited in Church *Report* 148] Soon other countries notified the companies that Tehran was no longer binding. [NYT 8-24-73:1] When on September 6, Yamani announced that the agreement was "dead," the news was already a month old, and his views had been coordinated with the other Persian Gulf producers. [NYT 9-8-73:39] The system shown in table 5.1 (which predated Tehran) remained in place. Yamani again repudiated nationalization [PIW 9-10-73:9], as he had for years. [Mikdashi Cleland & Seymour 1970, pp. 211–233] Yamani wished to preserve the integrated structure and prevent the emergence of a crude oil market where governments sold to oil companies. Events were to prove him right.

The new departure was the acceleration of tax increases initiated by Yamani's August 4 warning of "a very large one." A special OPEC meeting on September 15–16 fixed October 8 for a meeting with the oil companies to set new price schedules. [NYT 9-16-73:18, 9-17-73:1] All that they had not settled was "the precise formula and mechanism." [PIW 9-24-73:2] But before the end of September, Aramco and Saudi Arabia had already agreed on "a substantial increase" in take per barrel and a speedup in participation, to 51 percent in 1982. [PIW 10-8-73:1]

Two days before the October 8 meeting, the October War began. [PIW 10-15-73:1] Days later, "one would hardly know there was a war going on." Spot prices were few and indecipherable, as they had been for months. The substance of the meeting had been prepared beforehand. [Ibid.] Government take on Arab Light was raised from $1.77 to $3.04. As one oilman said, "They are no longer even going through the charade of negotiations." [NYT 10-19-73:61] There was no more formal price action until end-December.

The Production Cutbacks

On October 17, the OAPEC (Organization of Arab Petroleum Exporting Countries) nations agreed on an immediate 5 percent production cut per month, "until the Israeli withdrawal is completed from the whole Arab territories occupied in June 1967, particularly Jerusalem, and the legal rights of the Palestinian people restored." [NYT 10-18-73:1; on Jerusalem, see also PIW 11-12-73:5] A few days later, Saudi Arabia (and Kuwait) cut production more sharply. [PIW 10-22-73:1, 10-29-73-1] They were "entirely aware that this program would be very difficult to administer but they are looking to Aramco to police it." They also began classifying consumer countries by degree of friendliness [Aramco cable 10-21-73, 7 Church 515]

On October 21, Ambassador Akins sent Aramco a message urging industry leaders to "use their contacts at highest level of USG [U.S. government] to hammer home point that oil restrictions are not going to be lifted unless political struggle is settled in manner satisfactory to Arabs." [Aramco cable 10-26-73, 7 Church 517] As he had in January, the ambassador appealed to private parties to change his own government's policy. Moreover, he must have told his own government that the Arabs would not desist until given satisfaction.[12] This proved to be far wrong. He had been making promises to the Saudis since 1967 that the U.S. government would meet all or most of their demands. This too proved wrong. He misinformed both sides. Disappointment and chagrin would account for his later anger at Kissinger. (See appendix 5A.)

Later the Arab producers further divided consuming countries into what eventually became four categories: (1) "most preferred" (Britain, France) with no set limits, (2) "preferred," 100 percent of September shipments, (3) "neutral," reduced shipments, and (4) the United States and the Netherlands, under "embargo." The embargo violated a 1933 treaty with the United States. The State Department knew this but was "reluctant to discuss the matter." [NYT 12-19-73:12] The embargo also violated a UN General Assembly declaration unanimously approved in 1970. "Government officials are aware of this issue too, but are equally reluctant to discuss it." [Ibid.]

The Extent and Impact of Production Cuts

Table 5.4 shows the pattern of cutbacks from October to December. January was no longer a month of cutback, for demand was insufficient. Over the three months October through December, total lost output was about 340 million barrels, which was less than the inventory buildup earlier in the year. Considering as well some additional output from other parts of the world, there was never any shortfall in supply. It was not loss of supply but fear of possible loss that drove up the price. Nobody knew how long the cutback would last or how much worse it would get. Additional cuts were scheduled.

Precautionary demand was driven by the fear of dearth. Oil might be only a small fraction of a buyer's total cost of operation, but without it, a factory, or a power plant, or a truck fleet would stop dead. The loss was so great that it paid to take out expensive insurance against even a minor probability. Panic aside, it made sense for refiners and users to pay outlandish prices for oil they did not need.[13]

Speculative demand included those seeing a quick turnover profit or crude oil buyers trying to buy sooner rather than later. But an additional factor may have been even more important: oil product prices were largely controlled by contract or government. Every buyer and seller at the much lower mainstream prices knew that if the production cuts continued, those prices would also rise. Moreover, OPEC had nearly doubled the per barrel tax in October and would again.

Thus, buyers and sellers could hold crude oil or products with little downside price risk. Their increased demand raised prices all the more. "The spot crude oil market dropped dead last week ... as sellers decided to hang on to every barrel." [PIW 10-22-73:7] Those with stocks of oil or products sold as little as possible. Some sought to buy for an immediate

Table 5.4
Arab oil producers' output cuts after September 1973 (mbd)

Month	Saudi Arabia	Kuwait	Oman	Iraq	Qatar	Abu Dhabi	Libya	Algeria	Total	Percentage Cut	Cut as percentage of noncommunist world consumption
September	8.29	3.24	0.30	2.11	0.62	1.36	2.25	1.12	19.29	0.0	0.0
October	7.72	2.82	0.30	1.80	0.57	1.36	2.46	1.09	18.12	6.0	2.5
November	5.89	2.37	0.30	2.15	0.48	1.17	1.77	0.95	15.07	21.9	9.1
December	6.40	2.34	0.30	2.16	0.46	1.03	1.77	0.91	15.36	20.4	8.5
January	7.26	2.58	0.30	1.82	0.52	1.22	2.03	1.14	16.87	12.5	5.2
Minimum production as a percentage of September	0.71	0.72	1.00	1.00	0.74	0.76	0.78	0.81	0.76	—	—

Source: PPS 8-75.
Note: Noncommunist world consumption taken as 46.4 mbd.

resale gain, others to hold for higher prices soon. Thus the effects were out of all proportion to a loss of at most 9 percent for a month.[14]

Not the amount of cutback or "shortfall" but the fear of dearth did the damage. The lesson was ignored. The International Energy Agency, created to mitigate or prevent another such disaster, set a threshold of 7 percent for declaring an emergency, below which nothing needed to be done. (The mistake was proved in 1979, when the cutback was also small and easily made good but again convulsed the market. Time-wasting debate continued about how great the reduction needed to be to trigger a ponderous machinery whose slow motion made it useless.)

The "Embargo"

The production cuts and higher prices were real. So were the mile-long gasoline lines that endure in memory. But the shortages were created entirely at home, the result of price controls and allocations.

The reality of a selective embargo was questioned from the start. [NYT 10-18-73:1; LM 10-19-73] A *New York Times* editorial [10-18-73] overlooked its own news story, but the *Wall Street Journal* was properly skeptical. [Editorial 10-19-73]

As predicted non-Arab producers did well by diverting oil to "embargoed" countries. The oil companies deserve much credit for doing the complex logistics of swapping customers [FEA/Church 1975], but not for the inevitable result: all areas suffering a similar relative loss of supply.

There are two independent measures of loss, which take account of lags between crude output and product consumption. The Federal Energy Administration (FEA) compared January–April consumption, 1973 and 1974. The Arab producers viewed the United States as the target and Japan as "odiously neutral". Adjusting for the expected increase, each lost 11 percent. "Friendly" Western Europe lost 18 percent. [FEA/Church 1975, p. 8][15] Another measure is the drawdown of crude oil and product inventories in nine large consuming countries, from the end of September 1973 to the end of March 1974. The reduction in the United States was 4 percent. Three countries, including Japan, drew down less; five drew down more, including Britain (11 percent) and France (12 percent). [CIA-IESR 1977, p. 20]

Since the "most preferred" French and British did worst, perhaps that proves it did not pay to let the Arab producers take them for granted. Or perhaps everyone suffered about the same loss and the variations are only noise in the statistics.

Years later, Yamani said that the embargo "did not really imply that we could reduce imports of oil to the United States.... The world is really just one market. So the embargo was more symbolic than anything else." A former UAE ambassador to Britain said: "There was no embargo.... It was a lie we wanted you to believe." [Robinson 1988, p. 96] Akins continued to believe it.[16]

Frightened Consuming Nations

The consuming nations tried hard to have "access" and "good relations," and to make friends with the tiger so he would go eat someone else. The *Wall Street Journal* wrote, "Arabs don't have to police their own boycotts. Sycophant nations are doing it for them." [11-6-73:14] "Aramco acts as cutting edge in Saudi Arabians' U.S. embargo," noted the *New York Times.* [11-4-73:C1] And Aramco refused to supply American armed forces, which particularly affected the Mediterranean fleet. [NYT 12-18-73:5; 7 Church 514; Church *Report* 150]

The European Economic Community (EEC) nations, for their part, violated their own constitution, the Treaty of Rome, which forbids barriers to the movement of goods. "The French and the British feel they have a privileged position in the Arab world and ... they fear the Arabs would interpret European cooperation as confrontation." [NYT 11-5-73:61] Hence they opposed any joint action by the EEC or by the OECD. [NYT 11-5-73:61] The British secretary for trade and industry said there was no need for rationing because he had "assurances from Arab states." [NYT 11-6-73:49]

Soon "the French were shaken" when Algeria announced a cut in deliveries to them, and "in Britain the government was slowly beginning to admit that the oil crisis ... can't be ignored simply because of Arab 'assurances.'" [OGJ 11-12-73:91] The cut in deliveries embarrassed the British government, "which ... had received assurances from Arab countries." [POPS 11-20-73] The oil companies informed the French government of a 10 to 15 percent cut in oil deliveries, which came "as an embarrassment to the French government. [Oil was going] to other countries in short supply, where, as a result, prices are higher. The Netherlands, for instance,... is getting more non-Arab oil, some of which could have been destined for France." [NYT 11-22-73:61; P. Simmonot in LM, same date] "But official France has been pleased with what it calls its 'privileged position' in the Arab world, and has taken the lead in persuading other European countries to endorse pro-Arab diplomacy as the best defense against the 'oil weapon.'" [NYT 11-24-73:1]

The French government was "convinced that its special ties with the Arabs should protect it from shortages imposed on 'neutral' countries." [PIW 11-26-73:7] It also hoped that its "special ties" would procure a disproportionate share of exports and contracts. President Georges Pompidou appeared to have talked of little else but French exports during a meeting with King Faisal in May 1973 [LM 5-15-73] but professed to be "shocked" that anyone would think exports had much to do with his Middle East policy. [LM 11-19-73] At the end of December, the French reported a "big deal for Saudi Arabian oil." [NYT 12-25-73:37]

At an EEC meeting on November 19, the Dutch appealed for help but were met with silence. [NYT 11-20-73:9] The next day, the Dutch warned they might restrict natural gas exports "unless oil moved freely within the Common Market," and they "emerged apparently satisfied." [NYT 11-21-73:12]

As for the United States, Yamani awaited the cold: "This winter, when there is a shortage of fuel in the United States and your people begin to suffer—the change will begin. Americans are not used to being uncomfortable." [NYT 11-7-73:11] Ambassador Akins "indicated that fuel supplies to the East Coast of the United States would become critically short this winter if Arab oil supplies were not increased 'in a matter of days.'" [NYT 11-10-73:12] This was a far-fetched notion. First, nothing could take effect within days. A voyage from the Persian Gulf to the U.S. East Coast takes a month. Moreover, electric power companies held about 58 days' supply of fuel oil at the end of 1973. [DOE: MER 8-89:81] If their supply were cut by, say, one-fifth (twice the actual case), stocks would last 288 days, plus the time gained by switching to coal or buying power from coal- or gas-burning plants. The Arab exporters secretly resolved in late November to cut off all heavy fuel shipments to the Caribbean and other transshipments points. [OGJ 12-10-73:48] But the U.S. government took comfort: "[A] very high official who is a policy maker in this area [said] ... he feels King Faisal ... at the last minute would prevent any serious economic harm from being done to this country because he is at heart a friend of the United States." [MP 11-25-73]

In Japan, the Arab producers' charge of "odious neutrality" must have struck MITI minister Nakasone very hard in view of his past disengagement from any common front of consumers and also because Japan had openly cooperated in the Arab boycott of Israel for years. [NYT 4-21-68:20] Japan now refused demands to break diplomatic relations with Israel, or to cease trade, or to give military aid to the Arabs "in return for a stable oil supply." In making these demands, the Arabs "appear to have been influenced by their success in recent years in enforcing a boycott [by Japan against Israel]." [NYT 11-9-73:13] Some Japanese began to see that those who try to please will be told to try harder, that "it is very easy to apply pressure to and manipulate Japan."[17] [Kimura 1986, p. 75] By March, when Japan was reclassified as friendly, the "embargo" and the production cutback were both long gone.

During December, Yamani and the Algerian minister Belaid Abdesalam toured Europe and conferred with EEC representatives in Copenhagen. At a special meeting with Dutch representatives, they asked for acceptance of their previous demands plus "a special gesture," both of which the Dutch refused. [NYT 12-2-73:12] They held a press conference that "reeked with ridicule and defiance of Europe." [OGJ 12-24-73:22] The British and French proposed resolutions, which the West Germans and Dutch opposed. [LM 12-18-73] The small powers regarded Pompidou and Prime Minister Edward Heath as not being "good Europeans" and were "convinced that M. Pompidou had underhandedly rigged the Copenhagen meeting in order to force their hand." [LM, "Les malaises de l'Europe," 12-18-73]

Italians asked why they were not recognized as friends, "although [Italy] had been trying so hard to earn that accolade ... [with] 'a new foreign policy' favoring the Arab countries in the interest of undiminished fuel supplies." [NYT 12-31-73:3]

The End of Production Cutbacks

A report on November 9, after only three weeks, "that the present cutbacks in oil output are the limit" [NYT 11-10-73:14], proved to be correct. On December 4, the Saudis canceled the month's scheduled additional 5 percent reduction. [WSJ 12-5-73:3] No reason was given. Perhaps it was continued high-level production by Iraq and Iran, which had rebuffed Arab demands for both lower production and expulsion of Iranian Jews. [NYT 12-18-73:61] On December 5, Yamani said production would be increased "step by step" with Israeli withdrawal from occupied Arab territory. [NYT 12-6-73:9]

It was becoming clear that the shortages were not nearly as acute as feared. [*Economist* 12-15-73:44] West German inventories had actually increased during the first half of December 1973, and a survey of seven West European ports showed increased unloading, although "information about actual flow of oil is a tightly guarded secret of the multinational oil corporations. These have managed to install a de facto system of oil sharing, which governments themselves have not been able to agree to, insuring that countries such as the Netherlands, on the Arabs' total embargo list, continue to get adequate supplies." [NYT 12-22-73:73]

By the end of the month, the diversion to the Netherlands was clear. Only the information blackout had delayed its recognition. [NYT 12-31-73:27] In the United States, "Tankers line up off East Coast, seeking dock space to unload." [NYT 12-29-73:1] There was no support for the story that the oil companies had deliberately kept tankers on the high seas, waiting for the expected price rise.

In mid-December, Yamani said on "Meet the Press":

> We were producing at a much higher rate than what we should for our economy. And that was a sacrifice on our part. This sacrifice should be appreciated by the whole international community so we can continue sacrificing, continue accumulating a surplus and losing money. We are depleting our natural resources.... If we have to produce more, then this is because of you, because we have to please you and help your economy." [PIW 12-17-73:1]

He did not say how often he had heard it from Americans that Saudi Arabia was subsidizing the world economy and was owed something in return.

On December 25, the Arab oil ministers met in Kuwait. Having canceled the additional December cutback, they now ordered a 10 percent increase for January—about 3 mbd, or 15 percent below the September peak. However, the meeting made even that cutback of no effect: "[T]hey decided to export to friendly countries according to their actual needs, even though this is more than the level of September 1973." Japan and Belgium were added to the "friendly" list. [NYT 12-26-73:65] But leakage and diversion of non-Arab oil would spread the newly unlimited increase from "friendly" countries to all. Perhaps most important, the amount supplied could not be sold in January.

There was no further mention of the original demands: Israeli withdrawal and Palestinian rights. Indeed, after the first week in December there are none in the regular periodical sources we use (*Economist*, NYT, OGJ, PE, PIW, WSJ). The announcement raising production limits never mentioned it. The Arab exporters had said that "the embargo would be lifted when ... the United States guaranteed Israel's acceptance [of] the principle of withdrawal from Arab lands ... [and] that production would be increased in steps parallel to the actual Israeli withdrawal." In fact, Israel has not accepted the principle.... Why then today's announcement? [NYT 12-26-73:1]

The Arab exporters' political demands had had an effective life of two months, from October 16 to early December. As for the embargo, Secretary Kissinger was to write years later: "The structure of the oil market was *so little understood* that the embargo became the principal focus of concern. Lifting it turned almost into an obsession for the next five months.... In fact, the Arab embargo was a symbolic gesture of limited practical importance." [Kissinger 1982, p. 873, emphasis added]

"So little understood"—by whom? Several persons, including a former OPEC secretary-general and me, had publicly explained why a selective embargo could not work. The 1967 attempt had been a notorious failure. But for five months—after another failure became evident, after the cutbacks ended, after supply was visibly in excess—the U.S. government was ruled by an obsession to cajole the Arab exporters into ending the nonexistent "embargo."

Prices during the Cutback: The Path to $7

The October 16 tax at $3.05 per barrel set the price floor. The market process could affect only the margin over tax-plus-cost. It is impossible to give any coherent account of the margin, or the price, after October 16. "The spot market dropped dead last week ... as sellers decided to hang on to every barrel they could." [PIW 10-22-73:3] Oil companies faced "an endless ratchet.... A full pass-along of the recent cost [i.e., tax] increases would boost market prices to a new plateau, presumably touching off another round of posting hikes—and so on." [PIW 10-29-73:7] The trade complained that "spot oil markets distort Europe's true pricing picture.... Refiners have avoided buying spot products because of high prices, and don't want to sell product because of the crude shortage." [PIW 12-3-73:5]

But even if we had a reliable sample of spot prices, their causation would be distorted by anticipations. In any market, precautionary and speculative demand added on top of consumption demand will send prices above the level required to balance supply with consumption demand. The temporary change is reversed, and usually soon. Indeed, the speculator's most important skill is to sense how soon the reversal is coming. A famous speculator (Bernard Baruch) explained his success: "I always sold out too soon."

The concern of the OPEC nations was to ratchet up the floor to prevent the down-wave from cancelling the up-wave and to make the price increases permanent. Since the OPEC action was anticipated, the price effect came before the cause. In December, Iran auctioned some crude oil at more than $17 per barrel, after exaggerated reports of an imminent unprecedented jump in government take. [PIW 12-17-73:1]

On December 22, Kissinger sent messages to all OPEC governments "warning strongly against another rise in prices." The Persian Gulf OPEC nations soon responded: the take on the official "marker" crude oil, Saudi Light 3, would rise from $3.04 to $7.01. The meeting "focused this time not on posted prices ... but on the actual governmental 'take' (taxes and royalties)." From the $7.01 tax was derived a posted price of $11.65. [PIW 12-31-73:1]

More "participation" had already been demanded. [PIW 11-19-73:1] The Kuwaiti minister's prediction of even higher prices in the spring [NYT 12-23-73:31] led to shocked surprise in the consuming countries. Kissinger later wrote, "We felt misled by weeks of exchanges with various Arab leaders. The huge rise in prices was made doubly wounding by being so totally unexpected.... Yamani, in conversations with Akins, predictably blamed Kuwait, Abu Dhabi, and Iraq ... as he had previously blamed Iran. [Nixon sent Faisal] ... an unusually stiff message." [Kissinger 1982, p. 890] Years later, Kissinger would call both Iran and Saudi Arabia "specially dexterous. Whoever we approached made a convincing case that the other was the culprit.... One was left wondering how prices could ever rise in the face of so much reluctance." [Kissinger 1982, p. 888]

But the shah had talked himself into myth as the price hawk. His "new concept"—that the price of oil should be equated to the cost of competing forms of energy [NYT 12-25-73:31; PIW 12-31-73:3]—was not so new; the US. State Department was ahead of him. But the $7.01 tax was based on that equality, as it was then estimated, and not on "the market situation."[18] It became overwhelmingly clear in 1974 how little the market situation explained the price.

Taxes Override Supply-Demand: First Half, 1974

We now trace the course of prices in 1974, first examining the balance of supply and demand to see whether prices would, under competition, be rising or falling and then looking at the actual course of prices, above all on new contracts. In earlier years, a contract price was typically a meaningless average of old and new deals. This was no longer true.

Supply and Demand Would Lower Prices

By the third week in January, there were "unmistakable signs that some buyers are getting increasingly wary ... particularly where prices are fixed beyond the immediate future." [PIW 1-21-74:3] In late January, spot prices "suddenly nosedive[d] around the world.... Tremendous quantities of oil suddenly have come into the market place—oil we didn't even know existed 10 days ago." [PIW 2-4-74:1]

February OPEC production was down by 9 percent [PE 11-74]. By March, when the Arab exporters declared West Germany and Italy "friendly" and rescinded the "embargo," the OPEC producers were openly worried about excess supply and agreed to study an Algerian proposal to prorate output. [PIW 3-25-74:3] By April, Saudi Arabia had 700,000 daily barrels of excess capacity. [PIW 4-15-74:1] Total OPEC shut-in capacity that month was later estimated at 4.5 mbd. [PIW 7-8-74:1] By August, it had widened to 6.3 mbd. [PIW 10-14-74:7]

In June, Middle East and African crude markets were "oversupplied" [PIW 6-17-74:2], and crude went "begging in world markets as oversupply persists." [PIW 6-24-74:1] By late August, the storage system was full, and production had to decline even more. A production control program would solve the problem, "but deep concern is expressed over OPEC's flexibility in 'fine tuning' output in line with seasonal demand variations." [PIW 8-26-74:1] The problem has never been solved.

Taxes Raise Prices

Had prices been ruled by supply and demand, there would have been a spectacular price collapse in early 1974 and a continued decline afterward. Nothing of the kind happened. In early May PIW reported that "crude prices drop as the spot market comes alive again." [PIW 5-6-74:1] But new contract prices continued to rise.

In 1973, the model of market behavior had been a ratchet (see figure 5.4): create a temporary shortage to drive up spot prices and then hold that level when the shortage ends. In 1974, however, it was back to the preratchet sequence; despite the huge and growing surplus, prices were not merely held but continued to be increased.

The mechanism outside Iran was participation. Governments were entitled to 25 percent of the oil, but they took only small amounts. The great bulk was sold back to the companies, at a price that was a percentage of

Table 5.5
Excise tax ("government take") on Saudi marker crude, 1973–1974

		1973		1974				1975
		Before October 16	October 16	January 1	March 31	June 10	November 10	
E	Government equity share	—	—	—	0.25	0.60	0.60	1.00
P	Posted price	3.01	5.12	11.65	11.65	11.65	11.25	
R	Royalty rate (% of posted price)	0.125	0.125	0.125	0.125	0.125	0.2	
T	"Income tax" rate	0.55	0.55	0.55	0.55	0.55	0.87	
C	Accounting cost	0.10	0.10	0.12	0.12	0.12	0.17	
B	Buyback price				10.83	10.83	10.66	10.66
X	Government take, \$/barrel = $[(1 - E)(P(R + T - RT) - TC)] + EB$	1.77	3.05	7.00	8.01	9.30	10.37	10.66

Sources: MEES; PIW; PPS; OGJ; NYT; WSJ.

the posted price and was far above tax-plus-cost. This increased the bottom-line government take per barrel. Table 5.5 restates table 5.1 under the new conditions. By changing lines E, P, R, T, C, or B, the final X could be increased to any desired number. The price history of 1974 is the gradual change in each of the six parameters to raise the final weighted take. An increase, from $7.00 to $10.66, in a glutted market is an even more impressive achievement than the $4.00 increase by the 1973 ratchet.

The agreement between Iran and the consortium made Iran in effect a free rider on the detail of table 5.5. It provided a single government take figure, with a "balancing margin" to equate it to whatever was being achieved on the western shore of the gulf. Finance Minister Amuzegar proposed that the producing companies charge buyers a price of $7.62 for the basic 34° crude oil (Iran Light, or Arab Light), composed of government take of $7.00 plus 12 cents cost plus a 50-cent company margin. But some other gulf governments were proposing an increase through participation, which would increase take to $7.60 [PIW 1-7-74:6] and make the price $8.10.

By the end of the month, some indicators of price trends began to appear. A large barter deal for Iranian oil was at only $7.45, but commercial crude buyers were "getting walloped with massive increases" imposed by the multinational companies, to around $8.32 on Arab Light, with "virtually no explanation from major suppliers" as to government take and company markup composing the price. [PIW 1-21-74:7] The companies said they were baffled: "'We simply don't know what to charge our customers because we don't know what we're expected to pay the governments.... So we have to guess at a price.' ... The companies have sold literally hundreds of millions of barrels of Middle East oil since October 16 without knowing to this day what margin of profit in the fourth quarter of last year." [PIW 2-4-74:1]

Kuwait held auctions for small amounts of participation oil that it had taken in kind. At first, prices ranged up to $20.00, but by the second week in February, only 25 percent of the oil offered was spoken for; bids were between $8.50 and $10.00, and all were finally rejected. [WSJ 2-14-74:2, 2-22-74:2; PIW 3-4-74:3] Disregarding the auction, Kuwait fixed its contract price on sales to Gulf and BP (table 5.5 line B) at a level that increased the take. This raised the balancing margin and total take in Iran. [PIW 3-11-74:1, 3] Saudi Arabia had just required 93 percent of posted price for buyback oil (line B). [NYT 2-25-74:39]

In early March, the level of government take was still noted as $7.00 by Kuwaiti oil officials. [NYT 3-5-74:43] But by the end of the month, transfer prices for 34° Arab Light had moved up to $9.25 to $9.30. [PIW 3-25-74:5]. Some oral accounts to me put government take at this time at $8. The OPEC secretary-general said it was "$7.50 or so." [WSJ 3-28-74:13]

Statesmen Posture

In late February, Secretary Kissinger left on his fourth Middle East trip in as many months and was well received. "Even King Faisal ... has treated Mr Kissinger well.... Saudi officials waived their normal visa requirements ... [to allow] Jewish newsmen in the party into the country. [They had refused to do this for a French group with Foreign Minister Jobert.]" [NYT 2-25-74:1] These reports indicate that the "embargo" was still atop the agenda. Nobody told the secretary what was happening in the market.

The Saudis had said in public that the price was too high, and King Faisal was reported as taking "very important steps" toward reducing the price. [NYT 1-28-74:1] He wrote personal letters to Persian Gulf producers, urging lower prices. [NYT 1-30-74:18]

In February, Secretary Kissinger convened a meeting of producer and consumer states "to promote long-term price and supply agreements." [WSJ 2-8-74:1] Such agreements would be a remedy. "Decades of inaction brought energy gap" headlined the *New York Times*. [2-10-74:1]

In early March, Yamani twice expressed "the hope that he may win support within OPEC for his desired reduction in the price of crude," but the shah of Iran insisted that the price would rise. [PIW 3-4-74:4; NYT 3-5-74:43]

Early the next month, the U.S. and Saudi governments announced a program of economic cooperation and weapons supply. [NYT 4-6-74:1] It appeared "precisely tailored to Saudi Arabia's position.... [Meeting] increases in world demand, beyond its own financial absorptive capacity, will depend on its ability to industrialize and diversify its economy." [PIW 4-15-74:1] The rationale was clear: only with special incentives would the Saudis expand output. At the UN, Secretary Kissinger "warned commodity producers against banding together to raise prices" and called for cooperation. "To maintain stable prices for raw materials ... an international group of experts [should] work with the United Nations in surveying resources and developing an early-warning system for scarcities and surpluses." [NYT 4-16-74:1] The world needed "an expanding supply of energy at an equitable price." [WSJ 4-16-74:3] In the real world there was an expanding flow of arms and goods to the oil producers. [NYT 4-25-74:1]

Rising Taxes Override Massive Surplus

In April, markets were weak [PIW 4-22-74:5, 7], but only in early May could it be said that "crude prices drop as *spot markets come alive again*." [PIW 5-6-74:1, emphasis added] A contract price for Arab light was at $9.50, about the same as in late March.

Late in May there were rumors of a Saudi crude oil auction in the summer. [PIW 5-27-74:1] Soon afterward another Kuwaiti auction was unsuccessful, and Kuwait warned buyers that failure to bid would shut them out in the future. [PIW 6-10-74:5] Not dismayed by weak demand, the OPEC nations were now discussing a proposal to jump the 55 percent tax rate, adding about $1 to the price; Europe was "gravely concerned." Saudi Arabia, however, would not "support any tax increases—at least not at this time." [Ibid.: 2]

The June Boost in Saudi Take

In June, Yamani called for a producer-consumer "'practical dialogue' on energy, raw materials and technology." [NYT 6-4-74:47] Shortly after, a milestone pact signed by the United States and Saudi Arabia created a joint economic commission and one on Saudi military needs. American officials

> hoped [it] ... would provide Saudi Arabia with incentives to increase her oil production, and serve as a model for economic cooperation between Washington and other Arab nations. American officials ... have made no secret of their hope that Saudi Arabia will take the lead in increasing production of oil ... and that way help bring about a drop in the world price. [NYT 6-9-74:1]

One account described the atmosphere among the Arabs at the Saudi Embassy as "close to euphoria last weekend.... There was a great throng of top American Government officials ... together with many private American industrialists and bankers. Said one American banker: 'Fantastic—imagine it, the great of the world coming to kowtow to the Arabs.'" [NYT 6-10-74:47]

The kowtow—knocking the head on the floor before the Presence—must be accompanied by rich gifts. The Saudis ensured it two days later, announcing that their ownership share of Aramco would rise from 25 to 60 percent and raising the take $1.30 by substituting participation oil for equity oil. [NYT 6-13-74:65; see table 5.5] Thanks to the "balancing margin" clause, Iran's increase "immediately followed the Saudi Arabia–

Aramco interim agreement on 60% participation." [PIW 6-17-74:1] At this point, spot prices were at the year's low. [Ibid.]

At the next OPEC meeting, at Quito in mid-June, an EEC message asking for no more tax and price increases drew "a frosty reaction." But Minister Yamani said: "We won't join them in increasing prices or taxes." [WSJ 6-17-74:6] The Saudis "disassociated" themselves from the meeting's increase of 2 percent in the royalty rate and let it be known that "the long-anticipated auction offering of a large volume of state oil" would be held before September. [NYT 6-18-74:1; PIW 6-24-74:8] The United States should "do more than it has until now to persuade Iran, an ally of the United States, to accept price restraint." [NYT 6-19-74:1]

It took two weeks for Washington to understand that the higher Saudi participation announced June 12 meant a higher tax and price. "'It's strange indeed,' notes one Washington energy official, 'that of all countries Saudi Arabia, which has been advocating a cut in world crude prices ... should now be the one to propose an approach that would have the net effect of increasing oil costs further, even though not designed for that purpose.'" [PIW 7-1-74:1]

The new participation steps also bothered the Aramco companies. They had not been disturbed by 100 percent Saudi ownership. They did not believe rumors about a Saudi desire for lower output, and they instead expected a big Aramco expansion, increasing their investment and profit. [PIW 6-17-74:1] "They would at least get the one thing that concerned them most: privileged access to long term crude supplies at special prices." But if they were to buy at official prices, there was nothing in it for them. [PIW 7-1-74:1]

The negotiations over the Aramco margin would drag on for years. Because the Saudis expected to expand output, the loss of experienced personnel and corporate memory would involve higher costs. That was the Aramco companies' bargaining asset. In contrast, Kuwait expected only to maintain production at the two operating fields and was therefore willing to allow BP and Gulf much less.

At midyear, Undersecretary of the Treasury Jack F. Bennett said OPEC excess capacity was 4 mbd, which put prices under pressure. Some in the producing countries were urging new sharp cutbacks, which should be regarded by all consuming countries as "an unfriendly act." At someone's urging, he later changed that phrase to "a counterproductive measure." [NYT 7-10-74:47] Actual production was estimated by oil companies at 2 to 3 mbd more than consumption [NYT 7-1-74:45] and by an Arab oil exporter at up to 4 mbd. [NYT 7-16-74:45]

In a long assessment that appeared in a *Wall Street Journal* report, the Saudis were widely seen as the good guys, the lonely fighters for lower prices:

> But a different picture emerges from conversations with sources close to the cartel and with key OPEC insiders. The cartel is banking on Saudi Arabia to prop up petroleum prices, not bring them down.... [T]heir concern about the world's economies hasn't overridden their desire for oil revenues.... OPEC's major problem right now [is] ... how to keep oil prices from crumbling....
>
> There was a surplus before the Arab oil embargo.... The surplus is even greater today....
>
> [When Saudi Arabia takes over Aramco] Sheikh Yamani has confided to others in OPEC that he will seek $10.83 a barrel.... The average [price] of Aramco crude is $9.35.... The Saudi settlement will quickly become the pattern for other Persian Gulf producers and, eventually, most other OPEC member nations....
>
> Some observers doubt that he can swing so high a price as $10.83 on such large volumes of oil. But who knows?... In any case, other OPEC officials wish Sheikh Yamani luck, for they expect that whatever price he negotiates will become the new benchmark—and floor—for new "market" prices established by the other OPEC countries.
>
> What isn't generally understood is the fact that, in a time of oil surplus, the cartel was able to make the increases stick only because of an artificial shortage created by Saudi Arabia's orchestration of the Arab oil embargo.... Saudi Arabia, the public advocate of lower oil prices, could singlehandedly bring down oil prices by stepping up production ... by at least 800,000 barrels a day.

But Treasury Secretary Simon disagreed. He expected "that oil prices will come down [because of] the decision by Saudi authorities announced during the Simon visit to Jiddah to auction off state-owned oil next month." [NYT 7-15-74:43]

The Saudi Crude Oil Auction

While waiting for the auction, various OPEC members were reported increasing royalty rates. [PIW 7-15-74:2] Kuwait had been disappointed with an auction held in early July, and so had other producers. [NYT 7-4-74:25; OGJ 7-15-74:28; PPS 8-74:293] Kuwait put it to the local producers Gulf and BP: although prevailing term contract prices were $9.50 to $10.00, they must pay $10.95 for participation crude (table 5.5, line B) or be cut off entirely. [WSJ 7-19-74:3]

The State Department scolded Gulf for giving in. [NYT 7-19-74:47] Saudi Arabia, where the weighted average take had been around $9.27 per barrel, was preparing to demand a single flat price and hinting it might go as high as $11.05. The expected increase would be matched automatically in Iran through the balancing margin clause. [PIW 7-22-74:1] In any case, "the question of which way oil prices will go will not really be determined until Saudi Arabia disposes of its participation oil." [NYT 7-20-74:41]

At the end of July, "Saudi Arabia's big crude oil auction [is] now expected soon.... [It] will be a real auction, without specifying any minimum price." [PIW 7-29-74:1] The offer would be 1.5 mbd, for at least four and possibly as long as sixteen months. [PIW 8-5-74:1] It conformed to the strategy sketched in the *Wall Street Journal*: if the auction price exceeded the current buyback price (table 5.5, line B), it would become the official price for all oil. "The scenario could actually lead to sharply higher prices for oil." [NYT 7-30-74:43]

August was the month for waiting. Japan advanced Iraq $250 million interest free for seven years, and then for eighteen years, at 4 percent, in addition to loans from commercial banks at below-market rates and technical assistance in various industrial projects. In return, Iraq was to supply 1.12 billion barrels over ten years. Price was not mentioned. [NYT 8-17-74:29]

At the UN World Population Conference, a resolution called for a New International Economic Order, "ensuring the equitable distribution of world resources, by eliminating the ... exploitation perpetrated by capitalistic multinational corporations." [WSJ 8-23-74:14]. This call echoed the April UN Special Session on Raw Materials.

In Washington there was worry that the arms sales to Iran were being matched by sales to its neighbors, making the whole gulf a more dangerous place. But the officials involved said it was to obtain "access" to oil, and they insisted that America's need for Persian Gulf oil gave them "little choice." [WSJ 8-24-74:1]

In mid-August, the Saudi auction was postponed. As the trade saw it, this was incidental. The important question was whether there would be a net increase in Saudi output. If the auction offerings were merely subtracted from Aramco's liftings, this would "negate the [Saudi] government's announced aim of lowering prices." But the world price of oil was harder than ever to figure because there were many types of open-market sales, which in turn differed from state oil deals. For any given company it depended on the respective proportions of equity oil, at government take-plus-cost, and of participation oil, at a higher price. [PIW 8-12-74:1,4] In late August, the average or blended government take at the Persian Gulf was around $9.50. [PIW 8-19-74:1] This confirms table 5.5, which shows the June 10 price at $9.30.

The surplus of refined products was becoming ever more burdensome, and product prices continued to drop in Europe. [PIW 8-19-74:3] Kuwait proposed joint output reduction at the next OPEC meeting. [NYT 8-26-74:45] The U.S. government's hopes were dimmed by the "mysterious postponement" of the Saudi auction. Since world production had apparently been greater than demand and storage tanks were full in important consuming countries, "the auction had been expected to start the way toward lower prices." [NYT 8-26-74:47] It did not occur to them that the auction was postponed because it might lead to lower prices. But they were hopeful for the next year, when Saudi authorities said production would rise from the current 8.5 mbd to 10 mbd, which would "exert at least a modest downward pressure on the world price." [Ibid.]

Reporters could see, even if high-ranking officials could not, that Saudi August output had been reduced by more than the combined cutbacks by Kuwait and Venezuela. [WSJ 8-26-74:3] Aramco blamed the cut on storms in the gulf. [WSJ 8-27-74:12] That was obviously untrue. The Saudis could have used their pipeline to the eastern Mediterranean, which had great excess capacity. [NYT 9-5-74:62] The Saudi preoccupation was clear: "An OPEC-decreed production control program could solve some of these problems for companies, but deep concern is expressed over OPEC's flexibility in 'fine tuning' output in line with seasonal demand variations. [But prices would not drop.] While short-term spot prices might decline, oil companies have a floor of high costs [tax]." [PIW 8-26-74:1]

On August 31, in a "setback for Washington strategy," Saudi Arabia agreed not to lower oil prices. "It doesn't want to wear a white hat, after all." [NYT 9-1-74:1] The weather was still blamed for reduced output, and the auction was still only "postponed." [WSJ 9-3-74:4] But two days later, Iraq announced an export reduction, and "Saudi Arabia, Algeria, Kuwait and Abu Dhabi have quietly agreed to do much the same in an effort to maintain current prices." [WSJ 9-5-74:2] Oil production controls were widely discussed in advance of the next OPEC meeting in mid-September. [PIW 9-9-74:3]

On September 9, King Faisal canceled the auction. Ambassador Akins "said it was too much to expect Saudi Arabia to single-handedly bring down oil prices." [NYT 9-10-74:53] In fact, the Saudis could have done so easily at any time, because they and others had much excess capacity. Saudi Arabia could have produced more or could have refused to join in the repeated output cuts. Then the others would have also desisted, and higher output would have brought down the price. The public record has shown the Saudis leading the price rise. The auction was designed to raise them further. When the hope disappeared, the auction was canceled.

The September 1974 OPEC Meeting

It was expected that the Saudis would have achieved complete ownership of Aramco by the next OPEC meeting, and a new flat price, 93 percent of posting, would be instituted. This "would mean a sharp but disguised increase in [transfer] oil prices, currently averaging around $9.50 a barrel." [WSJ 9-11-74:42]

Transfer prices at the Persian Gulf had been rising steadily because of increasing participation. (Elsewhere the picture was mixed.) [PIW 9-2-74:2] An Iranian minister said most Iran crude was being sold at $9.25, give or take 50 cents. [PIW 9-9-74:1] The Aramco negotiations were dragging on because of Saudi refusal to grant the companies preferential prices on long term contracts. The companies were eager to settle because it was not clear how much they owed for oil delivered since January 1. [WSJ 9-11-74:42]

At the September meeting, OPEC members agreed to raise the royalty on equity oil from 14.5 to 16.67 percent of posting and to study output reductions. Federal energy administrator John C. Sawhill called this "economic blackmail," but OPEC sources explained that higher taxes did not mean higher prices; let the companies pay! [NYT 9-14-74:1; WSJ 9-16-74:3] As in June, Saudi Arabia "disassociated itself" from the action but two days later said it was raising the average cost of its oil by approximately the amount of the royalty increase, which varied among countries. [NYT 9-16-74:54] In addition, "the chief proponent of lower crude-oil prices [Saudi Arabia] ... may have, in fact secretly increased the price" by raising the price of buyback oil from 93 percent to 94.864 percent of the posted price. [NYT 9-17-74:45]

Minister Yamani said he had told the Aramco companies in June that more tax increases were on the way. [NYT 9-18-74:55] The companies had just paid $1.9 billion to comply with retroactive tax increases. A rough estimate, based on January–August production, would be about 90 cents per barrel. [PIW 9-23-74:1]

Empty Barrels in Washington

There was now belated recognition that OPEC measures under discussion or already taken "will lead to escalation of world oil prices of such dimensions that leading industrial nations are now, for the first time, beginning to talk openly of a direct confrontation. The situation is getting 'very, very serious,' says one top insider." [PIW 9-23-74:1] President Ford sent a "bluntly" worded "warning to oil nations." Secretary Simon "took an even harder line." Others in Washington said there was no "justification" for further price increases. [Ibid., 1, 7] The next week Washington

> pushed hard on its new "tough line" in a carefully orchestrated public campaign to bring down "exorbitant" world oil prices.... No one can yet foresee what may come of the blunt "gloves off" challenge by President Ford and two of his key cabinet officers. Through Secretaries Kissinger and Simon, the U.S. Government had counted heavily—and publicly—on Saudi Arabia's implied promises to help bring prices down by an oil auction and other measures. Since the Saudis didn't act, Washington expected that OPEC at least wouldn't increase prices further. But OPEC did, and the final straw was that it set the stage for still further escalation. [PIW 9-30-74:1]

Arab countries professed to see a "threat" in President Ford's speech. [NYT 9-20-74:5] But Yamani insisted "that his country was the chief friend of the oil consuming nations although his Government has raised the price.... [H]is country had no intention of cutting production to help keep prices high." [NYT 9-21-74:37] President Ford, "in a harsh and threatening speech," warned the OPEC nations "of possible retaliation." [WSJ 9-24-74:1]

All this tough talk was designed to cover up "that the United States' campaign to bring about price reductions has failed, and failed conspicuously." Federal energy administrator Sawhill said accurately that "the United States does not have a policy ... directed at halting and reversing the rise in world prices." [NYT 9-25-74:3] Senator Henry Jackson praised Sawhill's candor [WSJ 9-30-74:3], Secretary Kissinger was angry [*Time* 11-11-74:61], and Sawhill's days in office became numbered.

Bluster was a substitute for resistance: the more words, the fewer acts; the more hostile the speeches, the more to be enjoyed by OPEC. They offered the thrill of battle array without the danger—what German soldiers call a *Blumenkorso*.

At the annual meeting of the International Monetary Fund and the World Bank, Robert McNamara said the "slow long-term decline in petroleum prices [had] called for correction." He did not explain why. The Indian finance minister spoke similarly: "This has been a fairly common attitude among the less developed countries, even though ... [they] have been hardest hit by the present situation." [NYT 10-1-74:1] A month later, at the UN General Assembly, nothing was said about reducing oil prices. "Instead, the diplomats said, they tend to applaud the redistribution of wealth."[19] [NYT 11-3-74:2]

More Saudi Reassurance and a New Price Rise

"Crude oil markets [were] in total confusion" in the wake of the September OPEC meeting because it was not clear what increase applied to what company. There was also confusion about the ascertainment of buy-back prices, equity crudes, and their respective weights. [PPS 10-74:362]

At an industry meeting, Minister Yamani said "his country could cover its financial needs by producing only 5.5 million barrels a day," but he "pledge[d] not to cut oil flow." Sawhill applauded, but the oil executives who were present dismissed the "assurance as formalistic and meaningless." [NYT 10-5-74:3] Shortly after Yamani "said emphatically that his country wouldn't cut back production to offset petroleum conservation measures in the consuming nations. Without saying so directly, Mr. Yamani seemed to agree with ... Sawhill that this would lead to price reductions for crude." [WSJ 10-7-74:8] But production was reduced later, and in March 1975 U.S. government sources complained that thereby the Saudis had "pulled the rug out from under them." [OGJ 3-17-75:67]

King Faisal now said there was "'not a moment to be lost' for America to force Israel into a 'full and prompt' withdrawal from all Arab lands," and he threatened another crisis, 'far more severe than the last one.' Ambassador Akins said that "the price would probably be '50% higher today' if the Saudis hadn't fought strongly within OPEC for price restraint." He added that "the oil price has nothing to do with peace" in the Middle East.[20] [PIW 10-7-74:1]

In October, Secretary Kissinger visited King Faisal, who "assured [him] that his country would use its influence to try to bring down the world price of oil." Moreover, he told reporters that the king had "outlined actions he had already taken to help drive the price of oil down.... This was the first time that the King had given oil-price assurances to Mr. Kissinger during the six meetings they have held since last November [1973]." [NYT 10-13-74:12] A *Times* reporter seemed a bit jaded: "It is a familiar scenario ... with the King cast as Good Guy and the Shah of Iran as Bad Guy." Ambassador Akins again said that the price would have been higher but for Saudi Arabia. If only one other major exporter would join them, then "maybe the price would come down." [NYT 10-16-74:59] That same day, the secretary was in Rabat, Morocco, expressing optimism [WSJ 10-16-74:12] But also that same day, the Saudis reneged on their plan and raised prices "in line with OPEC." They ordered Aramco to pay the additional taxes approved at the June OPEC meeting, those from which they had "disassociated" themselves. The payments were retroactive to June 1. [WSJ 10-16-74:18] The Saudis notified Aramco that they would "apply the last two OPEC price formulas to the hilt.... This is contrary to earlier word from the Saudis." [OGJ 10-21-74:76] These increases were matched by the balancing margin in Iran, which also was to receive a large retroactive payment. [PIW 11-11-74:1]

Former Undersecretary of State George W. Ball [*Business Week*, 10-21-74:25] proposed to enlist the cooperation of the OPEC countries by giving them a disguised price increase: inducements to invest in the United States, over and above the return available to other investors. By the end of October, the OPEC nations seemed to have agreed to abolish posted prices and taxes, in favor of complete ownership and a single price to all buyers. "Saudi Arabia plans to announce shortly a modest reduction in the price of her oil and the freezing of the price at the new level for a year." But actual take (royalties and taxes) per barrel was not reduced. [NYT 10-29-74:1] The price would be $10.35, but the Aramco companies would get a 50-cent allowance. Thus, Saudi take on the marker Arab light crude would be $9.85 on sales to Aramco, which presumably would be its chief but not exclusive channel. [PIW 11-4-74:1] This unsettled detail was to drag on for years.

As this was being discussed, Secretary Kissinger was in Riyadh, where "King Faisal pledged again tonight that Saudi Arabia would try to keep oil prices at the current level and if possible reduce them, at least symbolically." The secretary said "he would like to express our gratification." [NYT 11-7-74:1] But that same day it was learned that Iran and Saudi Arabia had already agreed, after "lengthy bilateral talks, on a higher price." [NYT 11-7-74:41]

The "price bombshell" delivered the next week caused the gulf multinationals again to cancel or suspend transfer prices and to inform affiliate customers "that the new price increases would be large." While posted price would be cut by 40 cents, there would be substantial increases in the tax rate, the royalty rate, and the buyback price. Weighted average take would be raised from $9.80 to $10.36. Yamani said the Saudis sought not to make money on a tax hike but "to take from the oil companies and give to the consumers." [PIW 11-18-74:1, 5] A month later, he was still denying that take had increased. [WSJ 12-13-74:21] Abu Dhabi, Qatar, and Kuwait announced their agreement, and Iran matched the increase through the balancing margin. [PIW 11-25-74:1] Secretary Simon, who had "predicted that the price of petroleum was soon going to fall by $2 to $3 a barrel because of the efforts of Saudi Arabia," said prices eventually would decrease. [WSJ 11-13-74:4]

Secretary Kissinger now announced "a vast new effort to deal with the world energy crisis and the deteriorating financial situation in many countries." His address was "billed by his aides as the most important he has made ... as Secretary of State." He spoke "in grim terms," advocating demand reduction, recycling OPEC revenues, and "dialogue." [NYT 11-15-74:1] Creation of the IEA was called a victory for the United States and a setback for France, which refused to join, alleging that the IEA was designed for confrontation, not dialogue. [NYT 11-16-74:41]

The 1974 stock buildup was almost as large as in 1973. [IPA 1977] It was involuntary, the result of not cutting back output or purchase quickly enough. As the year drew to an end, the massive surplus of productive capacity was still growing and was now over 10 mbd.[21] Although this situation "put considerable downward pressure on petroleum product prices in many parts of the world ... unfortunately for the consumer, price action by the oil exporting countries has put a solid and very high floor under product prices." [NYT 12-9-74:45]

But the crude oil price was not yet clear because government take was not. Haggling continued to the year end on the Aramco buyout and take. All oil was to be sold at the buyback price, which would become a single market price, but the companies would receive a producing fee. In Kuwait, the price for the lower-grade oil would be $10.37, and the fee 22 cents, a generous estimate of total development and operating cost in a country where only capacity maintenance in two great fields was needed and few "or no exploration opportunities left." (This appeared to be a safe prediction in view of Kuwait's small size, yet it has proved to be wrong.)

Saudi Arabia, unlike Kuwait, had a variety of field, some complex producing problems, "implementation of major new development programs, and a great deal of exploration to be done, both offshore and in remote inland areas." The fee would be between 30 and 50 cents, with some additional exploration incentive to encourage companies to assume risks in finding oil. [PIW 12-9-74:1; WSJ 12-19-74:32, 12-23-74:7; OGJ 12-23-74:15] Compensation to the Aramco companies would be $2 billion [PIW 12-16-74:1], or twenty-two days' production at the current price and output levels. At year end, the volume of liftings by the resident companies, which were no longer concession owners, was still unsettled. [PIW 12-23-74:1] Of course, the resident companies wanted as much as possible, since even the low fee was very profitable.

The U.S. administration continued to chase after its grand design: a worldwide price agreement setting a floor between $7 and $11 a barrel, to "give OPEC an incentive to cut prices soon in return for long-term market commitments." [WSJ 12-13-74:10; NYT 12-30-74:9] The nature of the commitments was not explained.

Even people with little sympathy for Secretary Kissinger's views in general echoed him on oil. Richard N. Gardner, deputy assistant secretary of state in the Kennedy and Johnson administrations, called for "'a great transcontinental bargain in which access to energy and other raw materials, which industrialized countries need, is traded for other kinds of access that developing countries need.' He points out that the Arab world will not be willing to exhaust its precious reserves of oil with a few years of full production so that the United States can continue its wasteful energy ways." And the columnist Anthony Lewis wrote: "We know now that the supply of oil is limited and that growing demand would probably push prices up before long in any event." [NYT 12-30-74] That was what Kissinger had said just a year ago. A distinguished international group said that oil had been "long underpriced and overused";[22] there was a need for "access" to oil and for a "special relationship" with the producers, especially the Saudis and the Iranians. Sober men professed to believe that the oil producers could be induced to produce more oil than was most profitable to them.[23] The general belief was now set in concrete. Henceforth, one hears countless times that oil is growing more scarce and "access" is vital.

The Period 1971–1974: A Review

I

In earlier chapters, we saw that the 1970 "low" price was well above long-run competitive levels set by investment requirements for continued expansion.

Over 1971–1974, there were two patterns of short-run price increase. The more dramatic was the ratchet. The fear of cutoff ignites precautionary demand; rising prices set off speculative demand; then the sellers, whose threats set off the excess demand in the first place, raise the excise tax in concert, to fix and hold the price at its new higher level.

This was the pattern during 1973. Worldwide stock buildup through the end of September, about 550 million barrels, actually exceeded the aggregate exporters' cutbacks in the last three months. But fear of more cuts and knowledge of more tax and price increases sent up the price. The excise tax price floor was raised on October 16 from $1.77 to $3.05 and effective January 1 to $7.01.

The other method of price increase was first followed in 1971–1972. Despite excessive supply and weak markets, excise taxes were raised in concert. Every seller had to increase price by roughly the same amount and, even more important, refrain from offering lower prices to move more. Then production was set by the amount demanded.

In 1974, there was a repeat of 1971–1972 but on a far bigger scale. By mid-January, supply was excessive, and by the end of 1974, storage tanks were full. For the first time, there was great excess producing capacity in world oil. Under competitive conditions, a price collapse would have been the mirror image of the 1973 price leap. Yet the excise tax was raised from $7.00 on January 1 to about $10.66 at the start of 1975. The governments now became the sellers at a price of nearly $11.00, and the companies were allowed a buying margin.

New contract prices rose along with the taxes. Open-market crude and product prices rose in anticipation of higher excise taxes, before the taxes were adopted. In 1974, the greater the excess capacity became, the higher the price was. This is absurd in a competitive market but made sense under monopoly. The output cuts, which increased excess capacity, were necessary for the price increases.

II

Between 1971 and 1974, there was no resource scarcity, nor was there any sign of an over- and underinvestment cycle. What happened was learning by doing. A new cartel gained experience and confidence by repeated success. It had come a long way from January 1971, when the shah was anxious, if not fearful. It felt free to threaten. The cartel's confidence was built up by continued success, as it sensed little resistance by companies and consuming governments.

There was no detailed strategy for the price explosion. The OPEC nations had a timetable, but like good commanders in war, they were quick to seize opportunities to surpass it. The annual increase in their revenue per barrel was tiny in the 1960s, small in 1970–1972, accelerating in the first half of 1973, and exploding in the second half. The 1974 price increase was almost as large, but it was even more impressive, for it was accomplished in the face of oversupply and excess capacity.

In 1971–1972, the tax and price leaders upward were Iran and Saudi Arabia. The Saudis became increasingly dominant in 1973 because they were larger and they were the leaders in the October production cutback. In 1974, the public record shows five examples of Iran as the passive beneficiary of Saudi and Kuwaiti price-raising initiatives. There are no contrary examples. But the legend endures: the shah as price hawk, the king as dove.

The Saudis talked repeatedly in favor of lower prices and repeatedly took the lead to raise prices. They promised and then canceled an auction; promised to maintain output, then cut it. Soft words and hard actions served them well.

III

The Arab oil production cutbacks were announced to achieve an Israeli withdrawal from the occupied territories, especially Jerusalem. The cutbacks were withdrawn after two months, with no demands met. The devotion of the Arab oil producers to the Palestinian cause was loud, not deep.[24]

The market worked. It distributed oil with rough equality among the consuming nations. The Arab oil "embargo" was a denial of "access" to oil and proved that "access" was a nonproblem. It gave the United States a truly "special relationship" with the Arab producers: their special enemy. Enemies did as well as friends, proving friendship and enmity to be irrelevant.

The British and French surpassed even the Americans in groveling, but it did them no good. They lost as much supply as others—or more—and paid no less. The lack of resistance by consumer nations, their fawning and constant anxiety to please, were an inducement to go faster, although we cannot say how much faster.

IV

The 1971 Tehran five-year agreements lasted about six months and were dead long before formal repudiation. Later assurances—of lower prices, of an auction, of maintaining output—were also broken.

Sovereign monopolists cannot be held to any bargain. In ordinary business life, an agreement is enforced by law and by competition: customers or suppliers can sue or go elsewhere, or both. But OPEC has suppressed competition, and governments are beyond any law. Nothing short of war compels them to keep any promises they make.

If promises have no value, threats should have no force. A threat is also a promise. The threat to damage our economy if we do not change our policy is also a promise not to damage us if we do change. If we cannot trust the promise, we should disregard the threat. Yielding to it, whether morally right or wrong, is useless.

Appendix 5A: The Ambassador and the Secretary

Ambassador Akins was forced to resign in August 1975 and left the State Department. "Much admired in the Arab world," he was viewed by some as "Saudi Arabia's ambassador to the United States"[25] His role in the Carter administration will be described in the next chapter.

Akins has alleged that Henry Kissinger approved the 1973 oil price hike in order to support Iran: "Attempting to hang the rap for oil prices on the Iranians ... was a poorly understood and highly successful ploy that continues to this day." ["'60 Minutes' v. Henry Kissinger" by Thomas J. Bray, WSJ 6-30-80]

The success continues. A recent Kissinger biography contends that the shah allegedly "pressed for the most radical price increase ever over the reluctance of the Saudis." [Isaacson 1992, p. 563] Akins has complained that the Kissinger memoirs ignore "King Faisal's repeated requests of Kissinger to put pressure on the Shah to help the Saudis restrain oil price increases." [Akins 1982] In fact, Kissinger mentions—with scorn, as hypocrisy—requests by both Saudis and Iranians.

Shortly before the "embargo" ended, Akins says he was instructed by the secretary "to issue an ultimatum" to the Saudis. "Although Akins had asked him not to, the Saudi foreign minister had shown the ultimatum to King Faisal, who would have 'broken diplomatic relations with the U.S.' if it had been delivered." [PONS 5-6-76:3] Akins gave no idea of the contents of the alleged "ultimatum."

He has accused Mr. Kissinger of lying. [Robinson 1988, p. 118] This sounds serious. (Perhaps it is not. One who differs with Akins is a "criminal fool" [PONS 11-24-76:1] or a "neo-imperialist." [Akins 1982b, p. 9]) But he does not say what Kissinger said, let alone why it was a falsehood.

Kissinger has sharpened the feud by ignoring it. Probably he never realized how ill informed Akins was on oil, but the reasons for dismissal seem fairly clear even on the public record and aside from the "ultimatum" incident.

Akins twice appealed to private persons to change U.S. policy. He misled Saudi Arabia with "promises" of what the U.S. government would do. Perhaps his inability to deliver on his "promises" was what made him so angry.

Akins advised that a selective boycott would be effective. Kissinger's "obsession" with this phantom—as he called it years later—made him waste time, energy, and political capital. Akins reported that the Arabs would never relax the cutbacks until their political demands were met; in fact, they desisted after two months, with no demands met.

Akins convinced Kissinger that the Saudis really wanted lower prices, for which Kissinger was repeatedly and publicly grateful. In fact, they were the leaders in repeatedly raising prices, especially in 1974, to the embarrassment of President Ford and Secretaries Kissinger and Simon.

A subordinate cannot keep his job if he causes a superior to make a fool of himself in public.

Notes

1. Sheik Yamani clearly stated the importance of the excise tax in a statement whose provenance is lost but must have been made around this time.

Realized (actual market) prices had only "the near cost of production" as the price floor. "However, working on the basis of the posted price, say $1.80 a barrel, the situation is very different. The company then pays the government half the difference between the cost of production (20c) and the posted price ($1.80) that is to say 80c. This means that the company has a minimum tax paid cost of $1 per barrel which sets an absolute floor to its price. It cannot go out into the market and dump its oil; it will have to sell at a price considerably above $1 in order to make a decent profit. This is why posted prices are absolutely vital for the producer governments."

In the same vein, he stated at a symposium in 1969: "In case of nationalization, the national oil companies of the producing countries will enter into competition on the world market and prices will fall to $1 per barrel in the Gulf." [recalled in OGJ 2-17-75:55]

2. In March 1972, I said, "Taxes and with them prices will be hiked again and again to some ill-defined ceiling set by eventual consumer resistance." [OGJ 3-13-72:26]

3. In 1988, I heard a high-ranking State Department official attribute the 1972 Iraq nationalization to the 1967 Arab-Israel war. He cited no evidence and ignored the government-company conflict, which had begun before the existence of Israel.

4. [PPS 4-72:118] The company and government needed to set four items. (1) Government owes company a certain sum each year to cover the amortized cost of the equity share it is obtaining. (2) Government loses the taxes formerly due on the share it now owns. (3) Company owes government the "price" of the oil that government now owns and "sells" to company. (4) Company owes government its pro-rata share of the year's profits. The subject of the negotiations was by what amount (3 + 4) should exceed (1 + 2).

5. In 1972, the average domestic wellhead price was 19 cents per thousand cubic feet (mcf). Algerian gas was landed at $1.38. [DOE, *Annual Energy Review 1990*, table 79]

6. Yet we read that Irwin was trying "to create a united front among the consuming nations."[Goodwin et al. 1981, p. 446] The author says eighty pages later: "In the absence of documents ... judgments about the origin of, and motives governing, State Department actions must be regarded as 'conjectural history' at best." [Ibid., p. 526 n. 96] That is no excuse for ignoring the public record.

7. "What incentives does [Saudi Arabia] have even to let its production remain at the present level? It certainly doesn't need the money.... Production of only 5.5 MBD would give the country all the income it needs, with a comfortable margin for reserves. In other words, there is no economic incentive to let production continue growing....

"Contrary to what some people have implied, Saudi Arabia has no moral or legal obligation to help us meet our energy needs at the expense of its own future generations by producing and exporting oil that might better be left in the ground. It has no obligation to deplete its own resources in order to provide another country with cheap energy or to bail out those who have failed to plan effectively for their own needs." [Briefing document reproduced at 7 Church 517ff., quotations from 527–528, 534–536]

8. Conversely, the pricing of natural gas in the United States at less than the cost of alternatives was proof that it was determined by competition. [Adelman 1962, p. 39]

9. I said in April 1973: "If the Arabs ever attempted to cut off the United States for political reasons, the non-Arab members of OPEC would simply divert shipments from non-American customers to American.... Whereupon the Arabs would ship more to Europe and Asia, and the net result would be simply a big confusing costly annoying swap of customers and no harm otherwise. If this is common sense, it is also the lesson of experience." [Reprinted in Adelman 1993, p. 492; also quoted in *Economist* 7-7:73: Survey:12]

10. In the summer of 1973, I spent six weeks in Europe interviewing officials in seven European countries, as I had done the previous year in Japan. The question was, What in their view was the problem of oil supply security, and what measures were proposed to deal with it? The report, completed in September, was not published, because an accurate summary would be: Everyone professed much concern over security, but no idea was worth reporting.

11. The correspondence quoted in the *Report* gives volume 9 of the *Hearings* as the source. I have been unable to find it and believe it is from a cable that, like several others, was retained in the subcommittee's files and quoted directly.

12. Later in November, King Faisal stated: "'Saudi Arabia will not change its attitude of suspending oil exports to some countries and cutting back quantities to other countries except after the fulfillment of these points collectively and in a manner acceptable to all Arabs—*no matter how long it takes.*'" [NYT 11-23-73:1, emphasis added]

13. Consider a taxicab operator who charged $1.00 per mile, and consumed one-tenth of a gallon of gasoline, costing him 3.5 cents. For lack of gasoline, he would lose the other 96.5 cents of revenue, with no offsetting savings. Insurance against a one-tenth probability of running out of fuel was worth nearly 10 cents, or three times its current price.

14. Verleger [1982] pioneered the analysis of short-term price changes and the effects of inventory buildup and drawdown. Spot price changes themselves reflected the expectation of official price changes, as well as fear of dearth and additional output reduction. Hence he has not been able, and it may be impossible, to incorporate them in any econometric relation. His demonstration that price changes in the 1970s could not be explained by changes in capacity utilization (pp. 39–45) did not prevent the Department of Energy and others from treating it as a basic relation. See chapter 8 below.

15. It is more accurate to compare 1974 consumption not with actual 1973 but with expected 1974 consumption, which we estimate by actual 1973, increased by the average consumption growth rate over the previous five years:

Area	Actual 1974 consumption (1973 = 100)	Expected 1974 consumption (1973 = 100)	Percentage consumption loss
United States	93	105	11.4
Japan	101	113	10.6
Western Europe	89	108	17.6

16. "Much more likely [than an indefinite Arab shutdown] would be a gradual resumption of exports to friendly countries in return for a supply of essential commodities." [Akins 1982b, p. 6]

17. History repeats itself. See "Japan placed in a dilemma by an offer from Saudis to guarantee oil supplies" [WSJ:A 8-28-88:10], which has an uncanny resemblance to the offers made to the United States in 1972.

In 1988, "Foreign Minister Yoshihoro Nakayama said oil importing countries must begin preparing now for a crude oil crisis. Japan, which imports 99.7% of its oil is consolidating its foreign relations to secure future oil supplies." [WO 11-88:13]

In 1990, the "Institute of Middle Eastern Economics, a Japanese think tank, called on Japan's government to take legislative steps to extend official development assistance to Persian Gulf nations. It urged the government to follow its own policy in securing Middle Eastern oil supplies as non-OPEC supplies are exhausted early next century." [OGJ 2-19-90:35]

18. According to Fadhil J. Al Chalabi, who was permanent undersecretary of oil in Iraq during 1973–1976, assistant secretary-general of OAPEC in 1976–1978, then OPEC deputy secretary-general, the price was fixed at the end of December 1973 to equate to the supply price of alternative energy sources.

"[T]he rationale behind the new price was not the market situation, as much as the decision to relate the government take to a level near the cost of available sources of energy. An amount of $7 per barrel was thought to represent, at the then prevailing conditions, an indication of the cost of alternative sources of energy to oil.... Accordingly the level of the new posted price was computed ... so as to produce a net government take of that amount, $7 per barrel." [Al Chalabi 1980, pp. 86–87]

19. During the "embargo," a letter to the *Economist* [12-15-73:4] claimed that Sheik Yamani, "a black man," was getting his own back from the detested white race. Sheik Yamani is one of the most photographed personalities of our time and obviously Caucasian. What the letter proves is that in matters of oil, some people cannot tell black from white. The next year, "Sweden's Minister of State, Carl Lidbom, gave broad backing to the third world countries, speaking approvingly of the success of the oil producers in organizing their cartel and having made a 'breakthrough in the massive dominance of the industrialized countries.'" [NYT 9-4-75:13]

20. This was inconsistent with his 1970 lecture or his 1972 speech that the Six-Day War created a sellers' market or his reports as ambassador. Two years later, out of office, he reverted: "Only with [Arab-Israel] peace can we have petroleum." [*Oil Daily* 10-28-76:1]

21. In January 1975, the U.S. Treasury estimated OPEC surplus producing capacity at 10.5 mbd, which breaks down in this way: Saudi Arabia 2.4, Libya 2.25, Kuwait 1.5, Abu Dhabi

1.2, Iran 1.0, Venezuela and Nigeria 0.5 each, Indonesia 0.35, Iraq 0.32, Algeria 0.2, others a total of 0.31. [OGJ 2-17-75:NL1]

22. "The OPEC countries, whose principal product has been for so long underpriced and overused ... need, and undoubtedly want, both the enduring goodwill and the continuing economic prosperity of the oil-importing countries." Farmanfarmaian et al. [1975].

23. On the history of the "Euro-Arab dialogue," see Al-Mani and Al-Shaikhly [1983]. Neither side had anything to offer that the oil market and other markets would not supply anyway, without their action.

24. An opposing view: "Should there be no movement toward peace, [defined as Israeli "withdrawal from the West Bank, Jerusalem, Golan, and Gaza"] we must assume that producers in the Arabian peninsula will eventually find it expedient to cut down, if not to cut off oil production. It they do not, they could face revolution or sabotage." [Akins 1982b, p. 6] But none of them stirred, even during the 1982 Israeli invasion of Lebanon or the repression of the intifada.

25. *Economist* [8-23-75:63]. Two appraisals are: "A Diplomatic Situation Where Oil and Hauteur Just Didn't Mix," *Washington Post*, 3-14-76; "How OPEC Came to Power," *Forbes*, 4-15-76.

6 OPEC at High Noon, 1974–1981

A consistent underestimation of potential supply and a consistent underestimation of the consumers' ability to adjust their demand ... led OPEC (and usually leads every other cartel) to overestimate its strength.

—Sheikh Ali al-Khalifa al-Sabah, former oil minister, Kuwait

Preview

When Saudi Arabia made the last price increase at the end of 1974, the OPEC nations started to manage the new market and to prepare the next great advance. To finish the expropriation of the oil companies was easy, but it forced the nations to replace the companies in limiting and dividing up output. The task was harder because the worldwide recession following the 1973–1974 price explosions reduced demand. As in 1972, there was excess capacity, but now on a much larger scale, and it did not disappear.

From the start, market division was difficult. To make it harder, imports of goods and services soon rose even faster than revenues, and the huge 1974 balance-of-payments surpluses turned into deficits. This made more acute the task of setting the ultimate price: how far and how fast could it be raised? The non-OPEC supply response to price was reassuringly small. The demand response to price seemed negligible. Cautious plans for price raising gave way to seize bright opportunities. In the hour of triumph, nobody could see it as disaster because the price could not be maintained.

A Confederate officer at Gettysburg, shown a clump of trees a mile away as the objective, said he was sure he could get there but not at all sure he could stay there. No such cool weighing of probabilities was done in OPEC, and even less by the authorities in the consuming countries, which could see only dearth.

Nationalization

The main unfinished business of early 1975 was to complete the nationalization of the oil companies. Compensation for the producing rights was derisory: original cost of tangible assets, depreciated. Otherwise the process varied from country to country. It was described for Saudi Arabia by Clifton Garvin, the CEO of Exxon, who was the chief Aramco negotiator:

> [By 1974] the things that had been signed in the '72 agreement were no longer being honored.... [New arrangements] were reasonably well agreed to in the late summer of 1976 ... [but] never signed....
>
> [Agreements] didn't seem to survive very long. In other words, if it was in the interests of the Saudi government to change, they changed.... We [also] tried to make the exploration arrangements work for awhile, but it became obvious in a very short period of time that they were not going to honor what those crude oil arrangements had anticipated. [IRS 1991 Tr. 87–88, 162–164]

The sovereigns could do whatever they pleased, with no regard to "inky blots and rotten parchment bonds."[1] As Garvin makes clear, it is impossible to put a precise date on the expropriation.

Discovery Value

In March 1976, the "final" Aramco terms included a fee for newly found oil: 6 cents per barrel, as produced. [PIW 3-29-76:1] Depending on the interest rate, this had an in-ground present value of between 1 and 2 cents per barrel.[2] This was an estimate of the value to the Saudis of an additional barrel and the cost (to Aramco) of finding a barrel. At this time a North Sea reserve was sold at about $1.60 per undeveloped barrel in the ground, from 80 to 160 times as much. [PIW 3-15-76:2] Oil in ground was worth very little because current reserves were overabundant and being depleted so slowly. The negligible value of discovery value or user cost confirms the extreme monopoly distortion. [Adelman 1993, chaps. 10, 12] Saudi Arabia and others impound enormous underground inventories. To use them to produce more would flood the market and depress prices.

The New Order

Outside the United States, the integrated companies had competed mostly in selling finished products in many local markets. Sellers were few and competition restrained; changes in market share were small and incremental. Each company produced what it could sell at the pump. Each OPEC government pressured its resident companies to produce more, appealing to the companies' home governments, especially the United States. The myth that the OPEC nations wanted to produce less than did the companies belongs with the legend that before 1973 the companies deliberately held down the price. [Askari 1990, p. 28; El Serafy, in Ahmad et al. 1989, p. 12; Heal and Chichilnisky 1991] That would have meant less profit and more political danger, in both host countries and home countries, especially the United States and Western Europe, from local coal and oil producers.

But starting around 1976, the companies became the buyers for their refining-marketing operations. Margins were thin, and differences of a few cents, which went straight to the bottom line, could make a company switch suppliers. No longer could OPEC governments fix a price floor by fixing per barrel taxes and letting the companies compete freely above the floor. They now had to fix prices in concert and to trust each other not to undermine those prices by trying too hard for additional sales. It was an endless exhausting struggle.

Setting Prices in the New Order

Contract and Spot Prices

Figure 6.1 summarizes the oil market changes from 1973 through 1992. It contains at most four series, at times only two. Three of them are the Persian Gulf FOB prices (verified as arms length by the U.S. government) of Saudi, Abu Dhabi, and Iranian crudes sold to American purchasers. The fourth series is the monthly average spot price of Arab light, the largest component of Saudi production. Spot prices are for single cargoes, each price set by separate bargaining. The spot price is a valuable market indicator because it captures and magnifies any momentary excess supply or demand. For example, suppose that the amount of oil demanded is 1 percent higher than the amount supplied and that short-run demand elasticity is −0.1. If the whole market were made up of spot transactions, the price increase would be nearly 11 percent. But let the 1 percent excess all be channeled into the spot market, which we assume is 10 percent of the total. If so, the spot price increase is 187 percent.[3]

Suppose now that the sellers all disregard contract prices and offer oil only at the current spot price. The spot price will settle at (or collapse to) about 11 percent above the old level. What can be done for a small segment cannot be done for a large one. Indeed, if, as in 1973 and 1974,

Crude oil prices, 1973–1993.

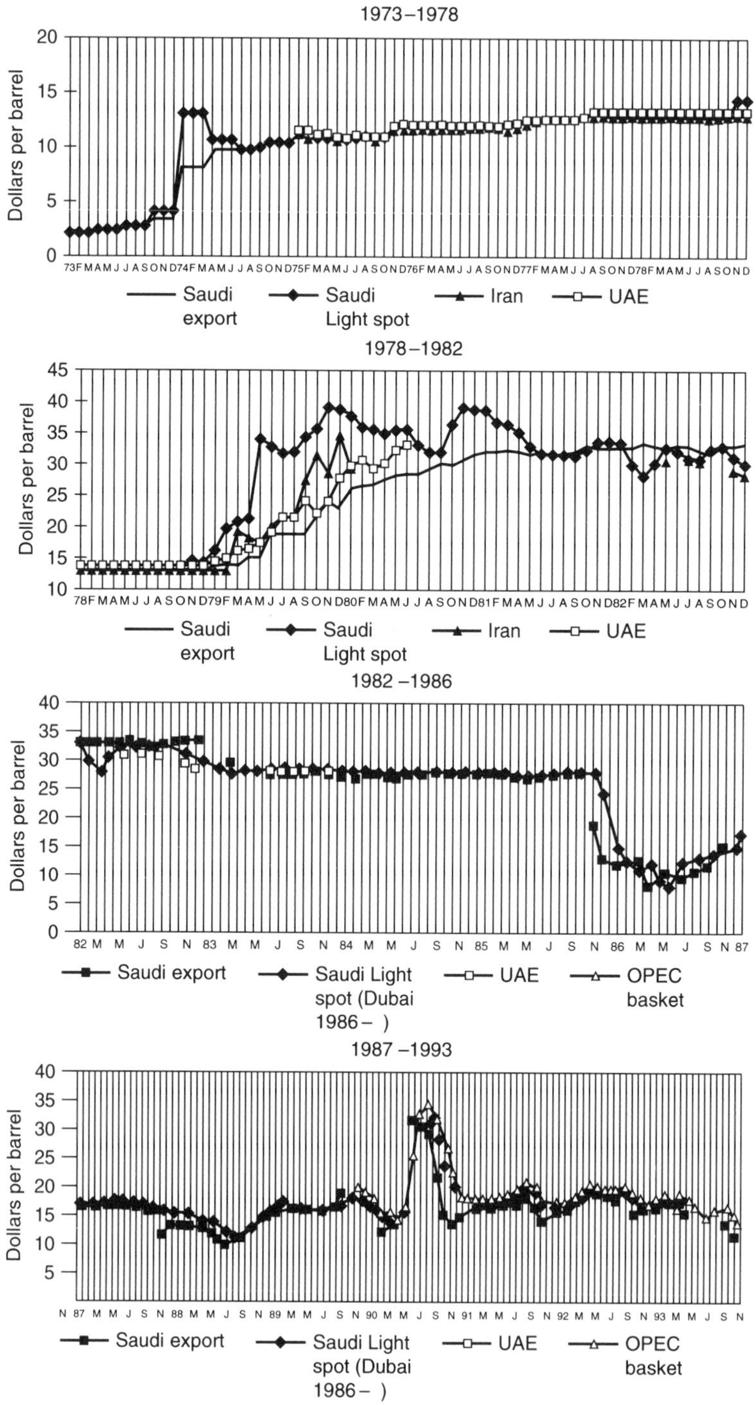

Figure 6.1
Sources: Spot prices from PIW, PMI; after 1986, Dubai Fateh substituted by them (and myself) for Arab Light. Export prices from DOE:MER. For 1973–1974 export prices from IEA, Crude Oil Prices 1973–1980 (Paris 1983). In months when Saudi export price is withheld, "Arab OPEC" (of which it is a large fraction) is substituted.

the spot market dwindles as a percentage of the total, the reported spot price may not be a real price. As figure 6.1 shows, even the Iranians, who in 1979 professed to sell only at spot, could never get very close to spot and overreached themselves three times. (There are no later comparisons because the hostage crisis ended sales to the United States in January 1980.)

Suppose the sellers now make the higher price official and simultaneously the transitory excess demand disappears. Demand is back to where it was. By the old standards, the price is now 11 percent "too high." But if the sellers all hold fast to the higher official prices and also cut back production, offering only what can be sold at the higher price, the spot price will be firm and equal to the official contract price, until the next change.

Thus, it is a mistake to regard spot prices as the "real market" prices and contract prices as merely catching up with them. Both spot and term contracts are real contracts with real prices. They are imperfect substitutes, like shorter- and longer-term interest rates. They are usually not equal, but they are strongly related.

The Sequence of Price Changes

The pattern first seen in 1973 has held over twenty years. The anticipation of higher official crude oil prices increases speculative demand, which raises the spot price, before the official price rises (similarly but more rarely in the other direction, until the 1980s).

Suppose an OPEC meeting is scheduled for December and is expected to raise the price, by 5 or by 10 percent, on January 1. An investment in buying a barrel at the previous price would earn the following:

Barrel bought on last day of:	June	July	Aug.	Sept.	Oct.	Nov.
	Annual Rate of Return (percentage)					
5 percent OPEC increase	10	12	16	22	34	80
10 percent OPEC increase	21	26	33	46	77	214

Of course, there was storage cost and the interest cost of carrying additional inventory until it was used up. There was also risk the OPEC meeting might not raise the price.

If those expecting an increase would bid for more crude oil and products, the increased demand would raise the spot price months before the meeting. The OPEC meeting, having provoked stock building and speculative price increases, would cite the spot price rise to justify raising official prices. It was merely "following the market."

Justifications serve chiefly for amusement, but econometricians must heed the old chestnut, *post hoc non propter hoc,* and not "explain" later official prices by earlier spot prices.[4]

After the expectation fulfilled itself, the higher price could hold only if the OPEC countries refrained from overproducing. This was particularly important just after the price increase, when the temporary additional demand turned into excess inventories. In late 1978, for example, the U.S. industry storage system held about 1,300 million barrels. Secondary distribution and consumer storage was about 500 million barrels. If primary holders tried to increase stocks by 5 percent, and downstream holders, who have a much larger range of operating flexibility, by 10 percent, this would increase the amount demanded by 110 million barrels: $(1300 \times .05) + (500 \times .10) = 110$. If the incremental buying was done during 60 days, it would be 1.8 mbd—over 11 percent of the total flow. This increment went mostly through spot channels.[5]

We now look to external developments: consumption and non-OPEC production.

Consumption, 1974–1979

The cartel price target was first set in December 1973 to equate to "the cost of alternative sources of energy."[6] The goal was restated in 1978–1979 by the OPEC Long-Term Price Policy Committee of oil ministers, chaired by Sheik Yamani.

For such a target, monopoly is a necessary condition. Only when oil-on-oil competition is suppressed can the price be set by the nearest alternative. But figure 6.2 shows what was overlooked: monopoly was not a sufficient condition. The true monopoly ceiling price was much lower than expected because consumers' response turned out much stronger than expected. But the effect took years to show itself (table 6.1).

The slowdown in oil consumption growth was due, before 1978, almost entirely to world economic recession and then very slow growth.[7] The price effect was negligible at first. A major report that summed up official opinion at mid-decade [CIA 1977] simply ignored price elasticity. Energy saving was vitally necessary but could be achieved only by disinterested political action. That was the consensus.

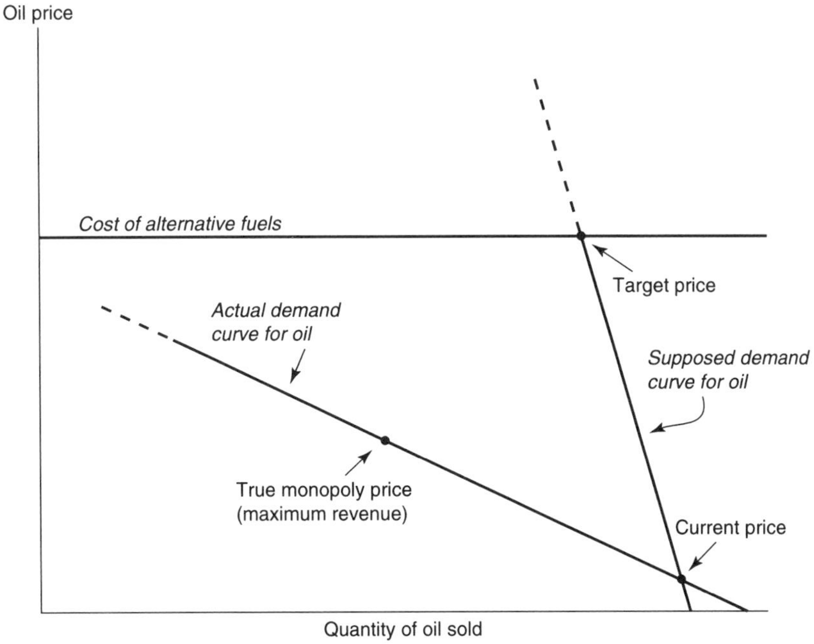

Figure 6.2
Long-run demand: supposed and real.

In fact, oil and energy consumption respond not to preachment but to price. The response is by investment. The faster the growth of the economy is, the more does its equipment reflect the new higher price level and conserve energy. Hence Japan showed the quickest adaptation.

As the price of a good rose, spending on it became a larger part of the firm's or household's budget, increasing the reward to substitution and the buyer's sensitivity to further increases. But as purchasers shed the more easily substitutable uses, their remaining demands were less price elastic. I suggest, without trying to prove it, that the first effect overbore the second.

But the elasticity of crude oil demand was only a fraction of the elasticity of product demand. In 1970 the Persian Gulf crude price was only 11 percent of the product price in Western Europe (table 6.2). All else being equal, a 500 percent increase in the crude price was only a 55 percent increase of the product price. But by the period 1978–1981 a crude price increase of "only" 156 percent raised product prices by 97 percent. Thus, the second price increase, which was proportionately much less than the first, had a much larger effect on the price of products and

Table 6.1
Market economies' consumption and OPEC exports, 1967–1993 (mbd)

Year	Market economies	Excluding OPEC	OPEC production	OPEC exports	OPEC share (%)
1967	30.8	30.0	16.9	16.1	53.8
1968	33.5	32.6	18.8	17.9	55.0
1969	36.6	35.6	20.9	19.9	56.0
1970	39.5	38.3	23.5	22.3	58.2
1971	41.6	40.4	25.4	24.2	59.9
1972	44.5	43.2	27.2	25.9	60.0
1973	47.8	46.3	31.0	29.5	63.7
1974	46.3	44.7	30.7	29.1	65.2
1975	45.2	43.5	27.2	25.5	58.7
1976	48.1	46.1	30.7	28.7	62.3
1977	49.4	47.2	31.3	29.1	61.6
1978	50.2	47.8	29.9	27.5	57.5
1979	51.3	48.7	31.0	28.4	58.3
1980	48.9	46.0	27.0	24.1	52.4
1981	47.2	44.2	22.7	19.7	44.6
1982	45.5	42.3	19.3	16.1	38.1
1983	45.3	41.8	17.8	14.3	34.3
1984	46.2	42.3	17.5	13.6	32.2
1985	46.2	42.3	16.4	12.5	29.5
1986	47.5	43.7	18.7	14.9	34.1
1987	48.5	44.7	18.3	14.5	32.4
1988	50.3	46.5	20.4	16.6	35.7
1989	51.5	47.6	22.6	18.7	39.3
1990	52.2	48.3	23.9	20.0	41.3
1990[a]	52.9	48.5	24.8	20.4	42.1
1991	54.0	49.6	24.9	20.5	41.4
1991[b]	66.2	61.8	24.9	20.5	33.2
1992	67.0	62.0	26.4	21.4	34.6
1993	67.1	61.9	27.1	22.0	35.5

Sources: BP; OPEC consumption from CIA, estimated for 1967–1972.
Note: Natural gas liquids included.
[a] 1990– , EIA, International Petroleum Statistical Report. Format change after July 1992.
[b] The former Soviet Union and China are now included with market economies.

Table 6.2
Consumer price and crude price, 1970–1990

Year	Saudi FOB	West European consumer	European tax	TR + REF + MKTG* margin	Saudi FOB share
1970	1.21	10.79	5.63	3.95	0.11
1971	1.69	12.50	6.03	4.78	0.14
1972	1.88	10.40	6.62	4.86	0.14
1973	3.25	13.65	7.54	5.16	0.20
1974	10.17	22.55	8.88	4.38	0.43
1975	10.87	27.10	10.97	5.44	0.40
1976	11.62	27.20	10.54	5.91	0.41
1977	12.38	29.20	12.08	7.78	0.38
1978	12.70	33.80	14.39	9.51	0.35
1979	17.63	46.00	16.89	12.49	0.37
1980	28.53	63.00	21.42	14.06	0.44
1981	32.48	65.00	20.80	12.18	0.50
1982	33.50	64.11	21.08	13.58	0.53
1983	28.03	62.37	21.26	8.97	0.44
1984	27.60	56.47	19.83	8.97	0.49
1985	22.04	57.57	21.03	14.50	0.38
1986	11.36	58.25	30.92	15.97	0.20
1987	15.12	66.59	36.77	14.70	0.23
1988	12.16	67.81	39.96	15.69	0.18
1989	16.29	69.74	39.53	13.92	0.23
1990	20.48	87.80	49.23	18.09	0.23

Source: IEA, tabulated and weighted by Centre for Global Energy Studies. Courtesy of Fadhil J. Al Chalabi.
*Transport, refining, and marketing.

generated a much stronger consumer response. In addition, as OPEC market share decreased, the impact of lower total oil sales on their sales was amplified.

Taking these two effects together, OPEC crude oil price elasticity was about 3.4 times as great in 1978 as in 1970. That is, conflating tables 6.1 and 6.2, the price elasticity of demand for OPEC crude oil went from 19 percent (.11/.58) to 61 percent (.35/.58) of the price elasticity of oil products. I think that product price elasticity had itself increased. Moreover, the second response was piled atop the first incomplete response to the first shock.

Crude oil price elasticity of demand that sellers could safely neglect in 1970 needed attention in 1978. It received none. The best informed among the sellers thought that "price elasticity of demand approaches zero." [Yamani 1978]

The response to the higher price level is measured roughly by the decline in oil consumption per unit of real gross domestic product.[8] (See table 6.3.) The reduction in oil product use per unit GNP, once underway, had to continue for a long time.[9]

Non-OPEC Output Growth: A Price Response?

Output from non-OPEC producers rose from 17 mbd in 1973 to nearly 19 mbd in 1978. It is widely believed that higher oil prices drew in more investment, which increased output.[10] In fact, practically all of the new supply of the 1970s was from fields in Alaska, Mexico, and the North Sea that had been discovered and committed to development before the price explosion.

Alaska North Slope

The Prudhoe Bay field was discovered in 1968. By the end of 1971, twenty-four development wells had been drilled, but activity had slowed, awaiting congressional approval for a pipeline to the port of Valdez. [AAPG 1972, p. 1175] In 1972, forty-six more development wells were completed, but "only 1 rig is being operated pending approval of the pipeline." [AAPG 1973, p. 1408] Had approval come quickly, the field would have been producing before 1973.

Mexico

The first discoveries in new areas were made in 1972, with depths typically at 13,000 feet or deeper. We use the alternative formulas of chapter 2 (table 6.4). Assuming the price in Mexico to be the same as in the United States in 1972, the rate of return is 168 percent per year. Assuming a 20 percent rate of return, the break-even price or cost is 63 cents per barrel. Allow as one will for errors in data or assumptions, the new Mexican fields were hugely profitable at pre-1974 prices. How much of the profit went into waste and payoffs is another story.

North Sea

In late 1972, the value of a barrel of oil landed in the United Kingdom was projected to be $3.37 by 1975–1976. [PE 11-72:421] Thus, a barrel

Table 6.3
Oil consumption and real GDP, major market economies, 1960–1993
(oil consumption in barrels per thousand dollars GDP)

Year	United States	Japan	Four EEC[a]	Total
1960	2.68	1.14	1.02	1.93
1961	2.67	1.25	1.14	1.96
1962	2.63	1.32	1.28	2.01
1963	2.61	1.56	1.42	2.07
1964	2.54	1.69	1.52	2.08
1965	2.50	1.89	1.65	2.13
1966	2.49	1.95	1.74	2.16
1967	2.52	1.90	1.80	2.19
1968	2.59	2.09	1.89	2.28
1969	2.65	2.27	2.05	2.39
1970	2.77	2.45	2.10	2.48
1971	2.77	2.54	2.14	2.51
1972	2.83	2.42	2.20	2.54
1973	2.82	2.58	2.18	2.56
1974	2.74	2.55	1.98	2.43
1975	2.70	2.25	1.87	2.33
1976	2.76	2.27	1.85	2.35
1977	2.76	2.36	1.80	2.35
1978	2.69	2.19	1.76	2.28
1979	2.59	2.22	1.77	2.23
1980	2.40	1.91	1.59	2.03
1981	2.19	1.79	1.46	1.86
1982	2.14	1.64	1.40	1.79
1983	2.05	1.53	1.35	1.71
1984	2.02	1.53	1.34	1.70
1985	1.92	1.40	1.28	1.61
1986	1.93			
1987	1.92			
1988	1.91			
1989	1.87			
1990	1.82			
1991	1.81			
1992	1.81			
1993	1.78			

Sources: MER; DOE/EIA 1990; Alan Heston and Robert Summers, "A New Set of International Comparisons of Real Product and Price Levels for 130 Countries, 1950–1985," *Review of Income and Wealth* (June 1988): 1–25.
Note: Real GDP is expressed in 1980 U.S. dollars. Estimates are based on purchasing power parity indexes obtained from the UN International Comparisons Project.
[a] France, Germany, Italy, and the United Kingdom.

Table 6.4
New fields in Mexico: Cost and return, 1972

1. Drilling cost per onshore well, depth between 15,500 and 20,000 feet (U.S. 1972)	$621,000
2. Allowance, equipment and other nondrilling costs	1.7
3. Adjustment, higher costs outside United States, Canada	2.2
4. Total investment per connected well (lines 1 × 2 × 3)	$2,375,000
5. Daily flow rate (barrels)	3,500
6. Investment per annual barrel (line 4/line 5/365)	$1.86

Assume Mexico price equal to U.S. price:
$i = 3.39/1.86 - .075 - .062 = 1.68$, or 168 percent return

Alternatively, assume a 20 percent rate of return:
$P = 1.86 \times (.075 + .062 + .20) = .63$, or 63 cents per barrel development/operating cost

Sources: AHKZ 1983, chap. 9; Joint Association Survey 1973, tables 2, 3; Pemex ML 1972, p. 2; Pemex DG 1973, pp. 13–14; See Appendix 7A for lines 2–4.
Note: New fields in later years [OGJ:WWO 1985:88] considered as Cretaceous production, Southern Zone, discovery data 1972 or later.

that was profitable at that price would be worth producing in 1972. Since our cost data are from later years, we need to adjust for the rapid rise in costs that followed soon after. A special inquiry was made by the British DOE. From autumn 1973 to spring 1975 development cost in the North Sea increased by 140 percent. [PE 8-76:316] We need to move to a 1978 basis for comparison. It is probably a considerable understatement to assume an increase of 200 percent from 1972 to 1978. Thus the 1972 break-even cost in 1978 dollars is overstated at $10.11 per barrel ($3.37 × 3.0).

We need to calculate the investment per daily barrel, which would be barely profitable at $10.11. We assume a 25 percent rate of return (before tax) and current operating cost at 5 percent of the investment for capacity. Then we can reckon, in 1978 dollars, a break-even investment per daily barrel of $10.11 × 365/.387, or $9,535.

Adelman and Ward [1980] calculated the needed 1978 investment per daily barrel by field and the peak production per field in the U.K. North Sea. The aggregate peak production is 3,982 tbd, which is greater than maximum production in any one year. Of this amount, only 250 tbd (Ninian field) cost more than the break-even amount and would not have been undertaken had not the price risen.

A good paradigm is in the Brent field, with cost overstated at $8,727 per daily barrel because it ignores the value of gas production. Brent was discovered in May 1971. In August, the owners, Shell and Esso, "startled oil explorers" by making a very high bid on an adjoining tract. In August 1972, Shell announced that the field would come onstream in 1975–1976. [PE 9-72:319] This was confirmed the next year. [PE 6-73:204]

Thus of the three new areas, all of the North Slope and Mexico and about 94 percent of the U.K. North Sea would have been developed even at the pre-1974 price levels. Possibly the North Sea would have been developed more slowly.

United States

There was a prompt investment response in the United States. In the six years 1973–1979, oil wells drilled rose from 10,200 in 1973 to 20,700 in 1979. In the next two years, they doubled again. Yet production fell. Even more important, reserves added in the mid-1970s were the poorest in decades. Ultra-high prices seemed to induce ultra-low outputs.

Some of this is to be expected. In the early stages of a frantic boom, there is hoarding of materials, machines, and manpower. Those with inadequate supplies of some factors hold them in the hope of getting more. Thereby, large amounts of the factors are locked up, held idle, or underused for a time. This helps explain why the absolute increase in U.S. oil well completions in the two years from 1979 to 1981 was twice as great as in the six years from 1973 to 1979.

But the question still remains why the reserve additions per new oil well drilled, or per dollar spent, were so abysmally poor between 1974 and 1980. The reason was probably price controls on oil and natural gas. Exotic high-cost hydrocarbons were uncontrolled and very profitable, and they spurred a huge overbidding for factors, which made ordinary price-controlled oil and gas very unprofitable.

Thus the 1973–1974 price increase contributed very little to additional supply for the remainder of the 1970s. Indeed, there may have been a net subtraction. In many countries, higher prices led to overtaxation and overregulation, discouraging supply, as with natural gas in Europe and oil and gas in North America. There was also an increase in risk factors, and therefore in minimum or "hurdle" rates of return, because the expropriation of the oil companies showed that a concession or contract would not be honored.[11] Even outside the Third World, there were widespread unilateral contract and concession revisions. And, of course, higher factor prices became inflated costs, hard to reduce.

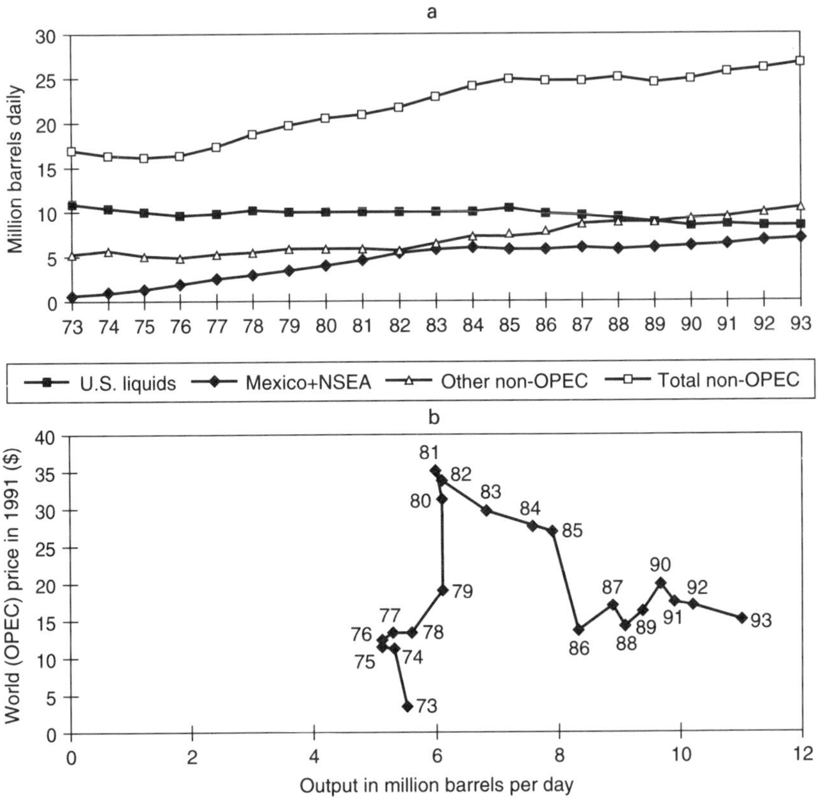

Figure 6.3
Oil output, (*a*) non-OPEC market economies output, 1973–1993, (*b*) output and price in the market economies, excluding OPEC, the United States, Mexico, and the North Sea. Sources: (*a*) PE, (*b*) DOE:MER and OPEC-SB.

Other Non-OPEC

The other non-OPEC group comprises many producers and is the best index of whatever underlying trend was at work (figure 6.3). Output increased mildly through 1981. Then as the world price (defined as OPEC average revenue per barrel) first declined and then crashed, production rose steadily and by 1992 was the largest of the three components.

Market Developments, 1974–1978

The changes in supply and demand just summarized meant a weak market. By the middle of 1974, OPEC excess producing capacity was about

Table 6.5
OPEC excess capacity, August 1974 (mbd)

Country	Capacity	Production	Percentage excess
Algeria	1.1	0.9	18.2
Ecuador	0.2	0.1	58.3
Iran	6.5	6.1	6.2
Iraq	2.5	1.8	29.6
Kuwait	3.8	2.0	47.4
Libya	3.0	1.4	53.3
Qatar	0.7	0.5	25.7
Saudi Arabia	9.7	8.5	12.4
UAE	2.3	1.9	15.9
Venezuela	3.5	2.9	18.3
Indonesia	1.5	1.5	0.0
Nigeria	2.4	2.4	0.0
Total	37.1	29.9	19.4

Source: PIW 10-14-74:7. The ultimate source was a study by Chase Manhattan Bank under a Federal Energy Administration contract. Arithmetical errors in source have been corrected.
Note: The FEA said capacity would increase to 41.8 mbd in 1975. Increase of 4.7 mbd per year implies 390 tbd addition per month.

20 percent (table 6.5). Previously the complaints of "surplus" referred to the underlying excess potential or to an occasional transitory excess caused by the fluctuations of demand or the need to bring on new supply in large increments that outran current demand.[12] Rapid consumption growth had rapidly liquidated any such excess. The 1974 surplus was expected to disappear in the same way. Instead, it grew.[13] Chronic excess capacity, which did not evaporate, was something new in world oil, although it had plagued the United States for many years of market-demand-prorationing, administered by a cartel of states. [Adelman 1964]

Operating under the New Conditions: "Sharing Losses"

The major discounters had become the national oil companies [NYT 2-18-75:1]. Those slow to respond were losers. Libyan production fell from the 1970 peak of 3.3 mbd to 1.6 mbd in 1974. Its freight advantage disappeared because tanker rates fell.[14] By the time the Libyans responded and cut prices, and even offered lower taxes to the remaining companies as an incentive for higher output [PE 7-75:242], exports were generally depressed, and they achieved only 2.1 mbd.

Abu Dhabi output fell by over a third from December 1974 to February 1975. After oratory about oil companies [NYT 3-3-75:41], officials cut the premium for their low-sulfur crude oil, and their March sales climbed back to the December level. The Saudis "made themselves a shock absorber" by cutting output [NYT 3-11-75:43], from 8.8 million bd in October 1974 to 5.65 million in March 1975. This "pulled the rug out from under" those in the U.S. government who had believed the Saudis' assurance that they would not cut output. [OGJ 3-17-75:67] By February 1975, noted one trade journal, "an extraordinary degree of preoccupation with the levels of production and exports is becoming increasingly evident among the oil exporting countries.... [D]espite the 'shut in and save' philosophy, uneasiness is appearing even in ... rich states ... over the continuing decline in oil offtake and what it might do to revenues, right now." [PIW 2-10-75:1]

Henceforth the chief topic at an OPEC oil ministers' meeting was "sharing the losses" of sales volume. [PIW 3-3-75:4] With past failures in mind, OPEC nations tried to avoid comprehensive market sharing agreements, but the OPEC secretariat said production sharing was urgent. [PIW 3-10-75:10] The June 1975 meeting announced a price increase for October. [NYT 6-12-75:1] This promptly raised spot prices: "The prospect of an OPEC price hike in October is breathing some life into the spot crude oil market in midsummer.... [But] storage costs some 20c to 30c per barrel per month, so a big OPEC hike would be necessary to make inventory-building worthwhile." [PIW 8-2-75:5]

The October 1975 price increase was very imprecise because OPEC used (for the last time) the fictitious "posted" prices, which were merely an interim calculation toward setting per barrel taxes. However, the Saudi bottom line—government take—increased from $10.16 to $11.18 per barrel, by more than most of the other governments' hike, which "found it expedient to pay some homage to the forces of the market by putting up their prices by less than the agreed percentage ... largely by trimming their differentials. But since departure from pricing norms tends to be anathema to all cartels, OPEC's Economic Commission has been trying to evolve an agreed scale of differentials.... It has not been an easy task." [PE 12-75:444]

"Those awkward differentials" [Ibid.] have bedeviled the cartel to this day. They reflect differences in transport cost and in quality—that is, the value of the products that can be refined from a given crude oil and the additional costs (for example, a light crude with a high wax content). Both differentials change frequently, in response to seasonal movements, inven-

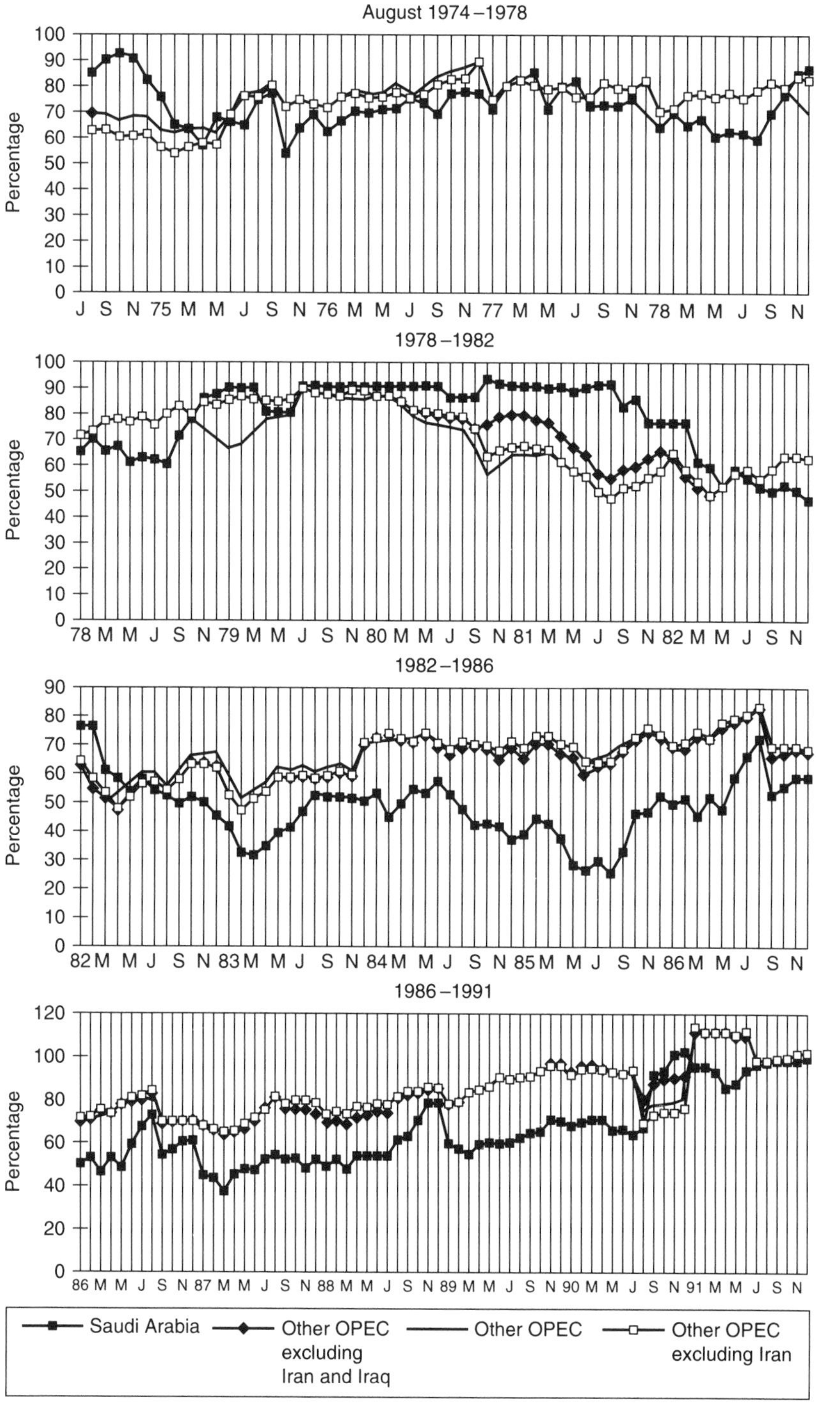

Figure 6.4
OPEC monthly capacity utilization, (*a*) August 1974–1978, (*b*) 1978–1982, (*c*) 1982–1986, (*d*) 1986–1991. Source: PIW.

tory changes, and consumer preferences. There are also slower changes in technology. A schedule of differentials that is right today will soon turn wrong. Some crudes will be underpriced; demand for them will increase, at the expense of the overpriced crudes. The division of the market is undermined; some sellers gain at the expense of others.

In mid-1975, a special report warned OPEC ministers of the menace of differentials as a means of price discounting. [PIW 7-7-75:1] Yet OPEC members headed in separate ways on differentials. [PIW 9-19-75:1] Over the years there have been repeated references to differentials as a divisive unsettled issue, but nothing permanent has been done.

Under the old regime, the integrated oil companies did not respond much to the differentials. Small differences in realizations among crudes could easily be absorbed in wide profit margins. This was no longer true after 1974, when the producer nations sold to the companies. The random movement of product prices and crude prices subverted the schedule of official FOB crude oil prices. (The end result appears in figure 6.4.)

Saudi Policy, 1975–1976

Saudi Arabia faced a never-ending adjustment. If it stuck with its official prices, it might lose business, perhaps suddenly, as happened to many. If it discounted too much, it might lose money it would otherwise earn; at worst it might bring down the whole price level.

In 1975, OPEC financial surpluses were rapidly fading. In early 1974, the U.S. Treasury had predicted a peak OPEC accumulation of $650 billion; the estimate was now scaled down to $200 billion to $250 billion. [PIW 5-26-75:5, 6-2-75:7] In 1974, Saudi Arabia did much better than average in its percentage of capacity used, but from early 1975, it restricted output more than the other partners. (The "hawkish" Iraqis, who had demanded a 30 percent price increase in 1975, actually cut prices in 1976 to gain more market share. [PE 9-76-338]) In early 1976, Saudi Arabia moved to reverse the trend.

During the Aramco negotiation of 1975 and 1976, the Saudis required the Aramco companies to lift a minimum of 7 mbd. As in 1973 and in 1975, the expectation of an official OPEC price increase in December 1976 generated speculative demand, which began pushing prices up in August: "OPEC's coming hike spurs early moves to stockpile crude ... Spot crude prices start rising as OPEC hike looms." [PIW 8-30-76:1] As in 1975, the explanation foreshadows the elaborate number crunching that came later, in the 1980s:

With a $1-plus a barrel OPEC boost all but inevitable, crude buyers can afford to pay out the 25¢ to 85¢ a barrel of added financing and storage cost involved in buying oil four months ahead of time.... Sometimes it backfires, though, as last May–June when OPEC-wide price hikes didn't materialize.... The specter of a big OPEC price increase Jan. 1 has snapped the crude oil market out of the summer doldrums as buyers rush to stockpile crude. [Ibid.]

Continued stockpiling in anticipation of the January OPEC price increase made the spot price continue to rise through October. [PIW 11-1-76:7] Then there was a pause.

In December 1976, the majority wanted a 10 percent price increase, with another 5 percent in July 1977. Saudi Arabia proposed only 5 percent for the whole year. Neither side would give in. The result was a 10 percent increase, Saudi Arabia in effect discounting by 5 percent. It was a simple uniform price cut, quickly proved more profitable than irregular discounts.[15] Right after the meeting, the speculative demand for oil dried up, and the other members were forced to scramble after the Saudis for sales.

The Payoff to Saudi Underpricing

During the fourth quarter of 1976, Saudi Arabia had produced 27.56 percent of OPEC output. During the first five months of 1977, it produced 29.85 percent. [PE 7-77:255] Thus a price 5 percent lower yielded sales 8.3 percent more, increasing revenues at their fellows' expense.

Of course, this could not last long. "Hawkish" Iran tried to discount indirectly, through barter deals. [PE 3-77:102] In July, the Saudis were ready to increase. [PE 7-77:255] The final result was a "compromise" increase of 10 percent to which all subscribed.

Some OPEC members were slow to suspect cheating. Nigeria "thought it had an agreement with Algeria and Libya in 1977 to raise prices in concert. By the time they realized what was happening ... they'd lost $1.5 billion in revenues." In less than six months, Nigeria lost nearly one-third of its sales because the price level was less than 2 percent too high. [NYT 9-17-79:D1]

The surplus was worse in 1977. "Oilmen hate short term gluts because they knock stock profits and undermine their increasingly frenzied predictions of long term scarcity." [*Economist* 8-27-77:88] "The first cautious glimmers of a slight price upturn are surfacing on the weak spot market ... mainly in anticipation of a possible OPEC hike next January.... With oil in oversupply worldwide, spot prices of Middle East crude have eroded steadily since the OPEC price split was resolved in early summer." [PIW 10-17-77:1] A week later, it was "a world awash in surpluses for now." [PIW 10-31-77] There was no "fourth-quarter stockpiling rush ... OPEC price rise prospects scarcely dent spot discounts." [PIW 11-14-77:1]

The spot market weakness was justified when the December 1977 OPEC price meeting took no action. "The ministers could not close their eyes to the over-supply situation evident in the markets for months past."[16] [PE 1-78:2]

Conclusion: An Effective Cartel

Between 1974 and 1977, sales were flat, and much excess capacity accumulated. With frequent or almost continuous contact over market shares and cheating, the OPEC members held together and even increased prices in the face of deficient demand. Secretary General Ali Jaidah looked back in 1980 with justifiable pride at the OPEC achievement in raising price despite a glut in 1974, 1975, and 1976. [PIW 9-15-80:1] However, the continuing inflation had more than offset the oil price increases.

On the Eve in 1978

It was generally believed in 1978 that OPEC still had unexerted price-raising power. In 1977, the mild drop in the consumption/GDP ratio had flattened or improved since 1975. Price elasticity near zero seemed at least compatible with the evidence.

In April 1978, OPEC appointed a Long Term Price Policy Committee of oil ministers, chaired by Yamani. Their report was not leaked until late in 1979 [PIW 12-31-79:3, 3-3-80:1] but agreement must have come much earlier. The long-term goal was restated more precisely: to raise the price to a level just below the cost of producing synthetic liquid fuels, supposed to be near $60.

But how fast to proceed? The committee proposed that they increase the inflation-adjusted price of oil each year by the OECD nations' growth rate. This would have been only a small percentage each year, and zero in recession, avoiding shocks to the world economy. Moreover, a slow advance would have afforded time to correct the committee's mistaken assumption about zero price elasticity. OPEC would have sensed the barrier to further increases. But within a year, the committee's recommendation was disregarded by its members. Why could they not follow their own good doctrine?

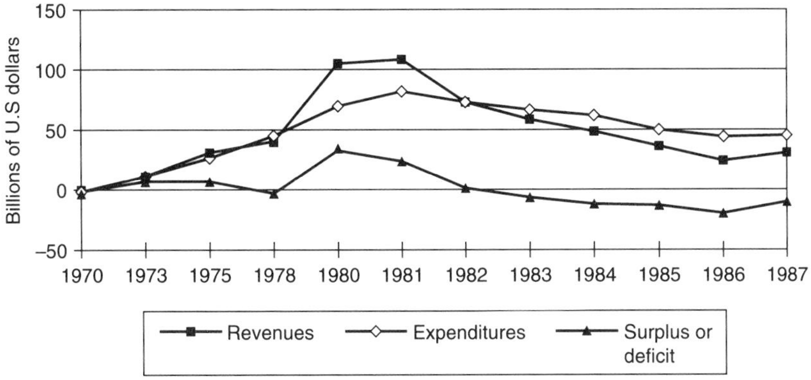

Figure 6.5
Saudi Arabia revenues and expenditures, 1970–1987.

The Menace of Deficits

One sees the picture most clearly in Saudi Arabia. From 1970 to 1978, revenues (nearly all from oil) rose twenty-one times, but expenditures rose thirty-one times, from $1.4 billion to $43.0 billion, and the Saudi budget was in deficit (figure 6.5). This was true of the whole Saudi economy. The 1974 current account surplus was washed away in four years by rising imports of goods and services (table 6.6). (The 1980–1981 surplus was almost twice as large as 1974, and it vanished in only three years.)

This rising tide of spending destroys the myth that the Saudis and other OPEC countries wished to keep oil in the ground and to produce only enough to meet their "needs." Their "needs" grew with their revenues. If the Saudis had really been producing more than they wished, then by mid-1974, and ever afterward, they would have served themselves, and won thanks from all other producers, by producing less to make room for them. In fact, they discounted prices and imposed heavy penalties on the oil companies for underlifting. In 1978, the eight-month average of Aramco production was 4.3 percent above the minimum but considered "too close to the line for long-run comfort." [PIW 10-16-78:5]

Short Horizons and High Discount Rates

We saw in chapter 2 why the cartel governments have shorter time horizons and higher effective discount rates than would private companies operating identical oil reservoirs. First, OPEC must bear all the fluctuations in world oil demand. Second, each must also cope with fluctuations

Table 6.6
Saudi Arabian current accounts, 1967–1992
(billions of U.S. dollars)

Year	Exports	Total receipts	Total goods and services imports	Current account
1967	1.5	1.7	1.5	0.1
1968	1.7	1.9	1.8	−0.1
1969	1.8	2.0	1.8	−0.1
1970	2.1	2.4	2.0	0.1
1971	2.6	2.9	1.6	1.0
1972	3.9	4.4	1.9	2.1
1973	7.5	8.3	4.9	2.5
1974	30.1	32.7	8.2	23.0
1975	27.3	30.6	12.5	14.4
1976	35.6	40.2	21.6	14.4
1977	40.4	46.5	29.1	12.0
1978	37.0	43.5	38.9	−2.1
1979	58.1	65.8	47.8	11.2
1980	100.7	112.0	59.3	42.8
1981	111.8	127.8	75.7	41.1
1982	73.9	92.9	75.5	7.6
1983	45.7	66.0	73.5	−17.0
1984	37.4	55.0	64.6	−18.4
1985	27.4	43.4	48.0	−12.9
1986	20.1	34.1	38.1	−11.8
1987	23.1	36.2	37.8	−9.8
1988	24.3	37.1	35.5	−7.3
1989	28.3	41.3	40.0	−9.2
1990	44.3	56.8	44.9	−4.3
1991	48.8	na	na	−25.7
1992	na	na	na	−13.4

Source: International Monetary Fund, International Financial Statistics (1990 and September 1991). The current account balance equals receipts, less net imports of goods and services, less "unrequited private transfers," which are largely remittances by foreign workers. For 1991, other components we use cannot be traced. The numbers for 1990–1992 are from 1992 OPEC *Statistical Bulletin*, table 7.

Note: The 1992 issue of *International Financial Statistics* leaves blank all items related to foreign trade and the current account for Saudi Arabia for 1991. Estimates of the 1993 current account deficit are between $11 billion and $15 billion. Making matters more complex is that Saudi Arabia is reported to have paid only about 20 percent of the estimated $65 billion owed to the United States and other countries for aid during the Gulf War. [9-93:20]. It is not clear whether the unpaid $52 billion has been included with the current accounts or whether it has been considered an import of capital. If the latter, it is many times the estimated $5 billion to $6 billion official reserves still in existence. Saudi individuals and firms are estimated to hold about $50 billion net assets abroad.

in sharing the OPEC totals. Third, the OPEC governments are peculiarly vulnerable to fluctuating income because oil exports provide the bulk of income and nearly all imports. The bulk of nonoil industries (particularly in the gulf countries) do not provide income; they absorb oil income and would vanish without it. [Gelb 1988] In Saudi Arabia, subsidies were 80 percent of revenues. Only 19 percent of the subsidies had "largely productive objectives." [Askari 1990, pp. 106, 112, 116] These were mostly in petrochemicals, whose rate of return was around zero.

In addition, most OPEC nations were vulnerable to unrest at home and attack from neighbors. A rational response was to grab what they could when they could. As had been said of them forty years earlier: "The future leaves them cold. They want money now."

The Washington Official Truth

Under the Nixon and Ford administrations, an assistant treasury secretary had said: "Breaking up OPEC would be detrimental." The State Department feared that disrupting OPEC would be "politically damaging" in the Middle East. There were "lush export markets"; a Brookings study estimated that income growth in the consuming countries was slowed very little by the oil price increases. [*Time* 1-19-76:54] Economists mostly thought otherwise.

The incoming Carter administration was much more articulate,[17] and the president had "regular and heavy correspondence with the Saudi leaders" [NYT 7-12-79:A3], which has not been made public. Carter believed that "Saudi Arabia is producing more oil than they want" [NYT 10-19-79:1], in accord with his adviser on Mideast oil matters, James E. Akins. One reporter noted that the "Carter Administration has gone all out to cultivate the Saudis. Akins ... served as a campaign advisor to Carter, and went to Riyadh after the election to talk oil prices with the Saudis." [Joseph Kraft, in *New Yorker*, 6-26-78]

As President-elect Carter's special envoy to the Persian Gulf nations, Akins stated in Kuwait that a 10 percent increase would be "acceptable." [PONS 11-24-76:1] The Carter administration, like the two previous administrations, desired "special relationships" with Saudi Arabia and Iran to induce them somehow to produce more than would suit their economic interest. The 1977 CIA report forecast an early oil deficit. The Soviet Union would turn from exporting over 1 mbd to the noncommunist world to importing 2.5 mbd by 1985. [CIA 1977] Akins had predicted in 1972 that the U.S. situation in 1976 could be "nothing short of desperate"; obviously, doomsday was now closer, and the president feared world oil reserves would be gone by the mid-1980s. [NYT 4-19-77:A24]

There was no room in their philosophy for a market process to equate the amount demanded with that supplied. Secretary of Energy James Schlesinger said the Carter Administration's strategy was "to continue to improve relations with Saudi Arabia, which has been the moderating influence." [WSJ 7-1-77:2] He warned of "a major economic and political crisis in the mid-1980s as the world's oil wells start to run dry and a physical scramble for energy develops." [NYT 10-6-77:67] The current oil glut would quickly disappear, and demand would exceed supply. [Schlesinger 1978, appended figure]

A prominent Brookings Institution economist, Arthur Okun, with close ties to the administration, thought strong action "essential if the nation is to have 'insurance against an energy *catastrophe*. The question is whether the American people will have enough maturity to act before the *crisis* hits.'" [Time 5-2-77:34, emphasis added] He did not explain "catastrophe" or "crisis." A Brookings study concluded that by the mid-1980s Saudi Arabia would be unable to repress further price increases; unless importing countries curbed energy consumption, oil prices would necessarily rise sharply. [OGJ 4-10-78:26] The undersecretary of state for economic affairs, Richard Cooper, believed that we lived in "an age of energy shortage" and that "economic expansion will sooner or later run into a shortage of oil at existing prices." [Cooper 1986, pp. 53, 66] He did not explain "shortage." He thought the 1973–1974 price explosion was caused by a surge in demand (which we saw had not happened). He made no reference to supply.

The Carter administration doctrine of worldwide scarcity and the need for cooperation with the Saudis and other producers is well stated in *Project Interdependence*. [CRS 1977] An article obviously reflecting official thinking argued that high oil prices were good for the United States. [*Washington Post* 7-10-77; see also *Forbes*, 3-20-78] In June 1978, Yamani said "key U.S. officials have told him ... that gradual price increases during the next several years are in the U.S.'s best interests." [WSJ 6-5-78:5]

The only place considered a source of future supply growth was the Middle East. A Carter official said: "It's hard to bare our teeth at the Arabs when we're groveling for their oil." [WSJ 10-21-77:OpEd] One does not see much with one's belly to the ground. A subcabinet member later boasted of "unprecedented closeness in the Middle East." [*Energy Daily*, 4-3-78] Oil imports from Mexico had to be restrained in order not to disturb "carefully cultivated relationships with the Middle East." [NYT 11-29-78:D1] Mexicans were not amused by the State Department, where "some even compare Mexico's breakneck expansion of oil sales to a derelict selling his blood in the morning to buy an afternoon bottle of wine." [WSJ 8-30-78:1]

Bilateral Relations

The "carefully cultivated relationships" with the Saudis and other mideast producers were taken very seriously. "During the 1970's, when oil prices were escalating, it was routine for the U.S. Treasury Secretary to be dispatched to major OPEC countries, chiefly Saudi Arabia and others in the Mideast to urge pricing moderation." [WSJ 7-25-88:3] The General Accounting Office noted:

> To achieve the U.S. objective of access to adequate supplies at 'reasonable' prices, the United States uses its *bilateral* relationships with friendly producers in an attempt to influence their pricing and production decisions. This is especially apparent with Saudi Arabia, with which, according to a Department of State official, the United States has a "very active" *bilateral* policy. Frequent visits by cabinet-level officials including the Secretaries of State, Treasury, Defense, and Energy, during the past several years illustrate this *bilateralism*. [GAO 1982, pp. 49–50, emphasis added]

The previously secret record of one of these visits, a meeting of the Secretary of Energy Schlesinger with Crown Prince (later King) Fahd of Saudi Arabia in January 1978, was recently released. [Schlesinger-Fahd 1978] Said Schlesinger, "[T]he outer limit of [Saudi Arab] feasible production is on the order of 14 million barrels per day." [Par. 21] In fact, in 1972 Aramco, with Saudi approval, had proposed expansion to 20 mbd. [PIW 10-9-72:3] In 1973, an Aramco vice president, in an internal communication, said, "You could go a lot higher than [20 million]." [7 Church 540] In January 1974, the chairman and a vice president had stated that Aramco could produce 20 mbd "on out through the end of the century without any problem at all." [NYT 1-29-74:39] Later the CEO of Chevron had testified Aramco could go in eight to ten years from around 8 mbd to beyond 20 million. [7 Church 453] In 1977, Aramco estimated that on the basis of current reserves it could produce 25 mbd. [PE 7-77:286] With current reserves at 150 billion barrels, 25 mbd would have maintained a 16.5 : 1 reserves-to-production ratio; the usual industry rule of thumb is 15 : 1.

"It is true, of course," Schlesinger said, that oil in the ground would be a better investment for Saudi Arabia than the alternatives which are now available for investment of its cash revenues." [Par. 15] Ostensibly he was trying to persuade the Saudis to invest more in oil production. One wonders, What would he say to *discourage* investment?

More to the point, the Department of Energy in 1977 had forecast world crude oil prices' increasing at a maximum of 3.5 percent per year, real. [DOE 1977, vol. II, pp. 8–9] Since the Saudi cost was negligible in relation to the price, that would be roughly the return on holding oil in the ground rather than selling it off. One tries to imagine an investment banker advising a client to make a risky investment for at most a 3.5 percent return.

The Washington view was the international consensus view, shared by everybody who was anybody. A large group of experts in March 1978 substantially agreed "on one key point: the world is presently heading toward a chronic tightness, or even severe shortage, of oil supply." [Rockefeller 1978, p. vii] The current surplus they called "ephemeral." They obviously believed that markets could not allocate. Hence, they feared "preemption" of oil supply "by the few nations better able to pay for it" and urged the importing and exporting governments to "develop a common approach to cope with a potential chronic energy supply crisis." In summary they wrote:

> The industrial countries must find incentives for oil exporters such as Saudi Arabia, the United Arab Emirates, Kuwait and Libya, which do not have an immediate need for all the resulting revenues for economic development, to produce nevertheless at the highest feasible levels. The report underlines the importance of U.S. relationships with Saudi Arabia and Iran. [Ibid., summary, pp. 2–3]

The market failure was so well known as to need no proof: a current "glut" only hid the basic "shortage" and a "scramble" for "preemption." Given the dominant attitudes in the consuming countries, the OPEC governments felt free to do as they wished.

The U.S.-Saudi Agreements

Early in the Carter administration, there was an agreement with the Saudis whereby they would provide "oil price moderation and sufficient oil supplies to meet the needs of the West and Japan. As a result, the Saudis say, they became the pricing moderates of OPEC constantly fighting for lower prices and agreeing to produce more crude so that tight supplies wouldn't force up prices." [WSJ 7-18-80:25] This seems to imply an awkward truth: that previously they had *not* been moderates. Then in early 1978 there was another and more specific agreement, providing that the Saudis would maintain output, and in return the United States would not buy oil for the Strategic Petroleum Reserve, a stockpile for use in a future shortage. It was mandated by a 1975 law, and the oil-producing nations openly disliked it. [NYT 2-29-80:D1]. Neither report of an agreement was denied at the time by any administration officials.[18]

The Second Price Explosion, 1978–1981

Spot prices kept sliding throughout the first half of 1978. "The crude supply surplus" was estimated at between 1 mbd and 3 mbd. [PIW 5-22-78:3] In May 1978, despite "a hot denial of any collective action on production controls" [ibid.:4], there was a secret OPEC decision to impose a production program; it was revealed only two years later. [NYT 9-18-80:D1; PIW 9-22-80:5, 3-29-82:S1]

Perhaps because of the secret May accord, spot prices strengthened. What may also have helped was a Saudi directive that Aramco produce more heavy and less light crude oil. If enforced, it would have cut Saudi production by something between 1 mbd and 2 mbd. [PIW 2-27-78:3, 5-22-78:3] This would help lift spot prices and hence "justify" an increase in official prices. [PIW 6-19-78:1, 6-26-78:1] But in July "no major price recovery [is] seen on [the] spot crude market." Buyer resistance seemed to be based on "substantial surpluses in producing capacity." But it was hoped that the expected price rise would stimulate some buying. [PIW 7-10-78:3]

The prospect changed in August. Anticipation of a higher official price at the forthcoming OPEC meeting was thought to be pushing spot prices up. [PIW 8-14-78:5] Shortly after occurred almost "a repeat of the summer 1976 crude oil stockpiling game: oil companies are lining up extra supplies to beat the almost certain OPEC price rise in January." But not all companies were doing so, since interest carrying cost was 8 to 10 cents, and leased storage charges 12 to 15 cents, per barrel per month. [PIW 9-4-78:1]

"The 'pre-OPEC' stockpiling rush" continued to raise prices in September. [PIW 9-11-78:1] In October, the market was quiet and balanced. Figure 6.1 shows that spot price was a good proxy for contract deliveries made for Saudi, Iranian, and the United Arab Emirates (mostly Abu Dhabi) crude oil, all staying close to each other.

The cartel was trying to repeat its success in 1976 and 1977 and raise prices in the face of overabundant supply, but it was more difficult: OPEC 1978 exports were down 6 percent from 1977, so the financial stress was greater. In September 1978, the OPEC secretary-general warned of the vanishing current account surplus. The role of residual supplier was definitely "unwelcome." [PIW 10-2-78:Sup]

Year-End Uncertainty

At the end of October, an Iranian oil workers' strike occurred. The supply uncertainty made spot prices rise while greatly narrowing the spot market.

> The losses could be "made up" if Saudi Arabia and Abu Dhabi eased their [output] restrictions, but not otherwise.... *The spot market has come to a sudden and complete stop....* Virtually no crude was being offered for spot sale after the export slash in Iran. *Suppliers are hanging on to every drop they have until it's clearer how long the Iranian supply disruption will continue.*" [PIW 11-6-78:1, 5, emphasis added]

November spot prices were higher. Kuwait said there would be no increase in output; rather, consumers should use up their "reserves and stockpiles." [PIW 11-13-79:3]

Then Iranian output resumed.[19] Spot prices dropped in early December [PIW 12-4-78:1], but everyone expected higher prices to be enacted at "OPEC's Momentous Meeting" [PE 12-78:498]. One assessment was for official OPEC prices to be raised about 15 percent, to "equate them" to spot. [PIW 12-11-78:3, 5]

Supply and Demand, 1978–1979

The price changes of 1978–1979 cannot be explained by insufficient supply. From the end of September 1978 to the end of July 1979, there was more than enough unused capacity available to take any strain (table 6.7). Moreover, in the last quarter of 1978, inventories actually rose contraseasonally. The world produced more than it consumed, when it would normally have been consuming more than it produced. In the first quarter of 1979, the inventory drawdown was actually less than in the sluggish first quarter of 1978. Before and after that quarter, production exceeded consumption.

The only reason for a price rise at this point was fear of possible loss of supply. Once again, precautionary demand for hoarding was added atop normal demand, for use. This raised the price, which added speculative demand. It was the same action as in previous years but now much larger.

All traders tried to buy and to "hang on to every drop" unless forced by contract to sell. The higher prices went, the higher they were expected to go. The more buyers wanted to buy, the less sellers wanted to sell.[20]

Year-End Stress

Fear and uncertainly had been promoted by the reluctance of Saudi Arabia and other gulf producers to expand output freely to the limits of capacity. In November, the United States kept its word and ceased to buy for the Strategic Petroleum Reserve. [PIW 3-26-79:7] The Saudis, nevertheless, continued to restrict production:

> Saudi Arabia has quietly ruled out any further immediate expansion of its petroleum output to replace supplies withdrawn from world oil markets by turbulence in Iran.... To fill in some of these supplies on the world market, the U.S. companies that account for the bulk of Saudi Arabia's oil output had asked the Saudi petroleum ministry for permission to remove temporarily some of the production constraints that restrict the flow of oil from Saudi Arabia. Neither the U.S. oil companies nor the Saudi government will comment on the development, but it was learned that the Saudi answer was negative. [WSJ 11-13-78:4]

With this assurance, the December 1978 OPEC meeting duly raised 1979 prices for each quarter, at an annual rate of 14.5 percent, as had been expected. They also agreed in principle to increase the premiums on light crudes, but in practice this would have required a highly detailed slate of differentials, on which, as usual, they could not agree. [PE 1-79:2; PIW 12-25-78:1] The loss of Iranian output was made up, and spot prices even decreased in December as the alarm eased.

The January Coup

In January, Iran was out of the world market, producing only for domestic consumption. For the first nineteen days, Saudi Arabia produced at 10.4 mbd, "still below the 'sustainable' ... cited by Oil Minister Yamani recently." [PIW 1-29-79:5; PIW 12-25-78:1] In mid-January, the head of OPEC said, "I think the supply and demand is already balanced." [NYT 1-19-79:D1] But during the second week in January, a report spread that Saudi Arabia was about to reduce production. Secretary Schlesinger said it could have "a very severe effect at this time, and another observer noted, "It's one thing for the Saudis to use the 8.5 million allowable in normal times to create a nice tight supply situation. It's another to use it to create a world crisis." [PIW 1-15-79, 1, 2]

Table 6.7
Production and capacity, Mideast OPEC, 1978–1979 (mbd)

	Sep.	Oct.	Nov.	Dec.	Jan.	Feb.	Mar.	Apr.	May	Jun.	Jul.
Iran output	6.05	5.49	3.49	2.37	0.40	0.76	2.22	4.13	4.11	3.94	3.77
Non-Iran											
Output	16.21	16.74	18.31	18.13	17.89	18.06	17.85	16.93	17.06	17.19	18.18
Capacity	22.35	22.35	22.35	22.35	22.19	22.04	21.89	21.74	21.59	21.44	21.29
Excess	6.13	5.61	4.04	4.21	4.31	3.98	4.05	4.81	4.53	4.25	3.11

Source: PIW 1-29-79:9, 2-4-80:9

During the third week of January, spot product markets, already under strain because of "refiners' unwillingness to let go of product supplies," were thrown into a panic by heavy buying by major oil companies. [PIW 1-22-79:3, 5] The companies already knew what would soon be made public: Saudi output was cut by 2 mbd on January 20, 1979.[21] A "spot price panic [was] now sweeping across the world market for OPEC crudes." [PIW 1-29-79:3]

> Exxon inaugurated a sweeping 10% worldwide cut [in deliveries]—about 400,000 bd. And other companies were expected to make added reductions.... Exxon ... carefully made no reference to developments in Saudi Arabia.... But customers feel the timing is no coincidence since it comes on the heels of the substantial cut in Saudi crude output in the latter part of January....
>
> The Saudi decision to scale down its production from a level well over 10-million bd has caused intense consternation in international oil circles and consuming countries.... "It's the difference between the current tight supply and an acute shortage," one observer notes. Aramco's production ... was ranging between 10.2 and 10.5 million bd in the first half of January.... Aramco had to slash its output substantially, and by late January ranged about 8 million bd. [PIW 2-5-79:1, 5]

Secretary of Commerce Juanita Kreps had been in Riyadh just before the cut, and "Saudi officials had 'assured' her that oil production would continue at a 10- to 10.5 mbd rate 'for some time to come.'" [PIW 2-5-79:6] But cut they did, by over 2 mbd. It went back to 9.5 in February, but only briefly.[22] Secretary Schlesinger said that the cut was "for technical reasons"—an inability to exceed 9.5 mbd—but he had not mentioned this a month earlier, and nobody even suggested it. [PIW 2-12-79:2] Saudi Arabia had produced 1 mbd more, and would soon do so again. Other estimates of capacity were 11.8 mbd, or 2 mbd higher. [PIW 2-20-78:10, 12-25-78:1; PE 8-78:333]

The Saudis' January production cut was an alarm bell that the market had been listening for for two months. Because supply was now more uncertain than ever, buyers were more avid and sellers more reluctant. The spot price kept rising through January and February, then paused in March. Governments kept raising official prices toward spot levels, "merely matching the market."

The Second Shock

By early March, Iran production revived, though at lower levels. Iran offered oil at spot prices but never obtained them. [PIW 3-5-79:1] Minister Yamani said Saudi Arabia would not increase output past 9.5 mbd and indeed "hoped" to decrease it to 8.5 mbd. [PIW 3-6-79:6] There was bewilderment over rising prices despite stagnant demand and adequate oil inventories. An oil executive explained: "When Iranian oil went off the market, OPEC tacitly agreed to limit production. It's much simpler to limit production so that price increases are automatic. The OPEC nations are acting the same way the Texas Railroad Commission did for 30 years." [NYT 3-18-79:E5]

In late March, spot prices decreased mildly. [PIW 3-26-79:1] The OPEC meeting on March 26 faced the danger that the return of Iranian output would make prices collapse, and Yamani accused Iraq of producing more than it had admitted. He declined to raise output to reduce the price increases. At the meetings, a general understanding was reached that "those countries that raised production during Iran's revolution will reduce output now. But there already are signs of differences over who should cut back and by how much." [WSJ 3-30-79:7]

During the first week in April, as spot prices slipped [PIW 4-9-79:3], the trade watched Saudi Arabia:

> So far, there's been no official cut in the Saudi volume, but there are hints it may be scaled down gradually as "market supply conditions warrant." Aside from Saudi Arabia, every OPEC nation is now scrambling to add extra charges.... The big question is whether the pricing free-for-all will continue or simmer down a bit now that spot crude oil prices continue to ebb....
>
> One sign of the spot price decline is the increasing nervousness among some crude oil traders with high-cost supply to resell. But one optimistic trader suggests, "If Saudi Arabia cuts production..." [PIW 4-9-79:1, 3]

The optimist was right. While Carter administration "officials said the Saudis, by keeping oil production at nearly 10 million barrels a day ... were making sacrifices to benefit the West" [NYT 4-3-79:A7], the Saudis had already cut from 9.5 to 8.5 mbd at the beginning of April. One supply specialist noted, "One must question the sincerity of Saudi desire to keep prices from escalating. If the Saudis really wanted to keep the lid on prices, they could have maintained their production at 9.5 million. That would have kept the pressure on the upper-tier price producers." [PIW 4-16-79:1]

With more perspective, in May 1979 the *Oil and Gas Journal* appraised the end-of-March meeting and its results, noting that Iraq was rebuked for again playing the free rider:

> Object of the cutbacks ... is to keep the market tight and let the pricing initiative remain with OPEC members. There was suspicion that the established Saudi

policy of moderation might mean a slow return to ... 8.5 million b/d.... But the Saudis were extremely prompt [to cut output].... Questions were also asked about the willingness of Iraq to ease back. Iraq tries to keep its production figures secret. And although the Baghdad regime admitted to increasing output from 2.5 million b/d to between 3–3.1 million b/d.... Yamani claimed that the actual figure was nearer 3.5 million b/d. [OGJ 5-21-79:35; see also PIW 4-2-79:1]

OPEC Brinkmanship

The trade was more sensitive, and the Saudis' renewed squeeze on output made spot prices surge even more in May than in January and February. Official prices took off in pursuit.[23] Secretary Schlesinger "stressed that since 1977 he has been warning [of an] inevitable deficiency in supply." [OGJ 6-4-79:55] While the spot price soared, the spot volume nearly vanished. In the third quarter of 1978, it was estimated at 2 to 3 mbd; in May 1979, some estimates were as low as 100,000 to 200,000 barrels daily. [PIW 5-14-79:1]

In July 1979 the Saudis raised output.[24] There was a price pause, and time for appraisal. Guido Brunner, the EEC commissioner for energy, said:

> Officials of the European Economic Community, who have just completed an initial exchange of views of members of OPEC, report that the overall objective of the exporters' organization is to keep exports running slightly less than world demand so high prices can be maintained.... A policy of maintaining a small crude supply deficit is economic brinkmanship. [OGJ 7-9-79:35]

The 1979 economic summit of the seven leading OECD nations adopted meaningless oil import "goals." [WSJ 7-2-79:1] They praised Saudi Arabia for saying that it would "never" permit prices to exceed $20. [NYT 6-22-79:D5] "Never" turned out to be three months. In October 1979, the Saudis raised their export price above $20 and kept increasing it. "Western consuming countries do not want oil prices to fall significantly, at least for now, because this would lessen the incentive to reduce imports and develop alternative supplies." [NYT 7-19-79:D5]

Another OPEC meeting was held at the end of June 1979, "to bring Saudi Arabia into closer proximity with the higher-priced OPEC majority." However, the resulting price schedule was so distorted, with such disparate prices for comparable crude oils, that oil company executives called it "a blueprint for chaos." [PIW 7-2-79:1] Saudi production was raised to 9.5 mbd at the beginning of July, and the spot price explosion stopped.[25] But official prices kept rising.

The Path of Prices

Toward the end of July 1979, inventories were close to "the peak of the oil glut last year" [WSJ 7-30-79:1], and some OPEC members were cutting back production to maintain prices. [NYT 7-31-79:D5] In September and October, a replay was developing: "Fear and uncertainty over the oil supply outlook are now the cement holding the spot market for OPEC crudes together.... Supplies around the world are ample, storage is rapidly filling.... Without the anxiety over OPEC's supply and price intentions, the spot crude market could collapse almost overnight." [PIW 10-1-79:1]

In October, an OPEC session raised the specter of an oil glut, and members called for production controls. A usual sign of price weakness was that delegates "again urged a 'dialogue' between producing and consuming countries." [OGJ 10-15-79:103] The United States had been reported to favor "global negotiations aimed at stabilizing the price and supply of oil." [NYT 9-25-79:A1]

Had Saudi Arabia raised output to its announced capacity of 10.8 mbd [PIW 2-4-80:9] at any time in 1979 and committed itself to stay there, it would have pricked the speculative bubble and brought prices down. Instead, they would promise to hold at 9.5 mbd for no more than three months. [PIW 10-1-79:12] During October and November 1979, as the OPEC nations continued to raise official prices, this reluctance to reassure markets about their production level generated a fresh scare over supply, and spot prices soared to a new high of $39 per barrel. Saudi Arabia did not immediately raise its official price, but in the past it had joined in price increases retroactively. Some insiders bet "that its anticipated January [1980] price boost might be made retroactive, and their question is, how far back? [Perhaps] ... even for the entire fourth quarter. Such a ploy would help soak up [Aramco] gains from fourth-quarter operations." [PIW 11-5-79:3]

After the last increase, in November 1979, there was a ten-month slide. When long-term policy was discussed at the December 1979 OPEC meetings, the worry was about too much oil, and some "production management" was suggested. One tool proposed was "an intra-OPEC financing facility that would provide a 'safety net' so that financially strapped members wouldn't have to produce their oil at a high rate to secure revenues. This would be almost tantamount to the perennial production programming idea." [PIW 12-31-79:4]

In early 1980, as in 1974, spot prices turned down, but official prices kept rising. In February, the talk was of "apportioning cuts" among Persian Gulf producers. [OGJ 2-25-80:31] In March, Minister Yamani said "the rising oil surplus exceeds the current production-cutting possibilities of OPEC." [WSJ 3-6-80:2] The "spot crude market shows symptoms of coming apart," and the peak was recognized as November 1979. [PIW 3-10-80:1, 3-17-80:1] Production cuts for supposed "technical reasons" were really for lack of customers. [Ibid.:5] But sellers continued to raise contract prices during the second quarter. [PIW 4-7-80:1]

In May 1980, another official Saudi price increase to $28 aligned it with some OPEC prices. [PIW 5-19-80:4] In June 1980, official contract prices were still being raised, but more slowly, because spot prices were weak. "Refiners are no longer willing to pay 'anything' to secure a long-term direct supply contract from OPEC producers." [PIW 6-9-80:3]

Although the Saudis kept raising their price, they kept it safely under their rivals'. As the others cut back output in response to shrinking demand, Aramco kept producing at a steady high rate. Figure 6.4 shows that the divergence of output rates kept widening for two years. In May 1980, Saudi Arabia said it planned to maintain the "temporary" 9.5 mbd rate through year-end. [PIW 5-12-80:3] By August 1980, official and spot prices had converged (figure 6.4), amid trade talk of burdensome stocks whose private carrying cost was reckoned at 15 percent interest plus $2.50 per barrel annualy. [PIW 8-18-80:3]

In September 1980, the OPEC deputy secretary-general, its senior pricing expert, expected the oil price "gradually" to approach $60 because that was the estimated cost of synthetic oil. [NYT 9-2-80:D3] OPEC secretary-general Ali M. Jaidah said history disproved "the common and widespread fallacy" that prices could not be raised in a time of glut; they had been, he noted, in 1974, 1975, and 1976. But to do it again, collective production policy was needed. He forecast that "the [Saudi official] marker price will soon be adjusted above its current level although the market is characterized by excess supplies." [PIW 9-15-80:1] This happened almost overnight. In late summer 1980, there was "a tacit gentlemen's agreement for a quasi-general OPEC production cut." It was "similar to the unofficial 1978 agreement to reduce output levels due to a previous surplus." [PIW 9-22-80:5; NYT 9-18-80:D1] Saudi Arabia now raised the official price, backdating it to August 1, and the Aramco companies were stuck with the bill for $728 million, since they had already sold the oil. [PIW 9-29-80:7] The Saudi official price now equaled the spot price. The outbreak of the

Iran-Iraq war in September 1980 again raised spot prices in October and November, although the gain of those two months was lost over the next seven.

The Iran-Iraq War

One cannot tell whether the success of the 1980 "gentlemen's agreement" for a production cut would have continued without the Iraqi invasion at the end of September.[26] The ostensible reasons do not concern us. The Iraqi government made much of the Battle of Qadisiyya in 637 A.D., when the Arabs had beaten "those insolent Persians." Saddam Hussein's uncle and mentor had written that God had made three mistakes: in creating flies, Jews, and Persians. These 1,350-year-old hatreds are still potent, but it was oil revenues that provided both the means and the motive for war. Billions had been spent on weapons. The Iranian Revolution, itself a reaction to the petrodollar flow, had weakened the Iranian armed forces and thus lowered the risk of the attack. The Arab-speaking province of Khuzistan (or Arabistan) contained most Iranian oil, nearly all within 150 miles of the Iraqi launching point. The return on Iraq's investment in aggression would have been very high.

The effect of war on the oil market was favorable. In mid-1978, the combined capacity of the two countries was 11 mbd and expected to go much higher. Two years later, it was down to 6 mbd and remained there to 1990.[27] Revolution and war are to the oil market as epidemics and famines to population control.

Said the supply experts, "'Everything has changed but nothing has changed.' It would take a five-month shutdown just to erase the existing surplus—and another year just to fully deplete all readily available world inventories." [PIW 9-29-80:6] There was also excess producing capacity.

Saudi Arabia immediately (starting October 1980) raised production from 9.5 to 10.4 mbd. It kept increasing the price but always stayed below other producers, which stayed below the spot price. The "Aramco advantage," which irked other companies, made it profitable for the Aramco partners to continue lifting the output. But there was a new and troubling note at the end of 1980: total OPEC exports were down by 17 percent since 1977.

> At the back of even militant Oil Ministers' minds is the longer term fall off in demand for OPEC crudes that is now seen increasingly widespread.... A market crunch between OPEC members themselves as soon as Iraq and Iran resume sub-

stantial exports is another preoccupation. Iran especially ... won't find it easy to persuade others to move over and make room in the market. Saudi Arabia carried the burden of a previous market shift in 1975.... But Yamani has indicated that Saudi Arabia is not ready to repeat anything along these lines until it gets concrete understanding for cooperation on long term OPEC strategy." [PIW 12-15-80:1, 2]

In January 1981, "Fear of soft market seems behind the latest Mideast [contract] price rise." [PIW 1-19-81:1] This apparent upside-down thinking makes good cartel sense. A deliberate contract price increase stops the rot in spot prices and raises the level from which, at worst, they will tend to decline. But by July 1981 spot prices were back at the lows of August and September 1980. From this month on, the market was unified; spot and contract prices, Saudi and non-Saudi, were approximately the same.[28]

The "Aramco advantage" was now gone. The Aramco companies had accumulated huge inventories, which they now began selling off at discounts of about $1 per barrel. [PIW 9-28-81:3] They had to reconsider whether their special relation was worth the drain of millions of dollars a year.

The most infatuated company was Mobil Oil, which in December 1980 announced participation in several Saudi industrial projects, in return for access to an additional 1.4 billion barrels over fifteen years (256 tbd). The deals were part of "Mobil's continuing efforts to cement its relationships with Saudi Arabia in order to guarantee access to crude oil." [NYT 12-10-80:D20] "Access" was mentioned three times in two paragraphs. In the next chapter we will see how the obsession delayed but did not prevent the Aramco companies' cutbacks of overpriced oil throughout 1982.

The Abiding Consensus: Permanent Shortage

When Secretary of Energy Schlesinger left office in August 1979, he set out the administration view:

> Oil, the fuel of choice that has driven the vast economic expansion since World War II, will no longer be available in increasing quantities to fuel the further growth of the world's economy. Prices will inevitably reflect the increasing pressures of demand against constrained supply.... [Even] forceful and intelligent corrective measures ... will serve only as palliatives.... Quite bluntly, unless we achieve the greater use of coal and nuclear power over the decade, this society may just not make it. [WSJ 8-23-79]

Even in the long run, supply could not grow. The consumer response to higher prices was not even mentioned.

According to a widely publicized report, reserve reassessment in Saudi Arabia had "led to decreasing reserve estimates" [reprinted PIW 4-23-79:S3], and it was said elsewhere that "shrinking reserves of oil" were worrying OPEC policy chiefs [PIW 9-24-79:1] (These stories continue to surface; they will never go away.)

The CIA reports of August 1979 and May 1980 deserve to be as well known as the 1977 report. In 1979, the CIA wrote, "the gas lines and rapid increases in oil prices during the first half of 1979 are but symptoms of the underlying oil supply problem—that is, the world can no longer count on increases in oil production to meet its energy needs." [CIA 1979, p. iii][29] And in 1980, it had this to say: "We believe that world oil production is probably at or near its peak.... Simply put, the expected decline in oil production is the result of a rapid exhaustion of accessible deposits of conventional crude oil.... Politically, the cardinal issue is how vicious the struggle for energy supplies will become." [CIA 1980]

In October 1979, James Akins said that soon "we shall enter a period of permanent oil shortage" and "semi-anarchy in energy, with the richest and strongest of the consumers making bilateral deals with OPEC." Exporting countries might refuse to sell oil for dollars. [PE 12-79:506]

The OPEC secretariat expected rising OPEC oil consumption to reduce exports. OPEC nations, they said, had a greater "obligation toward satisfying their domestic requirements than meeting external demand." [PIW 9-8-80:1] In mid-1980, the International Energy Agency (IEA) issued yet another warning (but now milder) of an approaching 1985 "gap," supply outrunning demand. [PIW 6-16-80:5, 7]

Summary and Appraisal of Price Actions, 1975–1981

Belief in Unexerted Price Power

Figure 6.2 showed the basic reason for higher prices: demand was expected to respond very little. With oil-on-oil competition suppressed, a monopoly need only look to the supply price of synthetic fuels. By 1978 or even 1981, aside from a trickle from Alberta oil sands, there was no sign of synthetic oil at any price. There appeared to be a long way to go upward.

The demand response seemed weak. In 1977 world consumption touched a new high. The response of non-OPEC supply was also feeble. The Alaska–Mexico–North Sea increase was not related to the price rise and would be limited. The United States was doing even worse than expected. Elsewhere, there was little action.

Consumer countries proclaimed the world in crisis for lack of oil. The United States was "groveling" for Middle East oil. When one's customers talk this way, it can be taken as the understated truth.

The Learning Period, 1974–1978

The OPEC nations expelled the companies and learned that while aggregate demand seemed very insensitive to price, an individual nation's demand could be shockingly sensitive and large amounts of revenue quickly lost by overpricing. From the first, relations were touchy between the other members and the leader, Saudi Arabia, which was vigilant but not always successful in protecting its share. (See Figure 6.4.) In 1975, the Saudis bore most of the cutback in output (table 6.8). But they were alert to the danger, and in 1976–1977 they underpriced—and obtained a record share of OPEC revenues. In 1978, OPEC made its first (secret) agreement on output sharing, followed by the proposed safety net agreement of 1979 and then the gentlemen's agreement in 1980. (The formal fixed quotas came in the 1980s.)

Spot markets tried to anticipate higher OPEC official prices. The sequence in time was: (1) higher expected OPEC price, (2) speculative demand, (3) higher spot prices, and (4) a meeting that raised the official OPEC prices. To the unwary, this proved that OPEC was merely following or ratifying "the market."

Financial Stress

OPEC nations were more dependent on oil than ever. They were committed to ever-larger expenditures at home and abroad, requiring ever-rising revenues. But revenues decreased in 1978 by 7 percent, and much more in real terms. Inflation in the OPEC countries was magnified by shortages in the specialized goods and services they imported. The foreign exchange surpluses of 1974 went negative by 1978 as import spending outstripped receipts.

The Events of 1978–1981

Unexerted monopoly power and financial stress promised a new price increase in 1979. The Long Term Price Policy Committee laid down some guidelines. But the Iranian Revolution was an opportunity not to be missed.[30]

Table 6.8
Saudi oil exports and revenues, 1970–1993

	Exports		Revenues	
Year	mbd	Percentage OPEC	Billion $U.S.	Percentage OPEC
1970	3.7	0.166	1.2	0.155
1971	4.7	0.194	2.1	0.179
1972	5.9	0.232	3.1	0.216
1973	7.8	0.261	5.1	0.224
1974	8.5	0.287	22.6	0.250
1975	7.1	0.278	25.7	0.271
1976	8.3	0.288	33.5	0.289
1977	9.0	0.306	38.6	0.312
1978	8.1	0.290	34.6	0.299
1979	9.2	0.319	57.5	0.295
1980	9.6	0.389	102.0	0.366
1981	9.8	0.471	113.2	0.448
1982	6.3	0.373	76.0	0.376
1983	4.5	0.292	46.1	0.287
1984	4.3	0.279	43.7	0.274
1985	2.9	0.188	27.0	0.203
1986	4.7	0.305	20.0	0.267
1987	3.8	0.238	23.0	0.237
1988[a]	4.5	0.254	21.0	0.238
1988[a]	4.2	0.249	20.2	0.232
1989	4.4	0.234	24.1	0.216
1990	5.8	0.289	40.1	0.274
1991	7.8	0.388	43.7	0.340
1992	7.9	0.371	47.6	0.348
1993				

Sources: PE 3-75, 6-81, 7-89; *OPEC Statistical Bulletin*.
[a]Because a new source is used in 1988 and later years, both sources are shown for comparison.

As Minister Yamani said in January 1981, "Prices increased in 1979–1980 as a result of another corrective action," as in 1973. [PIW 3-9-81:S1] A corrective action is conscious and deliberate. Their short horizons, their budget and foreign exchange deficits, and the worldwide delusion of scarcity were the devils that made them do it.

There would have been no spot price increases in 1979 had not Saudi Arabia deliberately refused to expand output, then cut production at crucial moments. The result was to panic the market. And even then, spot prices would have later sunk or collapsed, as happened in 1991, had not the whole group ratcheted up official prices and restrained production in concert.

The Saudi output cuts of January 1979 and April 1979 were an attempt to fine-tune the market with coarse instruments and achieve a tight supply situation. They did not plan a panic but made one and exploited it. As in earlier years, they broke agreements with the U.S. government, knowing it would publicly approve everything they did. But whether they would have done just as they did, even without its approval and groveling, is more than we know.

Appendix 6A: Saudi Arabia and the Strategic Petroleum Reserve

The United States stopped purchases for the Strategic Petroleum Reserve (SPR) in November 1978. SPR imports ceased by September 1979. [MER, various issues] By then, the world market was in growing surplus. But the United States would not resume purchase without prior Saudi approval. [NYT 2-22-80:D1] The Saudis urged delay, which they said was "necessary to help bolster the power of pro-American Saudi officials and offset the efforts of militant pro-Arab groups seeking to gain control of the kingdom." [NYT 2-29-80:D1] It was said in the United States that many Saudi "senior officials are anxious to cut exports as soon as possible to placate conservative Moslems, who regard high oil production as an unsound concession to the U.S." [WSJ 2-29-80:6] This political inside talk will always find believers.

The administration then proposed an agreement that "could give the Saudis an unprecedented degree of control over the use of America's strategic petroleum reserve." The Saudis rejected it. [WSJ 2-29-80:6] A DOE spokesman "said the United States would not go ahead with the decision to resume buying until an agreement with [oil] producers, including Saudi Arabia, was reached." [NYT 3-4-80:D9] When the royal assent was refused, the secretary of energy said resumption would be delayed. [NYT

3-6-80:D1] A month later, the administration let it be known that buying would resume in July 1981 at the rate of 100 tbd. [NYT 4-22-80:D9] Filling the SPR, whose size was reduced by a fourth, would take nineteen years. The belief in the United States was that Saudi opposition stemmed from its concern that "'filling the reserve is going to undercut the power of the pro-American element in the Saudi royal family,' one analyst of Arab affairs said here, adding 'There is going to be a great deal of fallout in the Arab world on this one.'" [NYT 5-23-80:A1]

There was no fallout. But DOE refused a Venezuelan offer to sell oil for the SPR "out of fear it might compromise Energy Secretary Charles Duncan's agreement with Saudi Arabia not to resume filling of the strategic reserve." [*Energy Daily* 6-6-80] Meanwhile, "The U.S. plays down the stockpiling resumption anyway, to avoid irking the Saudis; officials issued no announcement of their action." [WSJ 8-7-80:1]

Congressional impatience was building. Hearings were only the tip of the iceberg. [OGJ 5-5-80:138] When stockpiling was resumed in principle, the special adviser on SPR to the secretary would not admit any Saudi pressure influenced DOE. Senator Bill Bradley said he was "less than candid." [*Energy Daily*, 9-23-80:1, 9-26-80:1] SPR imports were finally resumed in September 1980, but at a very low level.

The secretary of energy was angered by a too-prompt receipt of some purchased oil because it might offend the Saudis. The DOE's denial of the discomfiture was believed by nobody. [WSJ 10-27-80:1] Later the GAO recommended faster filling of the SPR: "Despite abundant supplies, the United States has refrained from bidding for oil in the open market. This is widely thought to reflect a sensitivity to the wishes of Saudi Arabia." [NYT 11-3-80:D1]

At about this time, Saudi Arabia was pressing the Carter administration to stop an antitrust inquiry into Middle East oil pricing. "State has already asked that parts of the probe be halted." [OGJ 12-24-79:NL2] The U.S. government did not stop the inquiry, but in response to Saudi objections the Department of Justice dropped its demands for Aramco documents that were in the United States and that they had the legal power to obtain.[31]

Twice in a year, the Saudis could veto the enforcement of U.S. law in the United States. Such was the "bilateral" relation.

Notes

1. The negotiations shed some light on cost. Previously Aramco had received 22 cents per barrel for discovery, development, and production. [PIW 7-14-75:1] At 1968–1969 prices,

this would have been about 12 cents, close to the estimates in chapters 2 and 3. Exxon [1976] estimated Middle East "exploration and production" cost at 10 cents in 1970 and 25 cents in 1975. (See also [PIW 6-23-75:7, 3-27-77:11].) In 1978 it was about 32 cents. [PIW 7-3-78:8]. (See also [PE 8-78:337].) Some of these and later increases are due to inefficiency under nationalization.

2. Assume the barrel would be produced at the Aramco decline rate of 3 percent per year. Hence the first year's payment would be $6 \times .03 = 0.18$ cent. The present value of a perpetual stream declining at 3 percent would be equal to $.18/(.03 + i)$, where i is the discount rate. If the discount rate is .05, the present value is 2.2 cents; if .10, the present value is 1.4 cents.

3. Let the ratio of later to earlier price be P, and let the ratio of later to earlier quantity be Q. Then $Q = P^E$, where E is the elasticity of demand and $P = Q^{1/E}$. Now let the quantity be reduced by either 1 percent, i.e., $Q = 0.99$, or 10 percent, i.e., $Q = 0.90$. Then the respective values of P are 1.11 or 2.87.

4. The work of Verleger [1982, p. 56] avoids this mistake by recognizing that control of supply drives the system through a highly complex inventory mechanism, which I simply illustrate, without trying to measure.

5. This paragraph is based on Burton [1982].

6. The U.S. State Department in 1972 expected that the price of oil "would rise by 1980 to a level equal to the cost of alternate sources of energy." [Akins 1973, p. 464] *Price* and *take* are used interchangeably in the OPEC papers, reflecting the negligible importance of production cost.

7. Throughout we exclude internal OPEC consumption, which had no market effect.

8. The variations in the oil-to-GDP or energy-to-GDP ratio among nations do not measure the variation in energy efficiency unless one controls for variations in climate, industrial structure, population density, relative factor prices and consumer prices, and so on.

9. But much of the price increase was canceled in later years. The unanswered question is, How reversible were the effects? As late as 1993, twelve years after the oil price began to recede, the U.S. oil-to-GDP ratio was still creeping down. See Watkins and Waverman [1986], Gately [1993].

10. This has even been incorporated into a cyclical model. [Petroleum Finance 1988] The muddy theory and irrelevant statistics are discussed elsewhere. [Adelman 1991]

11. It is easily proved that if i = the normal discount rate, r = the rate allowing for risk of sudden loss, and p = probability of loss in any one year, then the present value of \$X in year t is: $PV(\$X_t) = \$X\ (1 - p)^t/(1 + i)^t = \$X/(1 + r)^t$, hence $r = (i + p)/(1 - p)$.

12. As Jack Hartshorn has pointed out in a letter, Aramco would add 10 percent to partners' forecasts for operational and seasonal fluctuations, and 10 percent for contingencies, a total 20 percent. This was cheap insurance, since the only cost was the actual investment. Taxes were on only production and sale. But capacity needed for insurance is not available for permanently expanded sales. Otherwise one must suppose the absurdity of the Aramco companies' building the spare capacity, being embarrassed by their own discounting on the basis of the spare capacity, expanding output, and then investing afresh in more embarrassing spare capacity.

13. The capacity estimates are from PIW. The CIA estimates for years confused technical and economic limits with policy limits in Saudi Arabia. PIW ceased to publish the estimates

after 1985 but resumed late in 1990. I have interpolated for the intervening years. The precision of these numbers is illusory, but they are useful indicators.

14. As we saw in chapter 4, the U.S. State Department in 1970 viewed the transitory freight advantage as proof that Libya deserved to collect permanently higher taxes.

15. One State Department economist pointed this out clearly. [Springer 1977] The document was declassified in 1990. Obviously, it had no effect on the views of those higher up.

16. On how the OPEC meeting promoted stockbuilding and speculative price increases, see also Wigel and Sandoval [1979].

17. President-elect Carter's economics task force, headed by Professor Lawrence Klein, had recommended an import quota auction system to undermine the cartel. [OGJ 1-3-77:31] It was not considered. (Detail is in "Oil Import Quota Auctions" [Adelman 1993, pp. 499–507].)

18. In contrast, a report of an alleged deal with Saudi Arabia on another subject was quickly denied. [NYT 7-12-79:A3]

19. The Khomeini regime ceased exports to Israel, which then tried to import crude oil from Europe. Both the British and the Norwegian governments refused to allow shipments. [OGJ 4-9-79:99]

20. The situation was worsened by the slow and lagging response of wholesale and retail prices of oil products. In part, the lag was institutional, in part the result of formal and informal price controls. Everyone in the chain was on notice that as the higher crude prices worked into the system, wholesale and retail product prices would increase. It was an almost riskless investment to buy every bit of oil available, whether for later use or for resale. Excess demand increased at both the crude and the products level.

21. The Saudis called the cut to 9.5 an "increase," since it would be 1 million more than their "normal" ceiling of 8.5 mbd. [WSJ 1-30-79:12]

22. In mid-February, Defense Secretary Brown "was plainly exultant about his meetings in Saudi Arabia." [NYT 2-13-79:A10] What he exulted over is unknown.

23. In April, the U.S. embassy in Saudi Arabia stated: "If we could do [what the Saudis wish,] they would give us all the oil we need and at good prices." [BW 4-9-79:99] They did not say what we needed to do, what the "good prices" were, or how long they would last. In May 1979, "Saudi Arabia is prepared to greatly increase its oil production 'but we will not be able to do so until peace is established in the Middle East.'" [WSJ 5-11-79:25] (This excluded the Camp David agreements, which they strongly opposed.) Nothing happened, but they increased output from 8.5 mbd to 9.5 in June and 10.4 mbd in July.

24. At the time, President Carter "could scarcely conceal his pleasure at the news from Riyadh: his 'friend' King Fahd had just agreed to push Saudi oil production toward [*sic*] 10 million barrels a day." [*Washington Post*, 12-14-83]

25. In July 1979, "Washington officials with close Saudi ties ... view the [July output increase] as giving [the United States] six months to achieve results, "particularly a solution to the problem of the Palestinians." They noted that OPEC would meet again in mid-December, "as the six-month period draws to a close." [NYT 7-10-79:A1] In August, "high Administration officials," sensing "new diplomatic openings" in the Middle East, asserted the need for "noticeable progress" toward Palestinian autonomy within the next ten weeks or else "the West is likely to face increased pressures on its oil supplies". [NYT 8-2-79:A1] Nothing more was heard of this.

26. The U.S. tilt toward Iraq was already underway. Secretary of Defense Brown asserted at that time that the regime "has changed. It has moderated its behavior." [*Economist*, 5-1-93:96]

27. Much Iran production had been quickly and permanently lost by the revolution. Nine-tenths of the foreign technicians had been sent packing early on [PIW 3-5-79:2; NYT 8-17-79:A1]. Native ones were lucky to be merely jailed. By January 1980, one oil industry executive estimated, less than 5 percent of the top Iranian technicians remained with the NIOC (National Iranian Oil Corporation). Expansion of capacity had depended on a huge gas injection scheme, which was now scrapped. [PIW 3-5-79:2, 9-24-79:S] In 1994, it was still being discussed, not planned.

28. As usual, in October "spot prices [were] rising on hopes for OPEC accord." [PIW 10-19-81:1] As expected, Saudi Arabia raised its marker price from $32 to $34, and the whole structure went up accordingly. There was even an accord on price differentials, for forty days, until the regular December meeting. [PIW 11-2-81:1]

29. In fact, the gasoline lines were unrelated to scarcity and reflected only price regulation and product allocation. [DOE 1987, p. 36] No other country suffered such difficulty.

30. An undersecretary of state for economic affairs characterized the price increases of 1973–1974 and 1978–1981 as "political events." [NYT 3-1-91:D2] Of course, the price fixers were political entities. But that merely made it easier for them to control the market.

31. I worked on the inquiry as an economic consultant to the Department of Justice. My opinion was and is that the decision in December 1983 to end the investigation was correct because there was no evidence of market power by the Aramco companies. Had I examined the documents, conceivably I might have changed my opinion.

7 The Cartel in Retreat, 1981–1986

OPEC at the Peak

OPEC had a grand illusion in 1981 that it still had much power in reserve and in time would raise the price of oil much higher. In fact, prices were already above what the traffic would bear. It was like someone who has walked out on a reef at low tide. When the water starts lapping about his feet, he finds it hard to accept that he must swim back most of the way he walked out.

By mid-1981, the second price upheaval was over. Figure 6.1 showed that prices that under stable conditions are close together and had been widely separated after 1978 were now rejoined.

Minister Yamani called the 1979–1980 increase "another corrective action," like the earlier one a long-delayed adjustment. [PIW 3-9-81 SS:1-4] So much for the notion that the price increase had come about inadvertently, despite the efforts of the moderates.

Prices were headed up and away. Figures 7.1 and 7.2 are the consensus forecasts tabulated by the International Energy Workshop and the Society of Petroleum Evaluation Engineers. Implicit in each forecast are two assumptions. First, the current price is the best estimate of the long-run competitive price. If the current price changes, that is simply a better initial estimate. Second, the price must increase from its current level. [Lynch 1992]

The Changed Market

From Reserve Price-Raising Power to Market Share-Price Trade-off

World opinion, including OPEC, was long guided by the vision sketched in figure 6.2. If the long-run demand for oil is highly price inelastic, a price

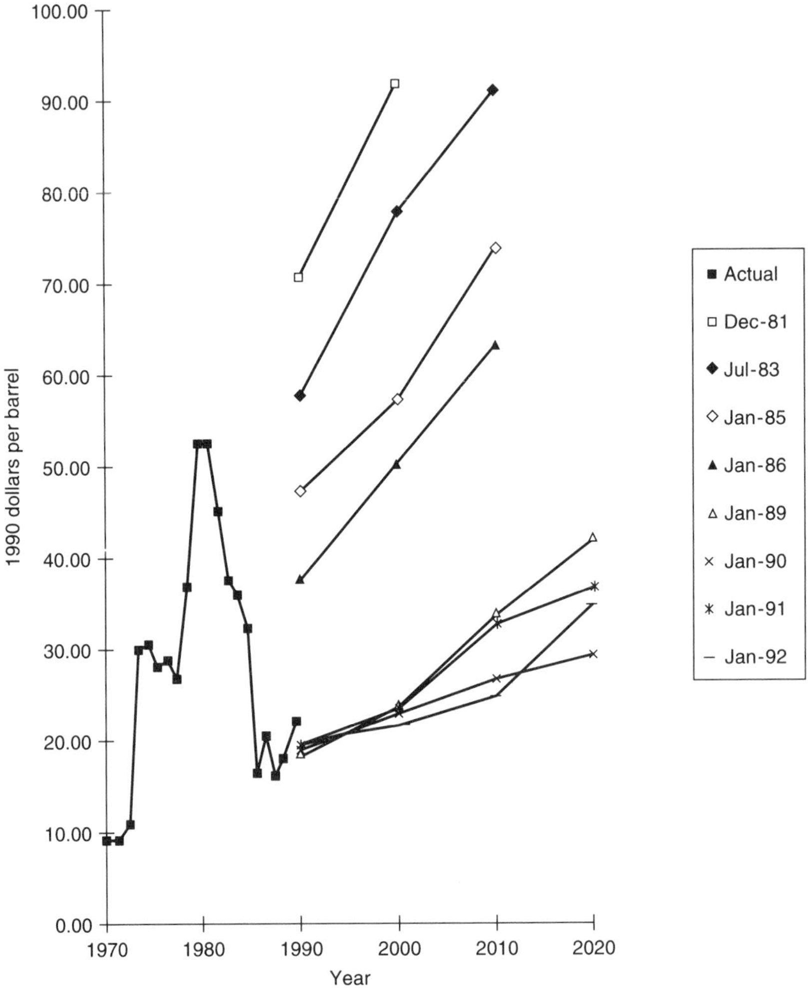

Figure 7.1
Crude oil prices, actual and successive International Energy Workshop polls.

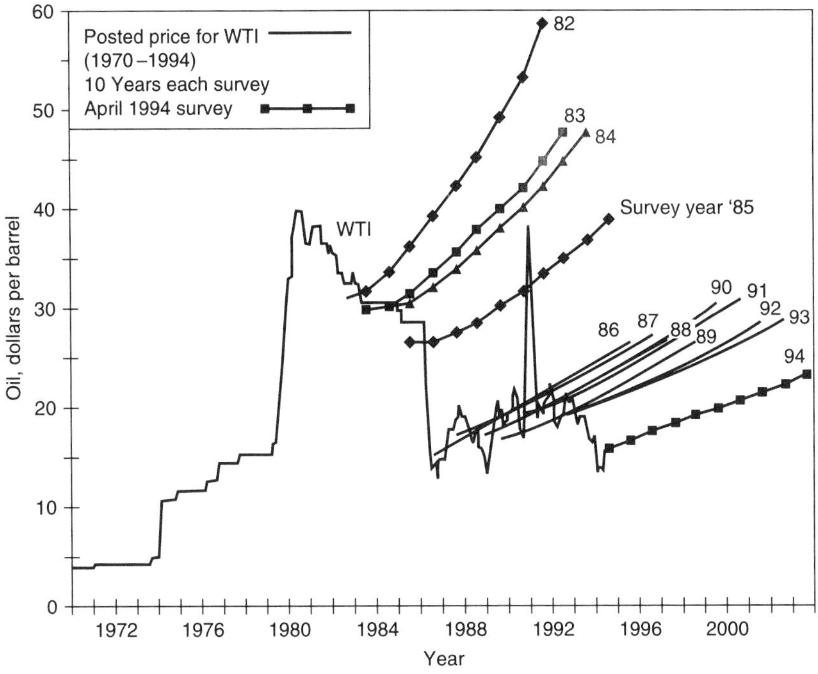

Figure 7.2
SPEE 1994 survey. Source: Society of Petroleum Evaluation Engineers.

just below the cost of producing synthetic crude oils in large amounts is a logical monopoly target. With OPEC competition repressed, and non-OPEC oil not expansible, the price could be set by the nearest alternative. This vision was the explicit guide to the first generally fixed price, on December 31, 1973. It was maintained in the 1979 report of the Long Term Price Policy Committee. OPEC saw a long way up to the ceiling and expected to raise the price again. [PIW 3-9-81:SS 1, 4]

A well-known Oxford economist and previous U.K. minister of energy estimated the ceiling at $90 to $100 per barrel (about $160 in 1992 prices). "The restraint by oil exporters is entirely political," in deference to the United States. [Balogh 1980] The price would exceed $80 ($140 in 1993 prices) because OPEC nations would reserve oil for domestic use rather than "deplete their dwindling asset to appease the appetites of the industrial nations." [Fesharaki 1980] An "optimistic" and "pessimistic" price, respectively, for 2000 A.D., would be $45 and $75 (in 1993 price levels, $76 and $126). [Yergin and Hillenbrand 1982] Bosworth and Lawrence [1982] explained that "the price of oil should rise as the market moves to more expensive sources." A monopolist will raise the price but increase it less. "Thus ... the monopolist is the conservationist's friend." [p. 121]

This chapter covers the loss of illusions. It was shocking when sales not only ceased to grow but actually diminished.[1] With the benefit of hindsight, we can see that the OPEC mission is to trade market share for a higher price. The question is when it pays to exchange and when it does not.

The Market Share Trap

Table 7.1 shows the relationship between market share and demand elasticity for total OPEC and for Saudi Arabia.[2] Readers can substitute their own assumptions. The column "notation" shows how to set up the table as a computer file and experiment with the parameters. The demand for crude oil is derived from the demand for oil products, whose long-run price elasticity was estimated by Griffin [1979] as around unity.[3] The price elasticity of crude oil is equal to the product elasticity, multiplied by the crude/products value relation, which in the OECD was only about 11 percent in 1970 but 53 percent by 1982. One could safely neglect price effects on demand in 1970 but not in 1980.

We take the initial price as unity and a hypothetical price increase of 50 percent. Following the Energy Modeling Forum [EMF 1992], we take the price elasticity of crude oil demand at .1 in one year, .26 in a decade. The initial OPEC market share is assumed at .37, as it was in 1981–1983 and again in 1989–1990, taking only the "market economies'" consumption as a base. It is lower today because of the integration of former Soviet Union (FSU) production into the world market. The Saudi market share is assumed at one-third of the OPEC share, which it has occasionally reached.

A 50 percent price hike increases non-OPEC revenues by 50 percent. All the curtailment of output is by OPEC, which gains 34 percent in revenues. If all OPEC members cut and nobody in OPEC cheats, the Saudis too gain 34 percent. If everybody cheats and maintains output, the Saudis gain only 1 percent. If the true elasticity is above −.10, the Saudis will lose substantial sales revenues on a price increase.

There is no escaping it: Saudi policy is dominated by their market share and the expected actions of their fellows. Experience confirmed that OPEC members are weak risk bearers. After 1981, nearly every member overspent its revenues. The cash squeeze made each one bitterly resist any reduction in volume or in price. Yet both were needed.[4]

Table 7.1
The market share trap after 1981: Result of an OPEC price increase

	Notation	50 Percent increase				10 Percent on products (long term)
		Crude oil, one year		Crude oil, five years		
Aggregate all sellers						
New price: ratio to old price	P	1.50		1.50		1.10
Price elasticity of demand	E	−0.10		−0.20		−1.00
New sales volume: ratio to old	$Q = P^E$	0.96		0.92		0.91
New revenues: ratio to old	$PQ = P^{(E+1)}$	1.44		1.38		1.00
Subgroup		(OPEC)	(Saudi)	(OPEC)	(Saudi)	Total
Initial market share	G	0.37	0.12	0.37	0.12	0.34
New sales volume: ratio to old	$Q(G) = Q - (1 - G)/G$	0.89	0.67	0.79	0.36	0.73
New revenues: ratio to old	$PQ(G) = P^{(E(G)+1)}$	1.34	1.01	1.18	0.54	
Subgroup elasticity of demand	$\ln PQ(G)/\ln E(G) + 1$	−0.28	−0.97	−0.58	−2.51	−3.23

Notes: If there is no cheating, Saudi sales and revenue are the same as total OPEC. If all others cheat, the result is as shown in column (Saudi).
Price elasticity from EMF (1992); five-year elasticity interpolated between −0.10 and ten-year −0.26.
The far right-hand panel shows the long-run effect of consuming-country 10 percent sales tax on oil product, assumed to be fully passed on.

Table 7.1 is not a model, but it does show what has happened since 1981: the interplay between what would be good for the group as a whole and what would serve the largest member. The three problems that OPEC needed to keep solving and resolving were price, total output, and output division.

Table 7.1 also shows the long-run result of consuming countries' fixing a 10 percent sales tax on oil products, assumed to be fully passed on. OPEC loses 27 percent of revenues. Nothing else was to be expected, yet when OPEC nations threatened to retaliate for tax increases with additional price increases, the threats were taken seriously in the consuming countries. (See chapter 8.)

Structural Market Changes in the 1980s

Disintegration and Spot and Term Prices

Outside Nigeria, Indonesia, and some Persian Gulf remnants, the multinational oil companies had been formally stripped of ownership in the 1970s. However, they had continued to produce and to market most or all of the host governments' production, compensated by a price preference or margin. This connection was jolted apart by the rapid increase in spot prices beginning in late 1978.

For over two years, as supply fears swelled the demand for hoarding and speculation and raised spot prices, the producing governments repeatedly raised their official prices in pursuit of spot. They disregarded contracts, openly or by various fictions. Some added various nonmoney elements to the price. It was often forbidden to resell to South Africa or Israel or the United States.

By June 1982, all OPEC state oil companies had discarded the restrictions. [PIW 6-28-82:1] They were ineffective and costly [PIW 7-26-82:1], but they had helped destroy the force of contracts.

The multinational companies ceased to sell large amounts of crude oil. In a few countries (Abu Dhabi, Nigeria, and Indonesia were the most important) they lifted oil for their own use. Contract terms grew short, as a result of quarterly price reviews and phase-outs and unilateral mid-quarter price changes. Security of supply was gone. "Japanese buyers in particular feel hard done by in their past term dealing with OPEC suppliers." [PIW 4-5-82:1]

Customer-supplier relationships continued because they saved transaction costs, of search and negotiation. But prices were not fixed far in the term contracts. They were smoothed-out spot prices, the difference depending on short-run market conditions.

Aramco became Saudi Aramco, and the four companies (Chevron, Exxon, Mobil, Texaco) became suppliers of special management and engineering skills. They bought crude oil at a small fixed differential from a contract price that was close to but not identical with spot.

Commoditization

With disintegration, short-term bargains, and price volatility, markets in crude oil and products were rapidly "commoditized." [Verleger 1987] As with other raw materials, the general and trade press was supplemented by many individual services reporting prices in great local detail, by place and even time of day. Traders came rapidly front and center. Some were independent, some affiliates, but all bought at spot from OPEC countries to resell to government, major or independent refiners.

Markets rapidly developed in the forward sale of physical volumes of crude oil, notably Brent in the North Sea. Even more important were the various futures markets. In 1978, futures trading began in middle distillates (heating and diesel oil). In the early 1980s, futures markets began in other products and in crude oils in 1983. [PE 6-83:227] By late 1990, there were "about 10 active futures contracts trading worldwide, with combined daily volume equivalent to over 150-million [paper] barrels a day." [PIW 1990:1]

Refiners and marketers, which depend on relatively thin margins between the purchase price and resale price, can lose all profit and run big losses when even relatively small price changes occur between the time they buy and the time they resell. They urgently need to hedge against unforeseen fluctuations and lock in the current price. Companies producing oil are at lesser but considerable risk because of fixed contract or tax payments. In late 1982, futures trading was still a novelty. [PIW 10-18-82:1] It soon became routine.

Buying or selling futures transfers the risk to speculators, who are expected to win some, lose some, but on balance make a profit. It seems to be agreed that these markets have worked well, and they provide liquidity and insurance to buyers and sellers. The futures markets now extend to three years from the date of the transaction, though the great bulk of the transactions are for eighteen months or less. Futures markets also provide information to the spot markets, which become more reliable indicators. (The important relation between "prompt" oil and future oil will be discussed later.)

There has also been a limited synthetic long-term market through swaps, a practice borrowed from the currency market. [PIW 1990; Arshi 1992] Here a bank or other "provider" promises a fixed price to a buyer or seller, for a period that may be as long as ten years but rarely exceeds three. The provider tries to make a bargain with a seller to offset every bargain made with a buyer, for example, promising a refiner a fixed price for its sales and an airline a fixed price for its jet fuel purchase. Thereby the provider avoids most of the risk and lays off the rest in the futures market.

In contrast to the futures market, where the "open interest" is published, not much is publicly known about the size of the swaps market. In theory, swaps could provide finance for investment in production. A variant of the swap is the simple preselling of oil, which is an extension of credit from buyer to seller. This has not amounted to much, partly because the credit standing of most OPEC governments remains poor. But efforts will undoubtedly continue as the OPEC countries try to find capital without capitalists.

Illusion: "The Market Sets the Price"

The wealth of quotations, and the constant interaction among them, give color to a half-truth: that the market sets the price. Of course, it does, but who controls the market? Under competition, nobody does, because nobody controls supply. Since the OPEC nations change supply up or down to set the price, they control the market. The various reporting systems discover the price, but they do not determine it.

Disintegration and Open Markets Promote Competition

The introduction of wide liquid markets has made it harder to maintain an above-competitive price. Producers can find a home for crude oil through the market, and refiners can find supply. When many parties can probe constantly for a slightly better deal, a discount by one seller is swiftly propagated throughout the market. Slightly lower prices on spot crude oil would become slightly lower prices on refined products, which would divert sales. OPEC producers that were threatened with losing market share would retaliate by discounting. Integrated private companies might prefer high prices to benefit their crude producing operations, but they were now a small, powerless segment. As a buyer, each looked for a lower price.

More than ever, the price had to be maintained by deliberate changes in supply. Thus, what was called the "clumsy cartel" had to intervene often and made the price more volatile. The transition was not complete until the mid-1980s.

Non-OPEC Supply

The increase in non-OPEC production was a shocking surprise. The powerful consensus had been and continues to be one of "limited resources," with oil supply running down everywhere except at the Persian Gulf.

On the Soviet Union, the famous 1977 CIA report stated the consensus. In Yamani's words, "All the available studies point to the fact that the Soviet Bloc will be definitely importing oil sometime in the eighties." [PIW 3-9-81:SS2] Instead, exports from the communist blocs to the market economies rose from 1.8 mbd in 1976 to 2.9 mbd in 1985. [BP 1985, p. 18]

The non-Soviet, non-OPEC countries had increased output from 17 mbd in 1973 to 18.7 mbd in 1978. An impressive multiauthor work forecast lower non-OPEC production, despite expected higher prices in the 1980s, because of the strong resource constraint. [EMF 6 1982] Instead prices declined steeply, yet non-Soviet non-OPEC output actually rose faster, to 25.9 mbd in 1985.

The United States, without Alaska

The discovery of large oil fields had dwindled after 1929, yet annual reserve additions increased for thirty years and then stabilized around 1960. The unit cost per barrel added decreased through 1972. It was mostly a one-time gain, the retreat from "market demand prorationing," a wasteful cost-raising cartel of the more important producing States. [Adelman 1964]

The price upheaval plus oil price controls and gas price controls were a misfortune. Oil wells vary greatly in cost. Even taking only depth as an independent variable, in the United States the most expensive tenth of all wells absorb half the capital expenditures. For deep or exotic high-cost oil or gas projects, which hard-working lawyers could winkle into a class exempt from price control, profits were huge. Investment in these projects boomed, raising costs and imposing bottlenecks. This depressed the development of known fields, which produced the crude oil whose price was under control. It was an attack on the very process that had kept the U.S. oil-gas industry in being—the continuing expansion of old fields.

The reserves added in the early 1970s were the lowest in decades. Production declined more slowly, from just under 9 mbd in 1973 to just under 7 mbd by 1980. Then price controls were abruptly ended at the start of 1981, and reserves-added and production rose, despite weakening prices, until the collapse of 1986. Gas prices were not decontrolled, but they had been considerably raised in early 1979. Thus, the effect of prices plus controls in the United States before 1986 was that production declined when prices rose and rose when prices fell.

The New Areas: Alaska, North Sea, and Mexico

We saw in chapter 6 that all three of these were profitable and would have been developed at prices existing before the first price explosion. In Mexico, the national oil company Petroleos Mexicanos (Pemex) was a gaudy exemplar of corruption and waste. The belief in ever-rising prices was a national disaster. With roughly 50 billion barrels in reserves and current prices exceeding $30 per barrel, it seemed safe to borrow $60 billion abroad. The result was a national financial crisis and massive disinvestment in oil production. Published money figures are indecipherable, but wells drilled in Mexico fell from 434 in 1980 to 103 in 1987. Moreover, many of the wells were uneconomic, drilled to provide local jobs and contracts. Mexican oil and macroeconomic policy was at first as prescribed by the school of Nicholas Kaldor at Cambridge (U.K.): still more doses of nationalization, barriers to imports, subsidies, and regulation. (For a good brief account, see [NYT 10-24-82:1F].)[5]

The policy was slowly reversed, especially after 1988, but revival and cleaning up the debris of long mismanagement is painful and slow. Pemex output has stayed around 3 mbd—a tribute to many unsung engineers and workers and an indication of how much fat there was to squeeze out. Thus, the higher price levels resulted in less, not more, Mexican supply in the world market.

In private enterprise countries, the booming prices led to drastic tax increases to capture the windfall rents. In Alaska, it is not clear whether they inhibited expansion. Alaska reached 1.2 mbd by 1978 and slowly expanded to just over 2 mbd, before declining to about 1.6 mbd in 1993.

In the North Sea, the largest and earliest discoveries had been off Norway, and as late as 1975 Norwegian output was six times British. But the U.K. expansion was far swifter, and one reason was a more flexible tax regime, particularly after 1979. British offshore production peaked at 2.6 mbd in 1984–1986 and then declined.

Other Non-OPEC Producers

Outside North America, the United Kingdom, and the communist countries are the "others" [EIA-IPSR]. (See table 7.2 for some of the largest.) Their output rose from 3.8 mbd in 1973 to 5.2 mbd in 1980, then to 7.5 mbd in 1985, and 11.3 mbd in 1994. Figure 6.3b showed the price-output relation. The obvious and largely true interpretation is that the supply curve kept shifting to the right, but to some extent there was an inverse relation: the higher prices led to even higher taxes, and vice versa.

Non-OPEC Investment Requirements

Table 7.2 shows the investment per daily barrel in the thirteen most important non-OPEC producers outside North America and Europe (those that by 1985 were at least 100,000 barrels daily). Their production rose from 3.8 mbd in 1981 to 7.0 mbd in 1992. Estimated expenditures include all development investment and most exploration. (See appendix 7A.) They exclude pipeline or tanker transport or gas processing, which vary greatly from place to place. Results can be replicated and checked from published sources.[6]

Return on new investment was high on new development projects and stayed high, even when the price (here measured by the average OPEC price on export sales to the United States) fell by half. Argentina, the worst performer, improved greatly after 1990, when its state company was privatized.[7] Ignorance of these low costs frustrated the otherwise well informed: "I just cannot understand how this low price can sustain investments in high cost oil areas.... Somebody somewhere must be losing his shirt." [Ali Jaidah in PIW 9-12-88:S3]

If costs and prices determined investment, we might expect with some confidence that output would keep rising for the group as a whole, but costs are less important than taxes.

Taxation of Oil Production

The ideal tax system induces the operator to invest as much with a tax as without it.[8] It captures all economic rents, leaving the investor just enough prospective profit to make the investment worthwhile. The

Table 7.2
Investment per daily barrel, non-OPEC producers 100 tbd or over ($)

	1981			1985			1990		
	Output tbd	Investment T$/IDB	Rank	Output tbd	Investment T$/IDB	Rank	Output tbd	Investment T$/IDB	Rank
Argentina	497	37.36	12	460	47.68	13	470	33.26	13
Brazil	213	18.33	9	563	8.59	7	653	8.93	7
Colombia	134	56.07	13	176	22.55	12	439	7.63	6
Peru	193	36.90	11	189	21.73	11	130	24.26	12
Trinidad	189	18.03	8	185	13.55	9	157	19.42	11
Angola	130	4.49	4	231	4.61	5	474	5.02	4
Cameroon[a]	86	23.34	10	190	5.18	6	155	10.21	9, 10
Congo	76	4.00	2	115	11.15	8	170	3.16	3
Egypt	627	4.66	5	895	2.69	3	893	6.80	5
Syria	172	NA	NA	176	NA	NA	388	6.72	NA
Oman	328	1.20	1	505	1.38	1	668	2.15	2
Yemen	NA	NA	NA	NA	NA	NA	194	0.93	[b]
Brunei	156	6.37	6	160	21.24	10	152	10.13	8
India	300	15.28	[b]	606	14.05	NA	656	NA	NA
Malaysia	258	4.27	3	444	1.59	2	621	2.08	1
Australia	394	13.17	7	575	4.47	4	578	10.21	9, 10
Total	3,753			5,470			6,798		
Number comparable		13			13			13	
Lower quartile		4.49			4.47			5.02	
Median		13.17			8.59			8.93	
Upper quartile		23.34			21.24			10.21	

[a] Offshore only
[b] Not available in a later or earlier year, hence rank eliminated.

higher the pretax rent is, the higher is the tax. With no rent, there is no tax. The ideal tax is progressive, not for equity but for efficiency.

Suiting a tax system to a mineral industry is particularly hard because of the difficulty of reckoning cost. In the United States, a rough but effective way to capture rent on new areas has been cash bonus bids. The oil company estimates how much it expects to earn, over and above the costs of exploration, development, and production, and offers as much of it as it thinks necessary to win the lease.

Bonus bidding is an efficient rent skimmer in the aggregate, with large individual gains and losses. [Mead 1986; Mead and Sorenson 1980] showed that the return on oil leases won by bidding was normal or even subnormal, but the excises and royalties (a private tax) were shown by Lohrenz et al. [1981] to lessen investment and production.

A cash bid system works only when the number of bidders is large enough to ensure independent (i.e., competitive) bidding. The industry in the United States was large and long established, and the public was correctly willing to assume competition. Elsewhere, with the partial exception of Canada, this was not the case.

The other vehicle for rent absorption, the resource rent tax, was worked out in concept by Garnaut and Clunies-Ross [1975]. Its practical difficulties are great. [Bradley 1986; Bradley and Watkins 1987] But Eckbo [1987] concluded at least tentatively "that a resource rent tax combined with cash bonus bidding would capture rent in a tax neutral fashion and behave robustly."

No country, however, has a tax system that even aims to capture rent and be neutral to investment. It does not help that the air is thick with mistrust of private especially foreign companies. Taxes are badly designed or hardly designed at all.

There are also hidden regressive taxes, like requiring crude oil or refined products to be sold at especially low prices; putting customs duties on machinery or materials; favoring or requiring local labor or suppliers; local graft and/or corruption;[9] and so on. The two price upheavals led to hasty ad hoc attempts to capture the suddenly increased profits. The net result would as likely as not discourage or preclude more investment.

When prices stagnated and then sagged after 1981, some nations began to learn that the price would not necessarily keep rising. The process is best seen in Canada, whose National Energy Program [1980] was practically one long drool over the revenues to be drawn from the constant price rise. Yet as prices fell, the program was soon scrapped and Canada moved toward freeing up prices not only of oil but natural gas.

Previously Canadian gas exports had been allowed only with a backup of twenty-five years' supply in the ground. Since investment in a large, unproductive inventory was unattractive, gas reserves failed to increase, which of course "proved" how scarce gas was. When border prices were nearly $5 per thousand cubic feet (mcf), the Canadian government withheld gas from export to the United States. It argued, with justice, that Canadian gas was no different from gas on which the United States was allowing nearly $10. But the $10 prices were only the temporary result of price regulation producing excess demand for gas and then channeling it into a few exempt classes. In only a few years, the border price was down to a little over $2. Canada had lost about 75 percent of the capital value of the gas it was "reserving."

It is impossible to summarize the changes that came about after 1981. The oil trade press mentioned many tax cuts and new areas being opened. By 1990, *Petroleum Intelligence Weekly* said oil companies were "spoiled for choice," with more targets than they could chase. [PIW 7-16-90:1]

But by looking at taxes as they were in the late 1980s and early 1990s, we see how little progress has been made. A comprehensive review concludes that "the great majority of the fiscal schemes do not perform very efficiently as extractors of economic rent." Only Australia had enacted a resource rent tax, and it coupled it with a highly regressive conventional royalty/production tax. [Kemp 1987, pp. 319–320] A more recent study of offshore exploitation of new fields in ten countries (United Kingdom, Norway, Denmark, Netherlands, Australia, China, Indonesia, Egypt, Nigeria, United States) concludes that the tax systems were highly complex, not well aimed at economic rents, and regressive. [Kemp Reading and Macdonald 1992, pp. 52–56] A more detailed study of Norway shows that its taxation aborts (often drastically) production, which is profitable pretax. [Lund 1992]

Regressive Taxes Increase Risk and Abort Discovery

Wood [1990] gives typical capital and operating expenditures, decline rates, and project lives for three hypothetical fields of 15, 50, and 350 million barrels. At his assumed price of $18 per barrel, we calculate that even the smallest highest-cost field would earn 40 percent before taxes, and the larger ones are much more profitable. Wood reduces a large number of known exploitation contracts in many countries to twenty types. The small field provides his minimum 15 percent return after tax in only four of the twenty cases. In only seven of the twenty is it worth develop-

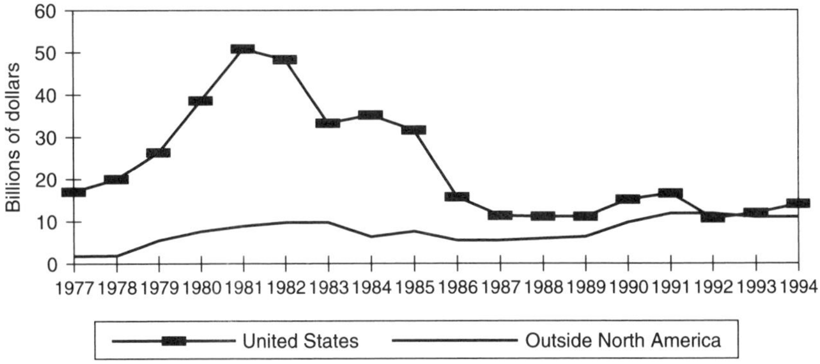

Figure 7.3
Expenditures for exploration and development by U.S. companies, United States and outside North America. Note: Transfer payments excluded. Source: OGJ annual tabulation.

ing even the middle-size field, which earns 60 percent before tax. The large field is always worth developing.[10]

The result in a typical new country is that none of these fields will ever be found. Exploration typically finds many more small than large fields. A small field, if taxed only on its net profits, would provide the company with funds, knowledge of local geology, and a stable beachhead for more exploration. But if it is known in advance that nothing small is worthwhile, the risk of complete loss is unacceptably high. *Large fields are not found because small fields are overtaxed.* (See also p. 271.)

Figure 7.3 tracks the U.S. totals fairly well and is probably a fair but not precise indicator of investment elsewhere. The totals increased through 1983, dropped and were stable over 1984–1989, and then rose moderately.

In summary, lower oil prices did not prevent, and may even have promoted, non-OPEC expansion in the 1980s, even after the 1986 crash. Costs were and are low even outside OPEC, and tax changes preempted price changes. National companies in non-OPEC countries like Mexico were another depressant.

The Path Downward, 1981–1982: Price and Market Share

Spot and Contract Prices

Figure 6.1 showed how Saudi Arabia discounted over 1978–1980. Its price is the solid line, FOB on shipments to the United States, and it should be compared with those from other Middle East countries. By

1981, the recorded contract price was a current price. Oil prices were now like interest rates, a family of imperfect substitutes, varying with the time period in the contract. Hence it is mistaken to regard the spot price as "the market" and to calculate the revenues that the leader, Saudi Arabia, could have earned had it sold everything at spot.

If we compare the Saudi export price not with the spot price but with the price charged by sellers that felt free to price as they wished, Saudi Arabia underpriced substantially through the end of 1980. Table 6.8 shows it was a great success. Its share of the rapidly rising OPEC revenues rose from 30 percent in 1978 to 45 percent in 1981, and only a minor part of the Saudi increase was due to the Iran shrinkage. The Saudi 1981 current account surplus was expected to be $50 billion. Foreign assets were nearly $100 billion and were expected to reach $145 billion by year-end. [NYT 3-24-81:A12] Today, its transactions with the IMF read strangely. "In the unaccustomed role of supplicant, the IMF has had to pay a high price to persuade Saudi Arabia to lend it ... around 16 billion dollars ... Because the IMF needed its money fast the terms have extracted are anything but concessionary." [*Economist* 4-4-81:81]

Haggling Invades the Market, 1981

At the end of March, Kuwait demanded, and Gulf, Shell, and BP refused, to pay Kuwait a premium on contract sales. [WSJ 4-1-81:2, 4-6-81:20] A Kuwait minister complained: "The companies have resorted to haggling." Kuwait cut its output rather than deign to haggle. "For the first time in years, OPEC seems genuinely worried that it has pushed oil prices too far too fast.... The time may have come to *slow down* the drive for higher and higher oil prices." [WSJ 4-13-81:21, emphasis added] To slow the rate of increase was not to stop increasing.

Then Arco notified Nigeria that it was ending two oil supply contracts. "This is the first time in several years that an oil company has willingly relinquished a guaranteed source of oil. [Previously] buyers have swallowed high prices to insure continued access to crude supplies." [WSJ 4-14-81:2] "I don't know how you could send a stronger signal that prices are too high." [NYT 4-15-81:D3] Before the month was out, Kuwait withdrew any premium on contract sales.[11] [WSJ 4-23-81:2, 4-29-81:4]

Shortly after, Minister Yamani said his country would not raise prices. "In the past, Sheik Yamani has made similar ... statements, only to end by raising prices." [NYT 4-20-81:D1] But Yamani suited the action to the word. "To the surprise of some observers," the Saudis kept production 1.5 mbd above "their preferred ceiling." [PE 5-81:186] By May, "the flood of

Table 7.3
OPEC quotas, 1982–1994 (tbd)

	March 1982	March 1983	October 1984	December 1986	November 1988	July 1990	Quarter 1991	September 1991	November 1992	Second quarter 1993	Second quarter 1994
Saudi Arabia[a]	7,650	5,000	4,353	4,348	4,524	5,750	8,034	8,600	8,030	8,150	8,150
Iran	1,200	2,400	2,300	2,369	2,640	3,350	3,217	3,300	3,340	3,340	3,600
Iraq	1,200	1,200	1,200	excl	2,640	3,350	excl	350	500	400	400
Kuwait	800	1,050	900	996	1,037	1,600	none	300	1,600	1,750	2,150
Qatar	300	300	280	299	312	396	399	400	360	364	378
UAE	1,000	1,100	950	948	988	1,590	2,320	2,300	2,160	2,161	2,161
Total Persion Gulf	12,150	11,050	9,983	8,960	12,141	16,036	13,970	15,250	15,990	16,165	16,839
Algeria	650	725	663	667	695	882	827	780	730	732	750
Gabon	150	150	137	159	166	212	285	290	280	281	287
Libya	750	1,100	990	996	1,037	1,320	1,425	1,500	1,350	1,408	1,390
Nigeria	1,300	1,300	1,300	1,301	1,355	1,720	1,840	1,840	1,780	1,780	1,865
Ecuador	200	200	183	221	230	273	273	290	293	—	—
Venezuela	1,500	1,675	1,555	1,571	1,636	2,070	2,235	2,300	2,260	2,257	2,359
Indonesia	1,300	1,300	1,189	1,190	1,240	1,460	1,443	1,400	1,310	1,317	1,330
Total OPEC[a]	18,000	17,500	16,000	15,065	18,500	23,973	22,298	23,650	23,993	23,940	24,820
Excluding Saudi Arabia	10,350	12,500	11,647	10,717	13,976	18,223	14,264	15,050	15,963	15,790	16,670

Sources: PIW, various issues, latest 12-7-92:7 and 5-3-93:4
Notes: August 1990 quotas identical with July 1990, except for exclusion of Kuwait and Iraq, hence not separately shown. November 1992 quotas are applied to second quarter 1993. Ecuador, having left OPEC, is included at recent maximum for comparability.
[a] Saudi Arabia was acting as swing supplier in 1982, 1983, and 1984. Hence the quotas stated here are those assumed at the meetings. Neutral Zone quota 0.3 mbd equally divided between Kuwait and Saudi Arabia
[b] Iraq refused any quota in 1986, hence excluded from quotas, and total OPEC.

Saudi oil is taking a stiff toll on the production rates of other countries, who have reduced output to maintain their official prices." [NYT 5-21-81:D1] "[A]ll other members are pressing them [Saudis] to slash production some two million to three million barrels a day [to] ... wipe out the oil glut.... [It] would send panicked purchasers into the market and would push prices up again." [WSJ 5-22-81:1] At the meeting, members demanded but Yamani refused to cut production. [NYT 5-26-81:1] Others did cut. There was a freeze on current prices, with the usual worry and inaction on differentials. [NYT 5-27-81:1]

The failed meeting had almost immediate results. The head of Exxon, who had recently made light of inventories, now said the company needed less oil and would not pay current prices. [WSJ 6-1-81:20] Other companies spoke similarly, but "in only the most guarded terms for fear of angering [OPEC] suppliers." [NYT 6-2-81:D1]

By early June 1981 there were numerous press reports of a "buyers' revolt [and] actual bargaining." Again, language mirrors life: "haggling" and "actual bargaining" were so unusual as to be remarked. Moreover, the "Aramco advantage" was gone. "The have-nots have turned into don't-wants."[PIW 6-8-81:3] For the next three months, spot and contract prices decreased, and "companies, in a tough mood that hasn't been seen in many years, are demanding that oil prices be shaved further." [WSJ 6-18-81:4]

The OPEC Long Term Strategy Committee, reported dormant for three years, was again convened. [WSJ 6-17-81:4] Expert reports to it recognized the drop in consumption and recommended a price policy "to keep price rises from forcing further conservation." [WSJ 6-19-81:6] This was the first acknowledgment of any significant price elasticity.

Expropriation had affected negotiations between companies and governments.

> Dozens of oil companies ... have suspended or phased out [deliveries].... After years of submission to the price demands of the oil exporters, the willingness of oil companies to force a showdown with at least some producers has come as quite a shock. "I have never seen contracts suspended on this scale," said the chief negotiator for an international oil company, who, like others, declined to be identified because of the sensitivity of the negotiations. The producing countries have responded to this effrontery with threats of blacklists, diplomatic protests and even economic sanctions against the oil companies' home countries—everything but a significant price cut.... [Previously] even in times of oversupply, like the present, the companies were willing to pay uneconomic prices just to maintain a long-standing relationship with a producing country. But times have changed.... *"We are no longer slaves to our capital investments in these countries."* [NYT 7-13-81:D1, emphasis added]

Thus in July 1981, the OPEC nations stood firm on prices and preferred to let liftings decrease. Some companies still felt "anxiety to retain friendly relations for the sake of future oil supplies." [WSJ 7-16-81:1] The OPEC nations demanded that Saudi Arabia either raise prices or cut production; either way, the non-Saudis would sell more at existing prices. As in April, the Saudis refused, continuing to produce at over 10 mbd. In mid-August, an emergency OPEC meeting decided nothing. The spot market practically suspended trading "in the apparent expectation that some [official contract] crude prices would be reduced." [NYT 8-25-81:D1]

On August 26, Nigeria made the first contract price cut, from $40 to $36. [NYT 8-27-81:1; WSJ 8-27-81:3] As differentials were viewed, it equated to the current official Saudi Light price of about $32. During the next week, other OPEC members cut prices also. Production, down by a third since 1978, was about 20 mbd, against 35 mbd capacity. [WSJ 9-11-81:6]

Saudi Arabia would not cut contract prices to the Aramco companies, which continued their high output. But Minister Yamani was worried over a price collapse, which he thought was invited by the intransigence of the other members. The current $32 was the downward limit. [WSJ 9-8-81:3]

In mid-September, informal negotiations again started. [NYT 9-13-81:22] The oil trade was apprehensive because it did not know how large inventories really were or the rate of drawdown in recent months; hence the industry did not know how much was being consumed. [NYT 9-15-81:D1]

Long-Run Comfort

The only formal model I know of that predicted the likely course of prices after 1981 was by Griffin, Daly, and Steele [1982]. It drew no attention. The consensus was upward. At the meeting of the Oxford Institute of Energy Studies, it was predicted that before the end of the century, six OPEC members would have exhausted their reserves and quit production. The world would be dependent on a few Persian Gulf nations with "small populations and therefore little incentive to produce much oil. Oil executives and the conference generally agreed.... Many top business and oil executives said they believe that the glut is a temporary phenomenon that will end when the world recession bottoms out." The head of the Oxford Institute, Robert Mabro, said it was wrong to "predict an end to the energy crisis. Forecasts of abundant oil supplies are as wrong as forecasts of the 1960's that saw limitless oil supplies." [WSJ 9-30-81:2]

James Akins, the former ambassador, predicted that if Airborne Warning and Control System (AWACS) aircraft were sold to Saudi Arabia, they would keep the price at $32. "Then sometime in the future (perhaps as long as two years) a gradual real increase up to the cost of production of synthetic liquid hydrocarbons." [Akins 1981, p. 9] (The aircraft were duly sold. The next month, the price was raised.)

While the sales of other OPEC countries fell by about 30 percent in the first eight months of 1981, Saudi liftings were steady. By September 1981, the Aramco partners had built huge inventories, which they now began selling off at discounts of about $1 per barrel. [PIW 9-28-81:3]

By mid-October, contract prices had been reduced by most OPEC countries. But an agreement was expected, raising the Saudi marker price from $32 to $34 and cutting Saudi production [WSJ 10-15-81:4; NYT 10-16-81:D1] "Spot prices [were] rising on hopes for OPEC accord." [PIW 10-19-81:1] In fact, Saudi Arabia raised its marker price, and the whole structure rose. [PIW 11-2-81:1] As ever, it made good cartel sense to raise the price in the face of lower demand. First came the expectation, then the effect, finally the cause.

OPEC expected the price line to hold. The business of the December 1981 meeting was "how to deal with the world oil oversupply." [NYT 12-9-81:D11] For once, the members did agree on a slate of differentials. [WSJ 12-11-81:6] But they then faced the "possibility that increases in the price ... are not in their own long-term interests." There was as yet no acknowledgment of any need for price reduction. The report of the Long Term Strategy Committee was put off for another year. [NYT 12-9-81:D1; WSJ 12-10-81:4]

When the price was raised, non-OPEC producers did not follow, and they cut price to move output, forcing OPEC to cut output to hold the price. [WSJ 11-9-81:2, 2-8-82:2, 3-3-82:1; NYT 11-21-81:33, 2-9-82:A1; PIW 2-15-82:3, 3-1-82:3] By February 1982, price leadership fell "momentarily in the hands of U.S. and North Sea crude sellers and their solution to a shrinking market is to reduce prices rather than lose volume." [PIW 2-15-82:3; also WSJ 3-2-82:6 (USA), 3-3-82:2(UK)]

Iran seems to have been the first open price cutter in OPEC. ([NYT 2-16-82:D12] noted its "second" price cut.) Oil executives thought the Saudis wanted a lower price [WSJ 2-25-82:2] This interpretation conflicts with the evidence. Yamani reprimanded Aramco partners for underlifting, without mentioning a price cut. [PIW 3-8-82:1] The Kuwaiti oil minister said the price could be held. [WSJ 3-9-82:3] Saudi Arabia and Kuwait threatened sanctions against companies that reduced their liftings in Nigeria, the most likely to cut prices. The threats accomplished nothing. [WSJ 3-26-82:2, 3-30-82:5, 4-1-82:4; NYT 3-26-82:21]

The Bank for International Settlements reported that by the third quarter of 1981, oil exporters were again net borrowers, as they had been in 1978. [NYT 2-15-82:D5; WSJ 2-16-82:31]

An emergency OPEC meeting in March 1982 showed "an unprecedented hostility toward Saudi Arabia" for not cutting output more. [WSJ 3-18-82:3] OPEC nations agreed to cut from 18.2 to 17.5 mbd, although it was unclear how they would apportion the cut. [NYT 3-21-82:1] But the Saudis finally had the commitment they wanted from the others, to share the reduction. It was viewed as a "new market-sharing approach instead of conventional price adjustment.... This is not OPEC's first fling at production planning, however, and previous efforts have been notably unsuccessful." [PIW 3-29-82:1]

By mid-May, spot prices were up and inventories down; "the crisis has passed," reported the *New York Times*. [5-17-82:D1] Even so, there was disagreement over whether to continue with production controls. [NYT 5-21-82:D1] The meeting reaffirmed the 17.5 mbd ceiling but did not want a permanent control mechanism. [PIW 5-24-82:3] Iran said it would not abide by any output ceiling and assailed Saudi Arabia for allegedly discounting, and for supporting Iraq. "When Saddam Hussein falls all of OPEC's problems will be solved. We will then see a much humbler Saudi Arabia." [WSJ 5-24-82:4] Yamani ruled out production planning. [PIW 5-31-82:3]

The market watch committee was unable to decide whether to change the ceiling, and "how any increase would be allocated among members." Some wanted to abolish prorationing, others to maintain it at least on a standby basis. [PIW 6-21-82:1] The emerging strategy, as Yamani had seen it months ago, was to "freeze or at least restrain nominal price and let inflation erode [it] in real terms." [Ibid., 6][12]

Prices, Policy, and the Long Term

Ministers Yamani and Al Khalifa Al-Sabah (Kuwait) now concede "that OPEC made a serious 'mistake' in letting prices rise so rapidly after the Iranian revolution." But they thought lower prices would be bad in the long term for everybody. "A break in the nominal price would only set the stage for another 'serious' shortage.... Prices must rise in the long run to avoid a basic resource crunch before the end of the century." [PIW 3-29-82:6] This, of course, was the consensus.[13]

The IEA warned that a slowdown in energy investment could pave the way for another oil price shock later in the decade. [OGJ 7-5-82:82] It warned as well of an oil supply shortage after 1985, "from 9- to 21-million b/d." [PIW 10-4-82:3] The group's chief economist, Herman Franssen, thought OPEC could hold the marker price. In April and in October, *Petroleum Intelligence Weekly* carried a special supplement by Robert Mabro, recalling that in the 1970s "OPEC was able to hold its reference price fixed at least in nominal terms in periods of slack demand, when an unchecked market normally would have led to a price collapse.... Failure to hold is to surrender ... the power over price.... The acid test of power is a slack market when strong competitive waves batter the price front." [PIW 4-19-82:SS1,2]

"Divisive" Differentials

Quality and location differentials could be used as a subterfuge or cover for deliberate discounting. Perhaps more important, at any moment a small competitive advantage for buyers could bring a big diversion of sales. Sellers who unexpectedly lost business had to discount or lose much revenue. By December 1981, "the divisive issue of differentials" was high on the agenda of the month's OPEC meeting. [PIW 12-7-81:1] The next August a new working group was formed to recommend action. [PIW 8-30-82:3] There was quick dissension, starting with the spread between Arab Light and Nigerian crudes. [PIW 9-13-82:3] The committee report was stalled [PIW 9-20-82:1], and the matter was referred to the full meeting [PIW 10-18-82:3], where nothing could be done. [PIW 12-13-82:1] The issue has never been resolved and remains an irritant to this day. So does the issue of what is or is not crude oil (e.g., condensate, very heavy oils.)

Managing OPEC Output

Production control was always the leading topic at OPEC meetings. Even good news was partly bad: an increased amount demanded could disrupt the division of the market. A year later, it would be said, "The disastrous pattern of Spring 1982 [was that] OPEC nearly succeeded in restoring market prices—only to have its production ceiling crumble the moment buyers showed some interest in increasing purchase volumes." [PIW 4-11-83:1] In July 1982, "The organization's own hopes are pinned on a gamble to hold firm for a few weeks in expectation that rising demand for OPEC oil by autumn will save the day, firming up prices and enabling discontented members to secure higher production quotas—if the revived Iran-Iraq war doesn't 'solve' the oversupply problem in an earlier and brutal fashion." [PIW 7-19-82:1]

But the market division fell apart in July. The Saudis threatened to cut their prices unless African producers raised theirs. Both Saudi Arabia and Venezuela threatened to disregard output limits if others did not stop overproducing. But others demanded that the Saudis cut production. Iran would no longer accept its quota and threatened "force."[13]

Yet the storm blew away with the reviving demand during the autumn and winter. In October 1982, the Gulf Cooperation Council's oil ministers warned that they would not tolerate "'irresponsible ... misguided' price cutting." But all they could do to punish offenders was to make even bigger cuts of their own. This might lead to a chain reaction "of downward spiralling prices all around." [PIW 10-25-82:1] Sources in OPEC and the companies said that retaliatory price cuts were "the only action that would catch the attention of Libya and Iran." [OGJ 10-25-82:91] But nobody dared to act.

Harbingers: Saudi Arabia's Overpricing

Previously Saudi Arabia had underpriced and been well rewarded by revenues' rising faster than total OPEC. But now it was overcharging. In 1982, on exports to the United States, the Saudi price rose by 63 cents, while the non-Saudi price fell by $3.05.[14]

The Saudis would permit the Aramco companies to cut only by 0.7 mbd; even so the companies "may yet have to take the oil they didn't take in February." [WSJ 3-1-82:2] The companies feared large losses on Saudi oil at $34, which was now some $4 to $5 above spot. But they were "even more worried about alienating the Saudis." [WSJ 3-4-82:3] Yamani had reprimanded them for underlifting. [PIW 3-8-82:1]

The Saudis cut "ceiling" production, which was now really a floor, from 8.5 mbd to 7.5, generally viewed as still too high. [WSJ 3-8-82:3] In May, Aramco partners said they were losing $5 per barrel. Texaco "continued to make these purchases ... upon the expectation of having continued access in the years ahead to Saudi Arabia's reserves." [PIW 5-3-82:2] But then they cut back abruptly, since Texaco's liftings for the whole year were half of the 1981 amount. [PIW 11-2-82:8] Mobil shed its illusions slowly. As we saw in chapter 6, it had paid heavily "to cement its relationships with Saudi Arabia in order to guarantee access to crude oil."

[NYT 12-10-80:D20] At a 1982 Mobil stockholders' meeting, president William Tavoulareas defended

> Mobil's currently unprofitable supply agreement with Saudi Arabia.... A temporary glut should not make us forget that oil is a dwindling resource, and the companies with access to it are the ones that will be able to supply their customers and maintain their earnings growth.... [Losses of hundreds of millions were] a price we must pay for a long-term arrangement with a country which owns a quarter of the free world's oil reserves and has proven itself to be a reliable supplier at moderate prices. [NYT 5-7-82:D4]

As late as October 1982, Mobil had "taken pains to lift its full allowable all year, apparently partly to maintain its standing with Riyadh." The loss was so heavy that Mobil corporate profits fell by half. [PIW 10-25-82:7]

By the end of 1982, the Aramco companies were "gagging on expensive, money-losing Saudi crude oil," on which they had lost about $2.5 billion in 1982. "Even formerly highly-prized rights to 'incentive' oil gained for participation in Saudi industrial projects have lost their attraction at full official prices." [PIW 12-20-82:1] At a cost of billions, the companies had finally learned that "access" was a mouthful of air. They cut liftings so far that Saudi output was at only 45 percent of capacity, other OPEC at 60 percent.

In December, the Saudis seemed "ready for a showdown on price cutting." A group composed of several former high OPEC officials and two others (Robert Mabro, Ian Seymour) warned that Saudi Arabia, although opposed to cutting official prices, might do so "if necessary to defend the market price by putting a halt to discounting." Pricing discipline was a must, and so was a new production quota system. [PIW 12-13-82:1]

The Slide Resumes in 1983

In January, the Aramco companies finally gave Yamani an "earnest warning": they would cut liftings even more than 50 percent unless the price was lowered. [WSJ 1-4-83:2; PIW 1-10-83:3] But a special OPEC meeting in Geneva took a week to agree on a trifling output cut, with no agreement on price levels or differentials. [WSJ 1-24-83:3] Yamani walked out of the meeting, calling it "a total failure." [WSJ 1-27-83:1]

Oil companies feared that if the price were not deliberately cut by $4, it would fall even more of itself. OPEC must "limit production levels for its 13 members." [NYT 2-2-83:A1] They also approached non-OPEC producers, notably the United Kingdom, Norway, and Mexico. "Nations in and outside of OPEC are joining in unprecedented bid to support oil prices." [WSJ 2-25-83:3] But no action was taken.

The sequence was now repeated. In January and February, non-OPEC price cuts were seen first in the United States. Lower prices on light crude oils from the North Sea forced down light African crudes, especially Nigerian, which in turn reduced the light-heavy margin and took sales away from Saudi Arabia and the others. [WSJ 2-1-83:3] By mid-February, the price of British oil was "equivalent to a $29 per barrel price for Saudi Arabia." [NYT 2-19-83:39] Nigeria then "clearly undercut the British" [NYT 2-21-83:D1] It equated to Saudi Light at $27. [NYT 2-21-83:A1] Yamani had threatened to cut from the $34 base, but he and his Persian Gulf allies were reluctant to do so. [NYT 2-21-83:A1] There followed a "hectic spate of mini-meetings." The Soviet Union "called on OPEC members to close ranks to halt plummeting oil prices." [NYT 2-23-83:D4]

Cooperation, Dialogue, and Interdependence

The Iran-Iraq war continued. France bought more oil and sold more weapons to Iraq "to prevent its military collapse and a destabilization of the region." [NYT 1-8-83:4]

Fesharaki urged that the marker price be increased to $36 and that production quotas be set with "a firm commitment to protect at all costs any member country that is pressured by the oil companies." [PIW 1-10-83:11] James Akins "derided analysts who predict sharp declines in the oil price. He said Saudi Arabia considers the 'correct price' of oil to be more, not less, than $34." [OGJ 4-4-83:NL1]

Opinion in the consuming countries was not too different. "American strategists have always found a silver lining in the stable pro-Western Gulf that high oil prices helped create." [NYT 3-6-83:E2] The OECD chief economist, Sylvia Ostry, thought lower oil prices were harmful. "The ensuing fall in OPEC demand for OECD exports would probably outweigh advantages of cheaper oil that would help further lower inflation rates." [PIW 1-3-83:8]

The World Bank's *World Development Report 1983* predicted that by the mid-1990s the real world oil price would be 20 percent above the 1981 peak—in 1993 prices, $64 per barrel. Daniel Yergin warned that by the late 1980s "the industrial world would once again rely increasingly on ... the Middle East. This, in itself, would bring us back to 1973." [NYT OpEd 7-19-83] The Department of Energy's 1982 *Annual Energy Outlook* said Saudi Arabia could meet its "needs" by producing only 3 mbd.

The Group of 30, "a high-level body of international economic and monetary experts and government officials led by ... the former managing director of the IMF," warned that lower oil prices would promote economic recovery, which would bring "a dramatic increase in the demand for oil by the end of the decade. Another interruption of oil supplies would then produce another oil shock, repeating the cycle of the 1970's." The group favored "tacit collaboration" between importer and exporter nations. The group itself had "helped to persuade OPEC to hold its marker price to $34 a barrel." [NYT "Economic Scene" 3-18-83; OGJ 5-2-83:98] They must really have believed this!

In the consuming countries there was a call for "Western aid to OPEC" to maintain the price:

> The West may see political advantages, too.... Stable prices would bolster Saudi Arabia and other ... friends of the West in a particularly volatile region. It would also improve North-South relations. In return, the West might get what it has sought in vain up to now—a long term agreement with the major producers for reasonable price increases in return for guaranteed supplies and an end to disruptive "oil shocks." According to Prof. Peter Odell ... the present glut offers "a golden chance to build new political bridges between North and South." [NYT 1-30-83:E2]

The Brandt Commission on International Development Issues also called for "a dialogue between major oil producing and oil consuming countries to consider arrangements beneficial to all parties, including safeguarding supplies to the poorest countries." [*IMF Survey*, 2-21-83][15]

An example of cooperation was Algerian gas. A trans-Mediterranean pipeline to Italy was built in 1981, but no gas was shipped until the Italians granted a higher price than the one agreed on. Finally the Italian cabinet approved a subsidy, "linked to improved economic relations [and] ... a tacit promise ... to buy more Italian goods." [WSJ 2-25-83:31]

Minister Yamani warned the consuming countries that if prices fell "you would see bankruptcies all over the world and mainly in the USA ... [and] the banking system come under serious pressure. So there is something to protect." [PIW 1-31-83:SS1]. "[C]ooperation between the producers and consumers ... has become more necessary now than ever before." [PIW 2-14-83:8; WSJ 2-11-83:2]

Yamani was, as usual, well informed about the United States. Until 1982, "most oil and gas loans were made on the premise that energy prices would continue to rise sharply.... The estimates had been for dramatic increases, [e.g.] ... at almost 10 percent a year, to $90 a barrel by

1990." [NYT 2-14-83:D1] But in 1982, banks were nervous over oil and gas loans. [WSJ 7-19-82:17]

Robert Mabro said that only coordination between OPEC and non-OPEC countries would work. Like Fesharaki, he suggested support for governments that were losing sales, which were "the special targets of oil companies in campaigns designed to lower oil prices." [PIW 2-28-83:1] In fact, there were no company "campaigns to lower prices." Since refining was a competitive industry, lower crude oil prices quickly became lower product prices. The integrated companies, like other producers, were damaged by the lower values of their reserves.

The March 1983 Production Allocation

The March OPEC meeting lasted nearly two weeks, preceded by three weeks of negotiations, "the longest negotiating effort ever undertaken by the 13 oil exporters." [WSJ 3-15-83:3] It was a success and put an end to the "hysterical talk of an oil price crash," as was noted a year later. [PIW 7-9-84:1]

The marker price was reduced from $34 to $29 per barrel, reversing the previous strategy of raise higher, the better to defend. But spot prices, and effective government selling prices, were already about $2 lower than $29, and there was widespread disbelief that the line could be held. [NYT 3-15-83:1, D22]

Total OPEC production was set at 17.5 mbd, 1 mbd below current output. Explicit quotas were fixed for all but Saudi Arabia. For the first time the Saudis accepted the role of "swing" or residual supplier, to make up the difference between 17.5 mbd and the total demanded at the new price. "Prices are floor prices ... quotas are ceilings." [PIW 3-21-83:9]

The marker price was above spot. "The growing opinion of many buyers that OPEC is the marginal supplier of last resort does not bode well in an environment that places little value on term contract arrangements." [PIW 3-14-83:3] In fact, the distinction between spot and contract was blurred. "Contracts now seldom cover more than one year," renewable if prices were satisfactory. Concern over "security of supply" had evaporated. [PIW 3-7-83:1]

The ministers feared the agreement could unravel; its very success was the result of that fear. [WSJ 3-22-83:1] For the first time, there was discussion of how far down oil prices could go in the short run before production began to be shut in. A PIW survey placed it in the neighborhood of $10. [PIW 4-4-83:4] Nearly 95 percent of U.K. North Sea output was said to have positive cash flow at a price of $5. [PIW 6-20-83:5]

As the OPEC members placed their chief reliance on production limits, they became aware of the deterioration in output data. The possible error was now as much as 1.8 mbd, or 10 percent of output.[16]

In mid-1983, spot prices actually rose close to official levels. [PIW 5-2-83:1, 7-11-83:1] An OPEC market monitoring committee expected to raise prices "in a gradual and orderly fashion after 1985 ... in the context of a global agreement with the industrial powers that will assure the world a reliable source of energy at reasonable prices.... Buyers must have confidence in us." There was already a production agreement, they thought, with Britain, Mexico, and Norway. [WSJ 4-13-83:31, 5-2-83:34] But two months later it was clear that only Mexico would cooperate. "Internal supply and price discipline will require sharp sacrifices for [OPEC] for the next two or three years, but [they] expect the oversupply position to ease later this decade." [PIW 9-19-83:3]

OPEC appointed a new ministerial committee on long-term strategy, again chaired by Yamani. They would need to recognize "the sensitive relationship between OPEC prices and demand levels for its oil. That was an element totally missing from the previous document." [PIW 7-18-83:5]

Efforts to Expand Capacity

Some OPEC governments were said now to be competing with each other for scarce company investment funds, but the return offered was derisory: "higher than the U.S. prime lending rate." [WSJ 4-7-83:33] Fear of expropriation precluded any substantial investment. [PIW 6-20-83:9] Despite the war, Iran and Iraq had made great efforts to maintain oil output. [PIW 1-17-83:6] The Iranian oil minister estimated "production costs" (presumably only lifting costs) as 5 to 20 cents. [PIW 10-17-83:9]

Iraq was perhaps the greatest underachiever. In January 1982, it announced a target of 6 mbd of production (its previous maximum had been only 3.5) with the help of foreign companies. [OGJ 1-4-82:73] That is precisely what they had said ten years ago and were to say ten years and two wars later. Iraq conducted a wildcat drilling program, with Mobil "providing technical services." [PIW 8-8-83:8] A year later, Iraq said it aimed at 7 mbd of exports by the 1990s. A pipeline across Saudi Arabia was quickly installed.[17] [PIW 3-26-84:3]

The Aramco companies were displeased by a new Saudi marketing outlet, Norbec, which bypassed them; they recalled that they had absorbed substantial losses for the sake of their "special relationship." [PIW 8-15-83:1] They believed that Norbec was "directed against the Aramco companies" [WSJ 8-28-83:15] The Saudis called it only a temporary entity [OGJ 9-5-83:NL1], but it survives today and is very important in short-term transactions and in managing the Saudis' floating storage, which also began at this time.

The Felt Need for Enforcement

OPEC was now "deeply troubled by the failure of demand to recover and their inability to trim output far enough,... giving thought to a series of really tough measures" against quota violators. [PIW 10-31-83:1] "The oil market is losing confidence in the ability of the $29 pricing structure to survive in the coming year unless there is a sustained demand upturn or dramatic new action by OPEC to rein in production.... It seems less risky to refrain from buying oil than to gamble on a market recovery." [PIW 11-7-83:1]

The advisory group, which a year earlier had issued a warning on the need for discipline [PIW 12-13-82:1], now called for an OPEC mechanism to allocate output and "a show of strength" toward non-OPEC exporters. In Mabro's words: "They too share the burden of defending oil prices." The panel said that the cheating of several OPEC members was masked by lack of production data and again urged a return to companies' buying under long-term contracts. As usual, lower oil prices were said to be bad for consumers. [WSJ 9-30-83:35]

A well-reasoned paper by the Iranian delegation urged lower output and restoring the $34 marker. They recognized a backward-bending supply curve: lower revenues increased the "propensity to export," hence a production surplus, a further weakening of prices, with "a chain reaction ending in a serious market glut." [PIW 11-21-83:5] The Iranian prime minister accused the Arab monarchies of "treason toward Muslims" in overproducing to reduce prices. [WSJ 12-6-83:6] James Akins said Israel had advocated joint U.S.-Israeli occupation of the oil fields but offered no proof. [OGJ 12-19-83:59]

The Mabro Proposal

In December, Robert Mabro proposed that OPEC members should "offer their traditional customers long-term contracts at official prices." Additional oil, prorated to members, would be sold at shorter term only through a central sales agency. It would match immediately all non-OPEC offers. There was a barely veiled threat: "If OPEC, God forbid, were to engage in a price war against non-OPEC producers it [the sales agency] could lower the price on incremental supplies without having to lose a penny on base-load sales, which would continue at official prices under long-term contracts. This is a formidable deterrent." [PIW 12-5-83:8]

Mabro was an important spokesman,[18] but the proposal had no chance of success. First, the companies could not sign long-term contracts at official prices. They knew that if spot prices went above contract, governments would raise contract prices; when spot prices fell, companies would be choked with unwanted oil billed at full contract price. Aramco's experience had shown that. Second, the threat to retaliate against price cutters was provincial. North Sea production plus Soviet exports were less than 6 mbd, but other non-OPEC production was over 17 mbd, of which 10 mbd was in the United States where producers were among the first to reduce prices. [OGJ 2-8-82:85, 2-22-82:67, 3-22-82:71]

Mabro ignored how competition works. When producers are many and independent, they cannot threaten or be threatened. To threaten them with punishment is to demand that they ward it off by acting in concert, which is exactly what they cannot do. Each competitor must take the price as an external fact and then adapt to it. Those unfamiliar with the economics of competition go astray because they can conceive of only a few actors reacting deliberately to each other. (The same lack of understanding was shown in 1990, blaming Kuwait for weak prices.) Unlike Mabro, Yamani understood the American scene. He pointed to damage inflicted by low prices on the U.S. oil industry and banks, which had political influence. If the U.S. government would help by setting a price-production policy, that might make a difference but not otherwise.

Thunder without Rain in 1984

Prices and Supply

Prices were mostly stable in 1984. The OPEC average declined only 86 cents, the Saudis' export price by 43 cents. Production data were still unreliable. [OGJ 12-26-83:Editorial; PIW 3-5-84:3] Members were "resentful that last year's [1983] big $5 a barrel cut failed to have the desired effect." [PIW 3-26-84:4] Nigeria was the weakest link financially, with foreign exchange reserves almost gone. After a coup at the start of January, the new junta reaffirmed Nigeria's ties to OPEC. [NYT 1-5-84:D10] Neighbors were helping the Iraq war effort, and the Americans were now definitely favoring it. [WSJ 1-6-84:24] Saudi Arabia had borne the brunt of export loss, but its foreign assets were still said to exceed $150 billion. [PIW 4-23-84:4] In mid-1984 *Petroleum Intelligence Weekly* wrote that

> hysterical talk of an oil price crash so prevalent in March 1983 has all but disappeared, even though producer nations are still fragmented, oil is in oversupply, and refining margins are poor.... The lessons of 1983 are now clearly perceived: OPEC output restraint, however unevenly applied, has a major strengthening effect on spot prices. [PIW 7-9-84:1]

A Fair Horizon

Some large companies were acquired (e.g., Gulf by Chevron) at prices that some thought extravagant. But it was widely believed that buying a barrel was cheaper than finding or developing one, as though there were no market to equate the two choices. Some analysts and oil companies, including Exxon and Amoco, disagreed. But Daniel Yergin explained: "An obvious reason for [mergers] is that the world is running out of oil." [NYT 2-9-84:D1] By the 1990s, he thought, the current surplus "may well have eroded, putting pressure once again on supplies." [NYT 4-29-84:OpEd]

Barter Deals

At an OPEC meeting in early July, Yamani "told reporters the recovery had already started" [WSJ 8-1-84:2], but then prices fell because Saudi Arabia increased output. A Saudi barter deal became known—of 34 million barrels for civilian jet aircraft, the oil reputedly valued at $27.10 instead of the posted $27.92.[19] Such deals were viewed as discounts. [*Economist* 7-21-84:62] "The market disruption is wholly unintentional, and does not reflect policy change." [PIW 7-23-84:3, 7-30-84:1, 8-6-84:3] But control of export volumes and prices was "increasingly eroded" by them. [PIW 2-13-84:1] Yamani was "understood to have resisted the deal as much as he could, but it was concluded above his head at the behest of senior members of the royal family without regard for state oil policy" to gain a $50 million commission. [*Financial Times* 8-1-84:10] The additional Saudi production may have gone into floating storage, not sales, and hence exerted no pressure on price. [NYT 8-21-84:D1] From this time, the distinction between Saudi production and sales became routine.

In early August 1984, the British government asked oil refiners "not to put pressure on the British National Oil Corporation (BNOC) to cut prices." [NYT 8-6-84:D4; WSJ 8-9-84:3] The London *Financial Times* opposed this collusion with a cartel "whose activities have been the source of many of the world's economic problems." [FT 8-13-84] But

> Washington, while not overtly supporting OPEC, is not fighting it either.... Lower prices would cut billions of dollars of windfall profits taxes. Cheaper oil would undercut coal. Development of wood, solar energy and synthetic fuels would suffer, as would programs for conservation and to produce domestic oil using new technology. Lower oil prices would lead to greater long-term dependence on imported oil and thus risk a future price shock. They also fear this would weaken Saudi Arabia, further destabilizing the Middle East. [NYT 8-19-84:E5]

The *Economist* [8-4-84:13] mentioned "commentators [who] bewail the 'danger' of much lower oil prices." Another key to world public opinion is in a statement by Pope John Paul II criticizing "imperialistic monopoly." Far from calling OPEC an example of monopoly, the pope "was unambiguous in supporting the call by third world countries for a redistribution of the world's wealth." [NYT 9-18-84:A9] A few months later, he "spoke sternly to the world's wealthy nations ... [and the International Monetary Fund about] ... the market for raw materials." [NYT 2-4-85:1]

France helped by renewing an oil supply contract with Iraq that the previous year "allowed Baghdad to overcome severe financial problems stemming from its war with Iran.... The deal was imposed on the state-controlled Elf Aquitaine and Total.... France has [also] agreed to refinance Iraqi debts totaling about $1.4 billion since May 1983." [NYT 8-20-84: D6]

Light Crudes Start a Decline

In September 1984, even as prices generally rose, Nigeria and Iran began discounting light crudes. [WSJ 9-26-84:42] In October, for the first time since the March 1983 accord, Abu Dhabi warned it would unilaterally cut the price on light crudes if it could not otherwise sell enough because of competition from Saudi floating storage. [WSJ 10-2-84:34] BNOC customers were no longer willing to accept the unchanged prices, which they had so recently approved. [WSJ 10-8-84:35]

Norway then reduced prices "in a break with British policy." [WSJ 10-16-84:2] BNOC followed. [WSJ 10-18-84:3] Within a day, Nigeria more than matched them. At an emergency meeting of OPEC ministers, both information and collusion were inadequate. An article in *Petroleum Intelligence Weekly* noted:

> The sudden fracturing in the $29 pricing edifice comes after an almost unbelievable series of miscalculations by both OPEC and non-OPEC producers....

Individual actions taken by producers ... can all be justified when viewed in isolation. But taken together in the context of a weak and volatile spot market, the moves domino [i.e., cumulate] far beyond the intentions of each producer. [PIW 10-22-84:1]

A few weeks later, the same publication wrote:

OPEC output must fall fast and far enough, in combination with rising seasonal demand, for the shaky $29 oil price to be reestablished.... The oil marketplace has not only lost confidence in OPEC intentions, but also in the industry's supply/demand appraisals.... A secret accord between Saudi Arabia and Kuwait is the real underpinning of last week's OPEC agreement ... [where] consensus was achieved more quickly than some had anticipated. [PIW 11-5-84:1]

Of course the Saudis and Kuwaitis wanted the accord kept secret, in order not to advertise that the two of them would bear the burden of additional cutbacks to starve the market into a price jump. As Yamani said, cutbacks could make spot prices rise "well above" official levels. [NYT 11-15-84:D30] But secrecy did not last out the week, and the known willingness to cut back did the Saudis immediate harm. Their December 1984 output was below 4 mbd.

Another meeting was held in December. Norway had begun openly to base its own prices on recent spot prices. [WSJ 12-12-84:4; FT 12-16-84:10] The fruitless quest for a schedule of differentials was resumed. [NYT 12-19-84:D17] Members first blamed Britain and Norway [WSJ 12-19-84:6], then each other. [WSJ 12-20-84:3] To regain credibility, they would retain an outside firm to audit production. [WSJ 12-21-84:2; 12-28-84:3] The previous month, after OPEC's quota cuts, "most forecasters ... were predicting that a yawning supply gap of 1 mbd was opening for November and December, but no such thing has materialized." [PIW 12-3-84:1] At year's end, the market was "closer to the brink now than in 1983,... OPEC is launching an *in extremis* bid to reestablish its credibility and ... role as manager and defender of the world price structure." [PIW 12-24-84:1] The actual prices cited in the desperate-sounding *Petroleum Intelligence Weekly* appraisal were around $27, only about 10 percent down in 21 months. But the potential for a crash was always there.

The Saudis Are Pushed Too Far, 1985

Emergency Meeting, Data Gaps, and Quivering Prices

In January, spot and futures prices continued to slip. BNOC followed the Norwegian example, abandoning official prices and selling by the current market. "OPEC's ministers repeatedly have threatened that such a cut by Britain would force the cartel into a price war" [WSJ 1-8-85:3], but the threat was empty. A spokesman for Britain's Ministry of Energy said that Britain would not reduce oil production to ease pressure on prices. [WSJ 1-15-85:5]

In January, at another emergency OPEC meeting,[20] Minister Yamani remarked that the $29 price was "not sacred" [NYT 1-29-85:D10], but nothing was done. [WSJ 1-31-85:3; NYT 1-31-85:A1] Still, the mere hope of production cuts caused a sharp jump in futures prices, regaining most of the ground lost since August 1984.

Saudi Arabia had suffered a sharp decrease in nonoil GDP, which was mostly a use of oil revenues, not an independent source of income. "The private sector ... is much smaller than it was even two years ago. Over the past 18 months, some 1,500 companies have either gone out of business or asked for emergency financing." Payments to American companies were much delayed. [NYT 2-18-85:D3]

OPEC retained a Dutch accounting firm to audit members' prices and production because "increasingly unreliable output statistics and imaginative pricing schemes ... plagued the group." [PIW 2-4-85:1] In July, an oversight committee, chaired by Yamani, was categorical: "Not one single OPEC country has from the beginning given the auditors sufficient access to date to enable unqualified positive endorsement to any member's complete statistics." [PIW 7-29-85:1] OPEC also hired a London-based group to track tankers to get reliable export statistics [PIW 8-12-85:1], but the auditors and trackers disappeared from view. A year later, the auditors' contract had lapsed. [PIW 4-14-86:4]

One indirect but useful data set on oil price evolution is the record of natural gas prices, which were tied by contract to oil (table 7.4). They were an appraisal of where oil prices were expected to be in the near or perhaps also distant future. In Europe, they peaked in 1982 and diverged greatly in 1985.

The Short- and Long-Run Picture

The anxiety of early 1985 was reflected in a poll of 125 forecasters. Nearly all thought that "oil prices, adjusted for inflation, [would] continue declining over the next several years, but [would] rise again in the 1990's." A dissenter not polled (this writer) thought the glut had existed for seventy years. "OPEC is on the verge of a dogfight." [NYT 1-30-85: D2]

Table 7.4
Natural gas prices, 1980–1992 ($/mmbtu)

	United States	Western European		Japan
		Low	High	
1980	4.42	3.00	3.70	5.01
1981	4.84	3.30	4.70	5.83
1982	4.94	4.10	5.20	5.74
1983	4.51	3.50	4.40	5.16
1984	4.08	3.50	4.20	4.90
1985	3.19	3.40	4.40	4.99
1986	2.53	3.20	3.60	3.98
1987	2.17	2.50	2.80	3.29
1988	2.00	1.90	2.50	3.22
1989	2.04	1.70	2.50	3.26
1990	2.03	1.80	2.50	3.60
1991	2.06	2.90	3.20	3.97
January 1992	NA	2.40	2.80	3.60

Source: Cedigaz annual report 1992.

The Norwegian oil minister, speaking in Kuwait, countered "accusations" by saying Norwegian output would level off at about 700 tbd. [NYT 3-25-85:D5] (By 1994, it was 2.5 mbd.) A high-level expert team called for "a new initiative between oil producers and users" [PIW 4-1-85:5], but Britain dissolved the national oil trading concern BNOC, in a move that was expected to make it harder to maintain prices. [WSJ 3-14-85:3]

OPEC officials have alternately pleaded with Britain to help prop up prices and threatened to initiate price wars or trade boycotts if it did not. The OPEC ministers have allies in the British Treasury.... The opposition Labour Party which created the oil corporation in 1975 condemned the announcement as the "final act of vandalism" against an organization set up to protect the oil industry and the security of supply to the oil industry." [NYT 3-14-85:D1]

A group of experts convened by the CIA in April expected some price decline, before increases in the 1990s. [WSJ 4-23-85:1] James Akins said: "There is no fear in OPEC of a price decline." In the long run, the price would rise to the cost of synthetics. [OGJ 4-15-85:32] Daniel Yergin expected that "market realities will again give way to geological realities—the concentration of oil reserves in OPEC and in the Middle East. And that will eventually put the era of surplus behind us." [NYT OpEd 7-8-85]

I "foresaw endless downward pressure on prices as far ahead as one can reasonably look." [PE 7-85:234]

The Saudi Dilemma: Retaliation Invites General Collapse

Spot and futures prices increased in March and April. But by May 1985, "three months of apparent restraint" had ended. "About the only thing that is propping oil prices is the Saudi willingness to take deep production cuts, which have pared output to a 17-year low." [WSJ 5-16-85:5] The Aramco companies' requested price cuts had been refused, "in the belief that any Saudi price cut would lead to a downward spiral in world oil prices." [*Financial Times* 5-25-85:2] Yamani called for official price cuts on heavy crude oils, in order to promote the sales of Saudi crude. [WSJ 5-30-85:14] It was the same old issue of price differentials. Saudi willingness to accommodate others' overproduction was "wearing thin." But Yamani had warned in similar terms in February, and nothing had happened. [PIW 6-3-85:1]

The Saudis called an emergency meeting in their summer capital, Taif. A senior OPEC official said, "It seems quite plausible that the Saudis are ready for a confrontation." [WSJ 6-3-85:3]. They were not; they only offered arguments. [WSJ 6-4-85:2, 6-10-85:26; NYT 6-10-85:D1] Yamani now viewed the price/production problem as more serious than what had produced the $5 marker price cut in early 1983. Production control was necessary but not sufficient; "marketing practices [were] ... undermining the pricing structure." [PIW 6-10-85:1]

A week later, he again warned that Saudi Arabia could not continue at 2.5 mbd production and demanded "at least" its quota of 4.35 mbd. Yet he would "oppose any reduction in the OPEC price level because it would only mean more discounting from a lower base." He also explained the process whereby a properly run OPEC would cope with demand fluctuations by the price-raising ratchet: "If you stick with the price ... then everybody will be selling a little bit below the quota.... Then prices on the spot market will be going up because supply is less. Then I think we would have to meet, increase the quota, and increase the [price] ceiling." [PIW 6-17-85:1, SS]

Acting as the swing producer cost the Saudis $2 billion in June alone. In 1983–1984 there had been a $22 billion drawdown of foreign assets, which now stood at $90 to $100 billion, of which only $65 billion was "reasonably liquid." [PIW 7-8-85:5] Presumably most of the nonliquid $25 to $35 billion was in "loans" to Iraq. We do not know how much of it found its way to Saddam Hussein's treasure trove, which enabled him to bear the embargo from 1990–1994. [NYT 7-27-92:A7, 7-31-92:A8]

By the end of June, "a growing number of oil analysts" expected a price as low as $20 by year-end. They tended "increasingly to doubt the threats of Sheik Yamani." [WSJ 6-28-85:1] OPEC members doubted also, or they felt simply unable to act together to accommodate Saudi Arabia to ward off disaster. Even the Aramco companies "abandoned their long-term contracts." [WSJ 6-28-85:13]

In July, the Saudis convened another meeting to insist on some "guaranteed minimum level" of production, said now to be 3.5 mbd, not the full 4.35 mbd quota. But nothing was done. [NYT 7-8-85:1]

> Saudi Arabia, increasingly desperate ... [is] prepared to increase their oil production from the current two million barrels a day to as much as nine million barrels a day by year's end unless other members ... agree to renounce all forms of discounting prices and cheating on production quotas.... [This was] the most blunt of Saudi threats that have been made in recent months.... Most OPEC members refused to accept production curbs, in part because many of them have come to disregard the Saudi threats as a bluff. [WSJ 7-10-85:2]

Japan, Sweden, and the Netherlands called for a dialogue with OPEC; others, including Britain, Norway, and the United States, rejected it.[21] [WSJ 7-10-85:2; NYT 7-10-85:D1] An OPEC consultative committee claimed that their "agreement in principle to end marketing 'malpractices' represents a significant advance in OPEC's slow campaign to defend oil prices." [PIW 7-15-85:3] This was a confession of impotence to rein in overproduction.

The consensus view of Saudi Arabia had been that the Western-schooled technocrats wanted less production in order to conserve the resource. In fact,

> More and more of these educated and increasingly vocal Saudis have come to feel that the country should forget OPEC, cut price and sell more oil.... Behind this view ... are serious concerns that Saudi Arabia's political and economic weight in the world are being undermined by its decline as an oil exporter. Arab diplomats note in interviews that Saudi Arabia's political clout within the Arab world, which rests solely on its wealth and its generosity in foreign aid, is waning. [WSJ 7-19-85:1]

The search continued for a "magic formula to squeeze quart-size oil supplies into the pint-pot of demand." The Saudis said they had been negotiating with both the old Aramco partners and others to move more Saudi oil. Their threats continued to be viewed as bluff. [NYT 7-22-85:D2]

"In OPEC there is growing dissatisfaction with its current strategy of defending oil prices through production restraint.... [Some want] head-to-head market confrontation ... matching but not undercutting non-OPEC prices. [PIW 7-22-85:1] Soon the Saudis would apply exactly this policy to their fellow members. Another special meeting in July came to nothing. Six members—Algeria, Ecuador, Iraq, Libya, Nigeria, UAE—refused to permit the production auditors to enter. [NYT 7-28-85:D1]

> [A] majority of OPEC delegates [were] convinced that a threat by Saudi Arabia to unilaterally boost production and touch off a price war was only a bluff.... The Saudis increasingly are viewed as the rich landlord in a slum. Resented for their wealth, the Saudis are seen as preying on poor members who can't afford to sacrifice much more of their revenue by cutting prices or trimming production.... On at least three occasions this year, the Saudis retreated just as their threats were beginning to bite.... The upshot is that most OPEC members now believe the Saudis are bluffing. [WSJ 7-26-85:3]

Perhaps because the Saudis' threats of retaliation were not feared, spot prices actually improved. Heavy attacks on the Iran export terminal at Kharg Island "may be Iraq's own way of 'making room' for its incremental exports this autumn, without weakening prices." These supply worries seemed to bolster the spot market, which approached $28, close to the marker. [PIW 8-26-85:2] However, Nigeria moved to reduce taxes on equity operations, which amounted to a price cut to those companies. [WSJ 8-26-85:2]

When the Saudis announced in August that their production would at some time double to quota level, 4.3 mbd [WSJ 8-1-85:4], nobody bothered to comment. Yamani stated in mid-September that this would happen by winter. "The Saudi move ... had been threatened but disbelieved for several months" [WSJ 9-16-85:1], but some still thought he was only bluffing for an October meeting. [NYT 9-16-85:D1]

Prices in the last quarter rose, propelled by new air attacks on Kharg Island, which halted Iran exports. At the same time, the Soviet Union, "apparently beset by production problems," temporarily stopped exports. And Saudi Arabia said it would accept no new customers for two weeks. An oil trader said: "In my experience, I've never seen a set of circumstances as bullish as this. Just everything is going right if you want to see prices rise." [NYT 9-27-85:D1; WSJ 9-27-85:33]

Everything continued right for several weeks as fear of supply losses drove up spot and futures prices. At another OPEC meeting in October, Saudi Arabia repeated that it had abandoned its role as swing producer [WSJ 10-7-85:3], but prices continued to rise.

Netbacks: The Saudis Prepare for Action

The Saudis now arranged, at first only with the former Aramco partners, for additional output to be priced at the spot product value of the barrel, less refining and transport cost—the FOB "netback value." "Predictably [it was] already drawing bitter criticism" from others. Some term buyers were demanding discounts for "security"—of outlet. [PIW 9-16-85:1,2] Under shaky collusion, assurance of outlet was worth more than assurance of supply.

The Saudis tried to separate incremental sales at netback value from "continuing sales at official prices," essentially the 1983 Mabro proposal. (p. 216) But customers resisted discrimination. The Japanese were outraged because "the new formula is for Atlantic destinations only." [PIW 9-23-85:1] Soon the Saudis were making netback sales to non-Aramco customers. They had already built Atlantic-area stockpiles, now to be drawn upon. [PIW 10-18-85:1]

By the end of November, all geographic discrimination was on the way out. The Saudis were preparing to sell generally at netback or at spot crude prices. More important, they were converting existing officially priced contracts into netback deals, and Japanese customers expected to be included. [PIW 11-25-85:1]

Oil companies had given up "trying to produce an internally consistent supply/demand balance for 1986." More than 1 mbd of supply was "in limbo" (without visible outlet) [PIW 11-25-85:1], yet the perception that inventories were low kept spot prices increasing.

At the end of November, Saudi Arabia suspended official prices. The avowed targets of what was called "shock therapy" were the United Kingdom, Norway, and the OPEC majority. "The idea is for a seasonal production pact with non-members, discreet enough to avoid embarrassing ... Britain and Norway, but concrete enough to persuade OPEC's maverick majority to start producing within their quotas." [PIW 12-2-85:1]

OPEC was said to be "preparing for a price war against rival producers outside of OPEC." [NYT 12-9-85:1] A special meeting declared their frustration by nonmembers' lack of restraint and called for "a fair share in the world oil market," whatever that meant. [WSJ 12-9-85:3] Within OPEC, there were differences

> over the degree of commitment to engage in a bruising battle for market share if non-OPEC exporters are unwilling to limit output. The hardline group, including

Saudi Arabia, asserts that the Geneva action is a concrete "first step" and not a bluff.... Yamani specifically singled out the UK as the "number one target".... [A committee would] examine ways of putting pressure on non-OPEC producers to restrain output, as well as look at alternate pricing formulas. [PIW 12-16-85:1]

"OPEC oil ministers have made it clear that Britain is their main target." [*Economist* 12-14-85:57] This target made no sense.[22]

The Ranks Collapse and Reform, 1986–1987

In 1985, the average "total OPEC" export price, less volatile than the spot price, peaked at $26.81 in April. As late as November, even with netback deals sprouting, it was $25.68. Between 1982 and 1984, the Saudi export price had stayed within 50 cents of "total OPEC." It is not available for January through November 1985, but in December it was down to $18.48, $5.00 less. It kept falling to the all-time low of $7.91 per barrel in May 1986.

The Netback as a Price Hedge

For buyers, a netback price was a costless hedge against price changes. Two months or more might elapse between the date of loading crude oil and the date of sale of the refined products. If the product price rose during this time, the crude oil price to the refiner was also raised by the netback contract. But if, as many expected and all feared, product prices declined in the interim, the refiner's crude cost would drop equally. The refiner need not fear that the slim margin might turn into a heavy loss. The crude oil seller now assumed all the risk.

Netbacks also had another, but more subtle, price-lowering effect. Previously, an autonomous drop in the product price level would make some high-cost refining unprofitable. The result would be lower output of refined products.[23] This slowed or stopped the product price decline. Lower product output in time became less crude purchased, depressing the price of crude. Thus the lower product price worked through to the crude level, gradually and not completely. Some of the lower product price was absorbed by the refining industry. But with netback deals there was no slowdown or buffer. The lower product price immediately became a lower crude price.

It was hoped that netback prices would "scare up cooperation" from non-OPEC producers. In January, Yamani "added to the havoc ... by speaking of a 'downward price spiral ... to less than $15 a barrel." [PIW 1-27-86:1] The result was that physical spot markets nearly dried up "because buyers and sellers are too far apart on prices and terms." [PIW 2-3-86:1]

Pain Inflicted

Some producers soon cried, "Enough!" Political pressures began to mount, but Arab Gulf producers show no willingness to ease pricing tensions by an early accord on lower production ceilings." Iran, Libya, and others called for an emergency OPEC session, but Saudi Arabia and Kuwait refused. "'Let them all stew,' says one high Gulf official." [PIW 2-3-86:1] Saudi Arabia and Kuwait were not moved by threats of terrorist violence against them. [WSJ 2-4-86:3]

OPEC wanted nonmember producers—Britain, Norway, the Soviet Union, Egypt, and Oman—to cut along with them. [WSJ 2-5-86:3] The largest non-OPEC producer, the United States, was rarely mentioned. [WSJ 2-5-86:3; PIW 2-10-86:1]

Britain was urged, and even expected by some, to join with OPEC because its high-cost oil would be shut down. Many believed that $15 was "below what's required to replace reserves" and that the persistence of such a price would "erode non-OPEC output enough by 1990 to put OPEC back in the driver's seat." [PIW 3-24-86:1] The belief remains strong today, but the number is lower.

In Saudi Arabia, unlike the rest of OPEC, lower prices meant immediate large revenue gains. "[T]he policy of maximizing production and pumping more money into the economy enjoys great popularity and will be hard to reverse." [WSJ 2-11-86:1] The Saudis and Kuwaitis stood fast against "a majority of OPEC members ... seeking ways to persuade [them] to call off the fight for market share." [PIW 3-3-86:3] So much for the legend that Saudi Arabia does not dare resist pressure from its partners.

The Persian Gulf producers said that "only cooperation between all producers inside and outside OPEC" could improve matters. [NYT 3-11-86:D11] But "even the Saudis [were] becoming concerned about the rapidity and depth of the current price plunge." [PIW 3-17-86:3] At home, some blamed Yamani for having previously cut Saudi production to prop up OPEC prices. Netbacks were said to be a repudiation of him. [WSJ 3-14-86:30] But such inside stories about a closed society should be put aside. The gulf producers were simply distrustful of their colleagues. "The Saudis believe the pain of reduced oil revenue should be extended through the summer to encourage adherence to any future OPEC limits on production." [WSJ 3-18-86:3]

The immediate objective was a $28 marker price. [NYT 3-22-86:35] The Iranian oil minister blamed the United States for trying to destroy OPEC "'because it is an organization belonging to the third world.'" He proposed cutting output from 17 to 13 mbd. But Sheik Otaiba of the UAE said he shared Yamani's view "that some countries had not suffered enough financially to assure strict production discipline." [NYT 3-23-86:9] After sitting for nine days, another meeting adjourned on March 24, having done nothing. Prices, which had stabilized since the start of the month, fell again. [WSJ 3-25-86:50]

Vice President Bush Speaks Out

On April 1, Vice President George Bush "said he would tell Saudi Arabia that the protection of American security interests requires action to stabilize the falling price of oil." He would "be selling very hard" on his forthcoming visit. But his was not "a price-setting mission." [NYT 4-2-86:1] He aimed only to persuade the Saudis to "stabilize—or even increase—the price of oil by cutting production." [NYT 4-3-86:A1; PIW 4-7-86:3] Hypocrisy so blatant was almost refreshing. It was scorned by many editorials. [NYT, 4-2-86; WSJ, 4-3-86]

Hypocrisy aside, what did Bush accomplish? The Saudis were producing more in order to inflict pain. When Bush said in April that their production was inflicting pain, the Saudis responded logically by producing more oil, and more pain, in May. At an OPEC meeting,

> many delegates saw these observations, later awkwardly retracted by the White House, as a sign that OPEC's drive to win market share by letting oil prices fall is reaching its goal. "It is clear the U.S. advocacy of free market pricing for oil has cracked," said an Arab delegate who asked not to be identified. "We must continue to push the price war further. The pain must go on to get real cooperation from non-OPEC members," he said. [WSJ 4-15-86:3]

What Bush accomplished was the contrary of what he intended. It was much like Undersecretary of State Irwin in Tehran in January 1971, telling the Persian Gulf producers how much harm they could inflict by cutting back output or by threatening to. As Karl Marx might have said the Irwin visit was tragedy, the Bush repetition was farce.

The solemn foolery of U.S.-Saudi oil "dialogue" continued throughout the 1980s, as it had through the 1970s. [*Washington Post*, 7-21-92:1, citing documents obtained under the Freedom of Information Act] Later, President Reagan's "top policy makers" claimed to have persuaded the Saudi monarchy to drive down oil prices in order to damage the Soviet economy. [Schweitzer 1994] This is pure assertion, supported by no evidence, contradicted by all of the public record reviewed here.

In March, OPEC had requested cutbacks by five non-OPEC producers [PIW 3-17-86:3], but it had not settled its own output. [PIW 3-24-86:1] By April they were hopeful over statements in the United States, Japan, and Norway, where there would shortly be a new and more sympathetic socialist government. [PIW 4-14-86:5] But in the end Norway and Britain not only failed to cooperate, they even lowered production taxes, offsetting some of the effect of the price decline on the producing companies. [PIW 3-24-86:1]

Netbacking Continues

As Saudi Arabian revenues rose, there was "popular support for the Saudi government's decision to pursue an aggressive oil-production policy despite the fall in oil prices."[24] [WSJ 4-8-86:34] Indeed, some of the royal family were reported as thinking Minister Yamani should have acted sooner to increase production. [NYT 4-13-86:F6]

At an April OPEC meeting, a majority favored cutting output to 16.7 mbd, but this was opposed by Libya, Algeria, and Iran, and the meeting did nothing.[25] [PIW 4-21-86:3] Yamani denied that any output ceiling had been set. Non-OPEC producers had first

> to trim a million barrels a day ... before OPEC tries to cut its own output when it meets ... in June. He said low oil prices will eliminate another 1 mbd of high-cost oil. Then OPEC might remove a third mbd or so. That, he said, would take care of an excess 3 mbd.... "But if we don't get anything from non-OPEC, nothing will happen." [WSJ 4-23-86:2]

In May, it became clear that other OPEC nations had kept up output and even undercut the Saudi netback prices. Saudi Arabia then escalated the conflict, offering a discount on every barrel of oil bought that month in excess of the previous month.

Futures prices had been rising but now fell again. As the trade saw it, the Saudis would do everything necessary to keep their market share. [WSJ 5-6-86:5] The previous summer's output of 2 mbd (exports between 1.0 and 1.5 mbd) was intolerable; "the country's prestige evaporated along with its oil revenues." Yamani said the "eventual" price goal was

$28, "but not in a year or two"; the interim goal was about $18. Iran's oil minister "roared: 'We will tell them it is a U.S.-Saudi conspiracy against the poor of the world.'" [WSJ 5-13-86:1]

By May, with no sign of the Persian Gulf alliance relenting [PIW 5-12-86:1], many were wondering how low the price could go, or was likely to go—two separate questions.[26] The new socialist government of Norway said it would consider limits on oil production, provided OPEC set a total and distributed it among members. [WSJ 5-15-86:7] A threatening development was that various consuming countries were raising taxes on oil products, thereby preempting OPEC revenues at the source. If this became more widespread, it would frustrate OPEC hopes for higher demand. [WSJ 5-28-86:39]

In June, an OPEC meeting heard an experts' report recommending more pressure on non-OPEC producers, none of which had cut production significantly. Many suggested that higher prices were also the responsibility of "bankers and nations interested in world financial stability." [WSJ 6-26-86:10] The immediate target was a price between $17 and $20, and ultimately $28, but members continued to resist production cuts, even as prices hovered around $13, with some Persian Gulf crudes below $10. [WSJ 6-27-86:4] By the end of the meeting, all but Iran, Libya, and Algeria were willing to cut production. [NYT 6-29-86:6] The meeting could not agree. [NYT 6-30-86:D1]

"A shrinking majority ... led by Saudi Arabia" thought it was impossible to parcel out a smaller total, hence impossible to cut total production. But "the price war eventually will eliminate much oil from non-OPEC producers, such as Britain and the U.S., because their oil is too expensive to produce." [WSJ 6-30-86:7] The Saudis raised production to 6 mbd at the beginning of July. [NYT 7-17-86:D16] For the first time, they actually exceeded their quota. It seemed to work.

The July-August Meeting

By the next meeting, late in July, the delegates had apparently learned: "The total volume of non-OPEC output shut in purely on the basis of production costs is likely to remain *surprisingly small.*" [PIW 7-21-86:1; emphasis added] In view of table 7.2, it is not at all surprising. Earlier in the year, the predominant industry view was for prices to be at $18 to $20 by year's end; now there were steep cuts in planned capital expenditures. [NYT 8-4-86:D1].

Some sales of Saudi Light were rumored to be at $6.08, although the reported spot price was $7.70. Yamani first proposed a production ceiling of 17.6 mbd, but the minority wanted a cut to 14.5 mbd and much higher prices. [NYT 7-29-86:D1, 7-30-86:D3] "Nearly all delegates appeared genuinely panicky over the prospect of OPEC oil prices sinking closer to $6 a barrel. The chief fear among delegates is that with prices so low, industrialized consumer countries [may] impose taxes and tariffs on oil, effectively holding consumption." [WSJ 7-30-86:3] This, of course, was what the *Economist* had advised in December 1985. It was apparent that there would be no help from non-OPEC governments. A week's deadlock showed the members' short time horizons:

> Partly, many experts say, the problem is OPEC's seeming inability to do without instant gratification.... The ministers are seeking to achieve cuts in their individual crude oil output to dry up a glut of oil on the world market and push prices to a level that would more than compensate for the cuts in volume they would have to accept. But while the ministers agree on the principle, they appear unable to trust the reckoning. "Two barrels is something I hold, it's real," a source close to delegates said. "One barrel for a higher price is a promise. So I hang on to what I have." [NYT 8-4-86:D1]

Distrust and suspicion were pervasive. Delegates interpreted conciliatory proposals as expressions of weakness. An explanation, widely believed outside OPEC, was a Saudi-Kuwaiti conspiracy to bankrupt Iran to let Iraq win the war. [Milton Viorst, NYT 8-5-86: OpEd] This seems far fetched; the pro-Iraq government in the United States opposed the low prices; but four years later the same author would find a Kuwaiti conspiracy to bankrupt Iraq.

A "temporary and fragile" production agreement was reached in August, covering only September and October. The new limit was set at 16.8 mbd, compared with a current output of 20.5 mbd. Saudi Arabia and Iran resumed their old quotas, others cut mildly, and Iraq was permitted to produce ad lib, which meant in effect its current output. [WSJ 8-5-86:3; NYT 8-6-86:D1, 8-7-86:D3; PIW 8-11-86:3] Therefore Iran no longer needed to insist that it must receive twice as much as Iraq.

The agreement turned the market around. By the end of August, the new OPEC cutbacks were taking hold, and buyers were being refused. [PIW 9-1-86:1] Norway said it would reduce exports by 10 percent, to help "stabilize oil prices at a higher level." [WSJ 9-11-86:4] China, the Soviet Union, Mexico, Egypt, Malaysia, Oman, and Angola each made a similar pledge. [NYT 10-3-86:D1] None of them kept it.

Yamani announced a six-year (1992) goal: 20 mbd of OPEC exports at $20 per barrel. [PIW 9-15-86:1] His forecast of exports was right on target, but $20.00, adjusted for the increase in GDP-IPD between 1986 and 1992, would have been $25.30 in 1992. The actual 1992 average export to the United States was $17.87.

In October, Saudi Arabia and Kuwait proposed to raise the Arab Light marker to $18. This would require further output cuts, ruled by a new quota scheme more favorable to them. [WSJ 10-7-86:7] Iran, however, favored the easier course of simply extending the August agreement. Temporizing was best because they could not tell, after another Iraqi raid on Kharg Island export terminal, whether they could export more. [NYT 10-10-86:D1]

Most members wanted to retain current quotas, the line of least resistance, but Kuwait demanded an increase in theirs. [WSJ 10-13-86:3] They and the Saudis proposed that quotas be based 50 percent on reserves, 20 percent on producing capacity, and 10 percent on population. [WSJ 10-14-86:2; NYT 10-14-86:D1, 10-18-86:35]

The Saudi cabinet had publicly stated they would not accept a quota renewal, but on October 18 they did, with the marker target set at $18 per barrel. [NYT 10-19-86:A1] Kuwait assented for the time being. [WSJ 10-22-86:2] Their minds had been concentrated by sharp declines in futures prices.

Yamani Dismissed

For Minister Yamani it was a last hurrah, since he was dismissed at the end of October. The press accounts were vague (and probably unreliable). Yamani had allegedly "opposed immediate steps to raise oil prices, [arguing] that raising oil prices is incompatible with the quest of many OPEC members, including Saudi Arabia, for larger market shares." King Fahd called the stated $18 to $20 target only a "first stage" of the advance to higher prices. [WSJ 11-11-86:2] Another account agrees and adds that the king demanded both a higher quota and a higher price. Yamani had opposed barter deals that enriched some princes but endangered prices. The king thought they could be kept secret. [PIW 11-24-86:1] Yamani was reported "under strict orders from King Fahd to refrain from any comments on the kingdom's oil policy if he wishes to retain his freedom of movement." [WSJ 1-11-87:A2] A later editorial in the *Economist* [6-27-87:13] thought Saudi Arabia was better off exporting more at a lower price. "Sheikh Yamani was sacked because he recognized this." (The editorial, incidentally, is an excellent brief summary of the Saudis' investment-expenditure plight.) Prices wobbled somewhat on Yamani's departure but then revived. Kuwait supported Saudi Arabia in seeking higher prices but also without suggesting how to cut production. [WSJ 11-12-86:3]

The American Petroleum Institute now suggested that the U.S. government establish a base price for oil through an import fee, which would take effect when the price fell below a certain level. This would penalize OPEC for lower prices and reward them for higher prices. Nothing came of it. The Reagan administration took no position officially. "U.S. officials root for higher oil prices, despite possible ill effects. They hope the OPEC meeting's outcome will raise prices into the high teens." [WSJ 12-19-86:1]

More Production Cuts and Stable Prices

It was widely reported that the Saudi endorsement of higher prices was part of a deal with Iran. [WSJ 11-14-86:12] Aside from the fact that the Iraq air force, rebuilt with Saudi-Kuwaiti money, had done heavy damage to Iran from August through October 1986, reports of a deal were at best superfluous. Both parties were on record as wanting higher prices, but they continued to disagree on concerted production control.

All oil analysts agreed that a substantial production cut was needed. [NYT 11-14-86:D1, D17] True, the OPEC pricing committee thought pricing discipline would of itself increase by $3 the value of a "basket" of OPEC crude oils, which now replaced Arab Light as the "marker." [NYT 11-15-86:1; WSJ 11-17-86:2] To nobody's surprise, it did not. And when King Fahd refused to cut production below quota, there was a renewed price decline. [NYT 11-25-86:D1]

Iran's deputy oil minister now reversed course, saying his country was willing to work with other members to cut production to bring the price up to $18. [NYT 11-26-86:D13] The December meeting should have been brief and rewarding. King Fahd, now in accord with Iran, had also conferred with the chiefs of state of Iraq, Libya, and Algeria. [WSJ 12-3-86:2] A senior delegate said: "The Iranians will run this meeting without doubt, and everyone will fall into line." The Saudis would make the biggest contribution, by cutting output from the current nearly 6 mbd to 4.3 mbd. [WSJ 12-8-86:3]

After eleven days of meeting, the delegates could not agree on production cuts. Iran had been willing in August to disregard Iraq because it was about to "control" Iraq's production by force; it no longer did. [NYT

12-14-86:3, 12-15-86:D12] But oil markets were encouraged, and spot prices rose, on a reported consensus that "quotas and cutbacks remain the only option." [NYT 12-15-86:D1] The next day, they "all but reached agreement on measures to boost oil prices by sharing production cuts. Saudi Arabia, with other members' support, urged OPEC to reduce production at a level below actual demand. Thus OPEC hopes to starve oil markets by next March to prepare the ground for yet another price boost." [WSJ 12-16-86:5]

But Iraq would accept nothing unless its production was equated to Iran. [WSJ 12-17-86:2] Its quota of 1.2 mbd had been set in 1984, when Iran had badly damaged Iraqi export capability. But it had been restored and was expected to rise. Iran proposed that Iraq be suspended for its recalcitrance, but this was not taken seriously. [NYT 12-18-86:D2] "Saudi Arabia, which supports Iraq's war effort with oil and substantial war loans, has tried to use its leverage to persuade Iraq to join the output accord. It has not succeeded so far." Yet "for an accord to push prices higher, there must be some assurance that Iraqi output will not surge." [NYT 12-19-86:33] Iran now relented, since "its need for firm and rising oil prices outweighed its desire to force Iraq into the pact." [NYT 12-20-86:35]

The new agreement emerged on December 21. Everyone was to cut back by about 5 percent, and first quarter 1987 output was set at 15.8 mbd. It fixed the prices of twenty-three crude oils, including the old marker (Arab Light) at $17.52 and the crude widely used as surrogate marker (Dubai Fateh) at $17.42. Iraq was still not limited by quota, but the assumption was that it would produce only 1.466 mbd, below the current 1.6 to 1.7 mbd. Members also agreed "to phase out all oil sales contracts that are based on a free-market pricing of oil." This meant the end of netback pricing.

As *Petroleum Intelligence Weekly* said, the production cuts would maintain the price through the winter, the most favorable season; but "buyers are still extremely reluctant to assume all of the commercial risk ... by accepting the new schedule of fixed prices." Both Saudi Arabia and Iran were losers "in their bid to be seen as effective leaders ... both overestimating their ability to influence Iraqi oil policy.... [This] may be highly damaging in the long run as [Iraq] expands export capacity in 1987–1988." [PIW 12-29-86:1] King Fahd now thought output reduction was "the only way to absorb the surplus ... the price must not be less than $18." [NYT 12-27-86:32]

Appraisal: OPEC Retreat, 1981–1986

OPEC awoke from the dream of moving the price up an inelastic demand curve toward the cost of synthetic liquid hydrocarbons. The true limit to price, set by customer substitution and non-OPEC output, was not far above but actually below the 1981 price. But OPEC members also believed the consensus view, often referred to. Since the price was surely going to rise, one need only hold ranks and tough it out. They waited for something to turn up—the demand for OPEC crude oil. (It did, after 1986. That is the subject of the next chapter.)

In 1981, they actually raised the price to $34 to conciliate all the members. They lowered it to $29 in 1983 but then were unable even to discuss a lower price. Nor could they agree on the production cuts needed to support any given price.

With the companies gone, the OPEC governments had to fix production and set down market shares in black and white. Any change in planned total output unlocked everyone's demands for a larger share of the market, and the whole deal had to be remade. Market share had always been the main topic at OPEC meetings before 1981. The previous chapter reviewed the secret agreement in 1978, aborted by the Iranian revolution. Even while still raising prices, at the end of 1979 they proposed a "safety net" for members in trouble, and in the summer of 1980, a "gentleman's agreement" on output, aborted by another accident: the Iran-Iraq war. In March 1982 they had made a loose allocation agreement, which was not well observed, and then the firm agreement of March 1983, which was not well observed either.

Each cartel member wants to shove the burden of curtailment on to others. When smaller cartel members cheat by producing more and shading the price, the largest producer fears retaliating, lest the whole arrangement crumble. The Saudis' share of OPEC exports fell from 47 percent in 1981 to 19 percent in 1985. The others must have known that the Saudis could not tolerate exports around 1 mbd, but they could not achieve an agreement to alleviate the Saudis' plight.

The only Saudi weapon was to threaten to cut prices. But after a while, nobody believed their threats, and by mid-1985, their partners were openly contemptuous. Indeed, even if another single member believed the threat, what was he to do? Unless all other members would cooperate, it was not worth any single member's time.

The Saudis finally put in place a mechanism whereby their price went down automatically as far down as forced to. There was never any Saudi price war. The Saudis would meet any price, not beat any. Once they were again respected, a new agreement was possible to cut production and raise prices.[27]

Between January and March 1983, a Saudi threat produced an agreement within two months. After mid-1985, it took them sixteen months of actual price competition, and a price collapse, to work out a new market-sharing agreement. They regained about one-third of the price ground lost. They have never been able (except briefly in wartime) to get much higher.

Appendix 7A: Investment Requirements and Rate of Return

The example uses table 7A.1. The start (line 1) is all wells drilled. This overestimates by including gas wells and exploratory wells. In the United States, it would overstate development outlays by about 40 percent.

Line 2 states the average depth per well. The next step (line 3) is to enter the Joint Association Survey table to find the U.S. drilling cost per well at that depth. But since drilling costs increase more than proportionately with depth, the average cost per well will be greater than the cost of the average well. Line 4 estimates the difference at 10 percent.

It is generally believed that drilling and associated costs are much higher outside the United States, mostly because of the dense network of service and supply industries. I have tested this belief by a regression

Table 7A.1
Investment per barrel of new capacity, 1990: Egypt

1. Wells drilled, all types	102
2. Average depth (feet)	9,755
3. Drilling cost per well, U.S. ($000)	913
4. Corrections: nonlinearity	1.1
5. Corrections: outside United States	2.2
6. Adjustments: nondrilling outlays	1.7
7. Total corrections and adjustments (lines 4 × 5 × 6)	4.1
8. Total outlays per well (lines 3 × 7) ($M)	3.8
9. Total oil production investment (lines 8 × 1) ($M)	383
10. Oil wells drilled	60
11. Average production per well (barrels/day)	932
12. New capacity (lines 10 × 11) (tbd)	56
13. Investment per daily barrel (lines 9/12) ($)	6,851

Sources: WO; Joint Association Survey 1990; Survey on Oil and Gas Expenditures.
Notes: Adjustments in lines 4 and 5 (and hence in 7) may be excessive. [DOE/EP 1993] Tab III-1 has 25 percent adjustment, equipment only, or roughly 16 percent overall.

study of rig time, a good proxy for expenditures. For thirty-six countries, I tabulated the number of rigs operating in 1990 and divided by the number of wells drilled. Rig days per well became the dependent variable. The explanatory variables were average depth of well, percentage offshore wells, and a dummy, zero for the United States and Canada and 1 for all others. The results were as follows (t-statistics in parentheses):

$$ln\ (\text{rig-time}) = .06 + 1.59\ ln\ (\text{depth}) - .66\ (\text{percent offshore}) + .81\ \text{dummy}$$

t-statistics: (.10) (4.53) (2.13) (1.84)

Adj RSQ = .48, F-stat = 11.6

The negative coefficient for offshore wells is correct in itself: drilling time is minimized in offshore wells. It points to a weakness of rig time as a measure of cost when not separated into onshore and offshore. But the only purpose of the regression is to hold the other factors constant and show that being outside North America will increase rig time, hence cost, by a factor of 2.2 (the exponent to the coefficient .81).

Therefore in table 7A.1 we multiply the average drilling cost per well by 1.1 (line 4), then by 2.2 (line 5). This may be a substantial overcorrection. It certainly is much larger than that made in a DOE study cited in the table. Finally (line 6) we use the U.S. proportion of nondrilling costs: lease equipment, enhanced recovery (gas and water injection, thermal and chemical methods, etc.), and overhead. Line 7 shows the total correction and adjustment, by 4.1. This is multiplied by the average drilling cost to give the total investment (line 9).

This investment is divided by the gross capacity increment (line 12), estimated as the number of newly drilled oil wells multiplied by the average production of all oil wells. I think newly drilled wells tend to be bigger producers than average old wells. If so, the true increment to capacity is greater than estimated, and the investment per unit is overstated. In areas, such as the United States, where the number of wells operating is very large, with a very wide distribution of well capacities, the method cannot be used.

Rate of Return

In table 7.2, the 1990 upper quartile value was $10,210 per daily barrel. That is, one-fourth of the group were more expensive, and three-fourths were less expensive. To calculate the rate of return, we use the break-even equations from chapter 2. First is the rate of return before tax:

$i = P/(K/Q) - a - c$

where P is the wellhead price, Q is the peak output, assumed to be in the initial year, after which it declines at a percent per year. K is the total of capital expenditures, assumed all to be spent in the year before production starts. In our example, $K/Q = \$10{,}210/365 = \$28$.

As explained in chapter 2, the levelized operating cost per barrel per year c is taken as .075 of the investment, the output-to-reserve ratio at a conventional 1/15, and the decline rate a at $.067 - .067^2$. We assume the price as \$15 per barrel.

$i = \$15/\$28 - .075 - .062 = .536 - .135 = .399$

This is 40 percent pretax. Assume now a fifty-fifty production sharing scheme, which amounts to the government's taking half the gross wellhead price P'. In that case:

$i' = \$7.50/28 - .137 = .131$

The pretax return is well worth investing for and could be split to benefit both parties. The posttax return is marginal.

Alternatively, we can solve for the break-even price at some defined discount rate (e.g., 20 percent):

$P = (a + i + c)(K/Q) = (.062 + .20 + .075) * (\$28) = \$9.58$

With factors often used as typical or conventional, a useful rule of thumb is that the break-even price is approximately one-thousandth of the investment per initial daily barrel.

Mexico, 1992–1993

Pemex was reported in early 1994 to have spent \$3 billion in 1992–1993, of which exploration and production accounted for about two-thirds, or \$2 billion. The objective was to maintain not raise reserves. [PIW 5-23-94:2] In fact, reserves decreased from the end of 1991 to the end of 1993 by 373 million barrels (mb). Aggregate output in those two years was 1,946 mb. [OGJ:WWO 1991, 1993] Hence gross reserve additions were 1,573 mb. Investment was therefore \$2 billion/1.573 billion = \$1.272/barrel in ground. But of the 1992–1993 wells drilled, 12.1 percent were gas producers. [WO:IOO 1993, 1994] Adjusting to exclude gas gives \$1.118/barrel in ground.

In the late 1970s, in the new Chiapas fields, the decline rate was about 7 percent. [Petroleos Mexicanos, *Generalidades del Proyecto de Construccion del Gasoducto Cactus-Reynosa*, July 1977, tables II, IV, pp. 19–27] If so, $K/Q = (K/R)/a = \$1.118/.07 = \$5.97$ per annual barrel, and \$5,830 per daily barrel. This is the investment under current conditions.

Pemex is generally considered very inefficient. On average, in 1990 it took 310 days to drill and complete a well, as against the 122 estimated from average depth and percentage of onshore wells. In 1993, Triton Engineering Co. drilled a Gulf of Campeche well in 127 days, compared with the usual 249. [OGJ 8-16-93:60] Hence the attainable investment per daily barrel was in the neighborhood of \$3,000, and break-even cost to return 20 percent was slightly less than \$3 per barrel.

Notes

1. I once speculated, on no firmer basis than the average lifetime of taxable business equipment, that the half-life of the consumption response was about nine years—that 7.7 percent of the response was felt the first year and thereafter 7.7 percent of what remained. It is hard to measure any such slow-acting process.

2. P is the ratio of the new to the old price; Q, the ratio of the new to the old amount demanded. If $Q = P^E$, ln $Q = E\ ln\ P$. Differentiating, $1/Q = E/P(dP/dQ)$ and $E = P/Q(dP/dQ)$, the conventional elasticity definition.

3. Kouris [1983] seems to indicate a modal value around 0.85. Dahl and Sterner [1991] estimate around unity for gasoline, the least elastic of oil products. CGES [1992] estimates product price elasticity at 0.85.

4. Table 7.1 is overly conservative because I set to zero the elasticity of supply in non-OPEC countries. As we see below, the non-OPEC supply response to higher prices in the 1970s and lower taxes in the 1980s was swamped by changes in taxation and regulation.

5. Kaldor thought oil prices should be kept high. "The very large expenditures incurred on prospecting and developing new fields since the war would not have been possible if oil had fluctuated in price in the same way as copper or tin.... Recently there was a threat of a collapse of oil prices due to reduced demand which (in my opinion) was rightly received with a great deal of misgiving, even by the large oil-importing countries: since they realized that in the long run they are likely to fare worse under a regime of fluctuating oil prices than under a regime of stable prices, even though the latter would be a relatively high one in terms of industrial goods." [Kaldor 1983, p. 34]

6. The measure is affected by changes in factor prices. The IPAA measure decreased by one-fourth between 1981 and 1985, then rose nearly 7 percent to 1990. The expectation [EMF 6, 1982] was that despite higher expected prices, output would be lower. A simple supply curve would be: marginal cost (equal to price) rises exponentially with the quantity of output, $MC = P = e^{aQ} - 1$. Then $a = ln((P + 1))/Q$. They expected P to increase and Q to decrease; the coefficient a had to rise. In fact, a decreased for all countries during the 1981–1990 period and for eleven of thirteen between 1985 and 1990. (Reserves added data are too unreliable for any use.)

7. After 1990, the number of wells drilled in Argentina declined by one-third to 1993; production rose by 34 percent to 1994. [WO-10]

8. The following account is based on [Bradley 1986; Eckbo 1987; Smith 1987; Bradley and Watkins 1987; Kemp 1987, including the postscript by Watkins; Kemp, Reading, and Macdonald 1992]

9. An oilman explained: "Graft is when you must pay someone in the Government to do what he should do without payment. Corruption is when you pay him to do what he should not do."

10. In a similar study [OGJ 9-17-90:29], P. J. Hoenmans, of Mobil Oil E&P, took a sample of twenty-five countries under 1989 published terms. In a hypothetical 50-mbd field, government take ranged from 35 to 90 percent. Only in six countries was profit after tax acceptable. In four countries, even a 200-mb field was unprofitable.

11. But even at this time, Belgium agreed to buy liquefied natural gas from Algeria at an initial price (subject to escalation) of $5 per mcf FOB. This was even called a "concessionary" (below-market) price, granted because "Belgium has a position we appreciate very much on the new international economic order." [WSJ 4-9-81:28]

12. On June 6, 1982, Israel invaded Lebanon. The siege of Beirut aroused much indignation worldwide, including in Israel. There was no suggestion of any oil action by Arab or Muslim oil producers. Iran and Libya produced above their ceilings. [WSJ 7-1-82:2] The Saudi ambassador to the United States "*hinted* ... that if the Israelis tried to enter west Beirut the Saudi Government *might* consider cutting off *some* oil shipments and transferring its dollar reserves out of American banks." [NYT 7-22-82:A8, emphasis added]

13. The foregoing paragraph is based on NYT 7-5-82:31, 7-8-82:D18, 7-9-82:D3, 7-9-82:35, 7-11-82:3, 7-12-82:D1, D3; WSJ 7-8-82:3, 7-12-82:2; PE 8-82:315.

14. Since Saudi production was 37 percent of the total and the price of all OPEC oil shipped to the United States fell by $1.69, we can set up an equation with one unknown: $.37(\$.63) + .63(\$x) = -\$1.69$, where x is the change in non-Saudi OPEC price. It is equal to minus $3.05.

15. My opinion, against the consensus, was that there was no scarcity but rather oversupply, repressed by a strong but "clumsy cartel," with whom agreement or accommodation was not good or bad, merely impossible. (See [Adelman 1993], especially chapters 19, 25, and 26.)

16. "It seems hard to believe nowadays that as recently as the early 1970's it was possible to assemble monthly production data (down to the last barrel) for all major oil fields, and often on a company-by-company or crude-by-crude basis.... Consumer nations should not be too quick to criticize OPEC for its output data shortcomings, since they have largely failed to provide complete current information on oil demand and inventories. The largest single discrepancy is in stocks, where assessments vary by as much as 700-million barrels—equivalent to almost 1.9 million bd over a year." [PIW 5-16-83:1]

17. With production down to 1 mbd and exports around 400 tbd, Iraq had told contractors: "'We can't pay so you get your governments to provide credits, and we will start paying you back in 1985.'... None of the major contractors left.... The companies wanted to keep in Baghdad's favor for the postwar period, and their own governments were willing to back them." [NYT 11-24-83:D1] They were not repaid in 1985, and the ploy was repeated, again successfully.

18. Robert Mabro's recently established Oxford Institute of Energy Studies was supported, among others, by OAPEC (not OPEC, since Iran refused support because of his supposed

partiality to the Arabs); the European Community; energy research groups in Japan, France, and Sweden; the UK secretary of energy; the Arab Banking Corporation; and other banks, corporations, consultant firms, and research institutes. Thus his proposals for stabilizing prices give some idea of what these groups were willing to support, or at least consider.

19. Such deals were nothing new. In 1981, the *Wall Street Journal* reported how various members of the Saudi royal family had cut themselves in as sales agents for oil sold through Petromin or (later) Norbec. "Commission payments—for oil export and for nearly every major development project inside the kingdom—are part of the glue that holds the 4,000 princes of the royal family together. Prominent princes have come to expect a big share of the kingdom's spoils.... What worries Western oilmen is the likelihood.... [of] a decline in the relatively orderly marketing operations of the four U.S. companies." [WSJ 5-1-81:1]

20. With nothing of substance to report, the veteran Youssef M. Ibrahim, who wrote at different times for the *New York Times* and the *Wall Street Journal*, and to whose work we and all other observers are greatly indebted, wrote an item on high living at the hotel: late suppers, and "a procession of $200- to $1,000-a night 'escorts' making their way to the elevators." [WSJ 1-25-85:1] Mr. Ibrahim was, very briefly, barred from the meetings.

21. There had been a general decline in the willingness of the industrial nations to support commodity prices. Secretary of State Kissinger had called in 1975 for a worldwide system of price support through North-South cooperation. An American official said: "We're fed up with it and now the Soviets are fed up with it." [WSJ 10-2-85:34]

22. In the meantime: "American congressmen are now making their umpteenth attempt to reduce the $200 billion budget deficits.... Congress could find salvation in the oil market.... The taxman could simply stand between the wellhead and the petrol pump, scooping every fall in the price of crude without raising the cost of American motoring.... The dismal science rarely gives politicians what they most want, painless solutions to their problems. It is doing so now.... If Congress does not go for a petrol tax soon, it will have missed its best chance of returning the world to cheap oil, slow inflation, low interest rates and rapid growth." [*Economist* 12-14-85:16] The chance was missed, then and later.

23. Oil refining, like oil production and almost all other industries, is an industry of increasing cost because it is an array, from lowest to highest cost, of units in each plant, and of plants in the industry. Therefore marginal cost exceeds average cost.

24. Official "liquid and semiliquid assets" had been estimated at about $110 billion in early 1983, $55 billion by the end of 1985, and due for another drop of $15 billion by the end of 1986. [WSJ 4-7-86:3] Another estimate put total assets at about $80 billion, of which bank deposits plus short-term investments were $50 billion. In addition, there were loans impossible to collect quickly (to the IMF or World Bank) or perhaps ever (to Iraq), and equity investments. But much could be done to conserve assets by cutting spending. [WSJ 4-8-86:34]

25. In April 1986 the United States bombed Tripoli in Libya. The merits of the attack are not our concern. But Italy and West Germany were reluctant to approve it or join in antiterrorist measures, because "Libya remains their largest supplier of crude oil." [NYT 4-14-86:A6]; Japan, because it is "a major importer of Middle Eastern oil." [NYT 4-29-86:6]

26. "M. A. Adelman of MIT warned ... that oil prices could sink to $5 and remain there for the next 10 years—though he assigns that a low probability. Prices are more likely to fluctuate between $5–$25, mainly in the $10–$15 range, he said." [PIW 5-5-86:1]

27. Thomas R. Stauffer later wrote [OGJ 5-21-94:105]: "Shaikh Yamani slashed oil prices in 1986 with three clearly articulated objectives: to reduce conservation; to stimulate global

economic growth; and to discourage non-OPEC energy supplies of all kinds." Yamani never reduced prices. His avowed aim was to stop others' price cuts. He devised a system whereby Saudi Arabia would meet prices but not beat them. By the time his rivals realized he was not bluffing, prices had greatly declined. There is no evidence that any of the "objectives" ever existed. It is certain that none was attained. The energy-to-GDP ratio continued a slow decline. "Global economic growth" continued sluggish. Non-OPEC energy supplies continued to grow.

8 Stagnation after 1986

After 1986, oil prices were relatively stable. Omitting late 1990, from 1987 to 1993 the monthly average Middle East spot price was $15.67 per barrel and the standard deviation only $1.78. In real terms, of course, this represented a continuing decline. There were many complaints of "unstable" prices after 1986, but *unstable* was a code word for *low*.

The stable price was very disappointing. Before 1986, OPEC nations had lived from month to month and season to season, always looking for an upturn in demand. After 1986, consumption in the market economies (excluding OPEC) increased strongly. Production declined in the United States. Mexico was static. The North Sea expanded only slowly. The Soviet Union fell apart. Elsewhere production crept up. In the aggregate, non-OPEC output barely increased from 1985 to 1992.

Accordingly, OPEC exports (including natural gas liquids) increased from 12.5 mbd in 1985 to 22.0 mbd in 1993, the highest since 1980. Excess capacity in OPEC had nearly vanished by 1989, yet the price did not rise.

The hard lesson was that cartels find it hard to cope with change, even favorable change, because division of revenues is a zero-sum game. Higher revenues mean a fresh contention over sharing the gain. The greater it is, the greater is the zeal to increase one's share. A higher price might profit the group, but without mutual trust, members cannot exploit a risky market opportunity.

Excess Capacity and Price Change

OPEC capacity was in excess after early 1974, and all attention was focused on the "surplus," as if it were some passing misfortune. It was expected that capacity would shrink, and prices would inevitably rise. In fact, excess capacity had declined greatly by 1986, yet the price crashed.

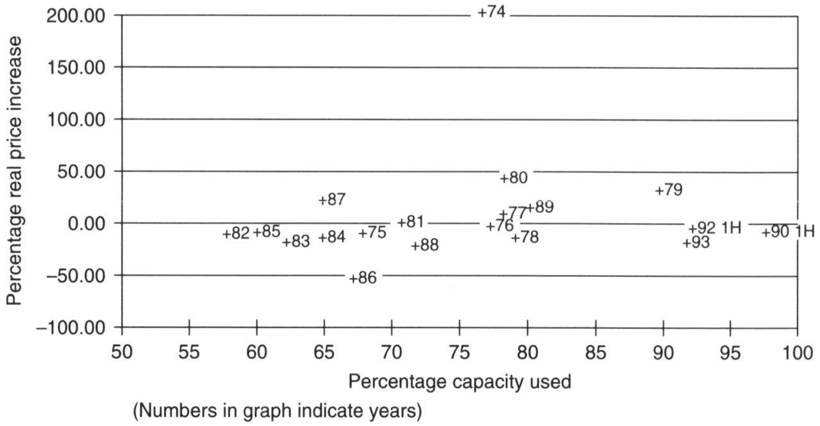

Figure 8.1
OPEC capacity used and real price change. Note: The numbers in the graph indicate years. Source: Table 1.

It kept declining afterward, yet the price did not rise. Shakespeare lends us a rule about excess capacity: "A little more than a little is by much too much." That is wrong under competition, right for a cartel.

Under competition, change in the percentage of capacity utilization is a good proxy for change in short-run marginal cost. High-level production calls into use the high-cost standby equipment, materials, and employees. It stops normal maintenance and downtime. The attempt to get more output than designed overloads the whole system. To produce an incremental unit makes every other unit more expensive. Thereby the cost of the incremental unit goes up, until the cost curve goes vertical. Hence, the capacity utilization curve—assuming good data—shows the price or price index under competition as the demand curve shifts up to the right or down to the left.[1] It is often used to measure inflationary pressure over the whole economy.

The sources and calculations of figure 8.1 are in table 8.1. The vertical axis shows not the price but the annual deflated price change. "OPEC is assumed to increase or decrease prices depending on whether capacity utilization is above or below the desired level [such as 80 percent]." [EMF 1992, p. 173]

The 1974 increase was in a class by itself. But we saw in chapter 5 that the production cuts had ceased in December 1973, and in January demand was less than output. Excess capacity was being calculated early in 1974, and the annual average capacity use was just under 80 percent. The price tripled in 1974 because the excise taxes were raised in concert.

Table 8.1
OPEC capacity used and real price change

Year	Average price[a]	Real increase (%)	Production	Capacity	% used
1973	3.45				
1974	11.33	202.17	28.69	37.07	77
1975	11.53	−7.07	26.41	38.61	68
1976	12.18	−0.68	30.50	39.13	78
1977	13.24	1.75	31.26	39.66	79
1978	13.22	−7.48	22.91	37.48	80
1979	19.24	34.03	30.82	34.19	90
1980	30.93	46.82	27.10	34.44	79
1981	34.56	1.55	22.73	31.86	71
1982	33.62	−8.41	18.83	32.20	58
1983	29.59	−15.40	17.74	28.23	63
1984	27.92	−9.61	17.72	27.13	65
1985	26.99	−6.81	16.45	27.48	60
1986	13.38	−51.72	18.67	27.68	67
1987[b]	17.12	24.03	18.21	27.88	65
1988[b]	14.30	−19.62	20.24	28.07	72
1989[b]	16.21	8.57	22.64	28.26	80
1989 1H[c]	16.71				
1990 1H[c]	16.36	−6.24	23.47	23.75	99
1992 1H[c]	16.39	−6.27	23.90	25.55	94
1992	17.58 (OPEC)				
1992	16.87 (DOE)		24.27	25.30	96
1993	14.87 (DOE)	−14.06	24.53	26.53	92

Sources: *OPEC Statistical Bulletin*; PIW; CEA, Economic Report of the President.
[a] *Formula: Average price = Value of petroleum exports/Exports of crude oil and refined product.*
[b] Interpolations.
[c] Prices for 1H1990, 1H1992, and 1993 estimated by applying ratio of prices "total OPEC" imports in MER. In 1974–1991, this correlated 0.99 with price calculated by formula.

The years of highest utilization were the first half of 1990, the first half of 1992, and the year 1993; in all of them, real prices decreased by comparison with the earlier period. (We avoid second-half 1990 and first-half 1991, during invasion and war.)

Excess Capacity as a Bargaining Tool

The curve of aggregate excess capacity does not explain price change. But individuals' excess capacity is part of the bargaining process. The cartel member producing to full capacity collects the higher prices but bears no burden of restriction. The burden falls on those with excess capacity. But their grievance is also their bargaining power. They demand higher quotas. If refused, they can expand output and damage the others. A cartel member that wants a larger quota must build excess capacity. Without it, he gets no respect. Hence, the announcements of OPEC capacity expansion are generally exaggerated for effect.

Members with excess capacity are often called price doves. The others, hawks, want the best of both worlds: a high price and the doves to carry the burden. If the doves stand firm and demand that the hawks share the cutback, an impasse weakens the price. When one or another side yields, there is an agreement. But there is always the danger of a price slide's getting out of hand.

Since 1986, despite constant pressure from the others, the Saudis have largely but not wholly held to their policy of the 1970s, which they unwisely left in the early 1980s: hold market share. Their constant theme has been: if you will not cut proportionately, neither will we. They sometimes made good their threats. The net result was a chronically weak price.

The Hopeful Start of 1987

The price fixed in December held through early November. Saudi Arabia, Kuwait, and Iraq had all notified customers they were back on fixed-price contracts. [WSJ 1-11-87:A2] Norway, the Soviet Union, Mexico, and Egypt announced their "cooperation." [WSJ 1-14-87:3] The Soviet Union cut back, but it had done this in the first quarter of the last three years, only to increase later. [NYT 1-23-87:D2]

The success resulted from "the simple combination of production cuts and fixed price notices." "Gaining term commitments from major companies to lift a significant proportion of crude needs at fixed prices is seen vital for OPEC success. For the buyer, the biggest stumbling block in the fixed price approach is a built-in lack of market responsiveness, but the special OPEC ministerial differentials committee may help on this score." [PIW 1-5-87:1]

A catalog of malpractices to be avoided was set forth, "drawn partly from OPEC documents." [PIW 1-19-87:7] And as prices improved, by "an OPEC-wide output cut of 1.5 mbd in the heart of winter," it was feared that the temptation to sell a little extra could bring spot crude and product prices down as the increment flowed through the refining system. [PIW 1-26-87:1]

Term Contracts, without Fixed Prices

A deal with the four Aramco partners sounded like old times: "a multiyear agreement ... at the official government prices." [WSJ 2-4-87:4] There was much pleased surprise in the industry [NYT 2-4-87:D1], but it was illusory; there was no penalty for underlifting. [PIW 2-9-87:1]

Months later, "the most fundamental problem is the seeming inability of producers to rebuild stable term supply relationships with primary contract customers." Fixed prices did not mean much because buyers were free to change volumes on short notice. Even the Aramco partners and the Japanese were reluctant to commit more than a month at a time. [PIW 5-4-87:3] Mobil Oil's former president waxed poetic over "Mobil's key link to the world's largest oil reserves beneath the deserts of Saudi Arabia" [WSJ 1-29-87:A3], but the chairman of Chevron spoke prose: "It is a long-term relationship, but one that returns nothing to our bottom line." [PIW 11-16-87:1]

OPEC producers "could restore value to the old fashioned term contract by keeping production low enough to push spot prices modestly over official levels." [PIW 5-4-87:3] A month later, fixed price contracts "for specific time periods are starting to make a comeback," although "arms-length crude sales by OPEC members at official prices still are a relatively small share of their total exports." [PIW 6-15-87:1]

The new Labor government of Norway promised cooperation. Japan, Sweden, and the Netherlands also feared another price collapse would make oil scarce. [WSJ 4-10-87:4]

The Saudis were again taking the brunt of the cutbacks. [WSJ 2-11-87:3; PIW 3-16-87:1] In March, they were down to 2.5 mbd, uncomfortably close to the August 1985 low point. [WSJ 3-20-87:14] As exports revived, they seemed increasingly committed to a $20 price that year and to $22 in 1988. Other members approved; Kuwait was opposed. [PIW 4-27-87:3]

Iraq's Expansion

"The wild card, of course is Iraq, which scorned the December agreement and is running some 430 tbd [29 percent] above its deemed allowable." [PIW 5-25-87:1] Its foreign debt was reported at $50 billion, half of it to Arab states—"a grim picture." [WSJ 2-12-87:29] Today more is known of how skilfully the $50 billion was used. By June, Iraq was still exempted from quota limits and planning new pipeline capacity for shipment through Turkey and also Saudi Arabia, which "appears to be trying to restrict oil exports from Iraq" [NYT 2-8-87:A7], through its territory to the Red Sea. But Iraq kept expanding and thumbing its nose, possibly because of better relations with the United States.[2] By mid-May 1987, the mood was upbeat:

> All 13 members of OPEC are impatient to reap the fruits of a year's discipline that successfully reversed the spectacular price collapse of 1986.... In April, the Saudi oil minister, Hisham Nazer, successfully extracted a promise from the U.S. to stop its "OPEC bashing," in the words of a senior U.S. administration official, a move that constitutes a major, but quiet reversal in Reagan administration policies. Furthermore, Mr. Nazer ... has persuaded major non-OPEC oil exporters ... and a large segment of the U.S. and international oil industry to join OPEC's drive for slow, steady price improvement.[3] [WSJ 5-18-87:2]

As usual, nobody explained what was meant by the "major reversal" of U.S. policy. But prices did not rise. Abu Dhabi, like Iraq, was producing over quota, and a "disjointed" differentials structure meant there was constant temptation for many to cheat. Only Saudi Arabia turned away customers. [PIW 6-8-87:3] Indeed, it was "resigned to acting as swing producer this year to support the $18 price, so long as volume cheating stays within narrow bounds." [PIW 6-29-87:1]

In the longer run, it was judged that Saudi integration into refining overseas would save it from the fate of being swing producer, which was logically correct. [Ibid.] Downstream integration is permanent built-in netbacking, the tactic that had brought the others around. "Kuwait continues to sell *its full production quota* regardless of OPEC's gyrations, disagreements and fights over the official price." [WSJ 6-25-87:1, emphasis added] But downstream integration of a small fraction of output was only a small help.

OPEC's 1987 midyear meeting had to face "a delicate market balance on the horizon." [PIW 6-29-87:SS1] The new production accord "set the stage for a hike to $20 in 1988—if discipline is preserved." As usual, the object was to starve the market: the ceiling "was deliberately pitched at a level below projected demand" to force up the price. But OPEC feared "creating upward price pressures that individual members will find difficult to resist, with the end result excess production and lower prices."[4] If so, "Saudi willingness to act as swing producer would again become crucial." [PIW 7-6-87:1] It was already crucial; the Saudis continued to refuse customers. [WSJ 7-17-87:5]

The Rosy Long-Term Outlook

The experts' consensus was as reassuring as ever. James Akins said in October 1986 that "OPEC stands a 95% chance of forging a long term production limitation agreement by yearend ... [and] such a pact has better than a 50% chance of lasting." [OGJ 10-6-86: NL1] Petroconsultants Ltd. made an analysis of non-OPEC output "based on reserves" and proved that a decline was "imminent and unstoppable ... well before the end of the decade." [OGJ 10-20-86:22] In 1994, Petroconsultants' "worldwide data base" showed there were 1100 billion barrels left in the earth [*Economist* 8-6-94:10], a figure no facts could refute.

By 1987 the consensus was that non-OPEC output had "finally declined" and would keep declining as fields dried up, with the world "becoming ever more dependent on OPEC oil." The trigger point for a crisis would be when production exceeded 80 percent of capacity, a point "approaching rapidly." There "will be a shift back to the Middle East," which would increasingly affect "the geopolitics of oil." [WSJ 8-2-87:1] "If supply and demand are left to the market, it will take only a decade for the Saudis to become the swing producer, able to control the world price by regulating the flow from their own wells." The price could approach $200 per barrel. [NYT 8-13-87] Such a price would console OPEC for its waning influence in the Third World. [NYT 8-2-87:E2]

Peter Odell of Rotterdam University thought OPEC should "be protected and cossetted, so in effect firmly incorporating into the Western system a group of countries which have hitherto enjoyed scant respect." [Odell 1987] The head of the Cambridge University Energy Research Group expected low prices to engender "a severe oil price increase that would make nonsense of what appeared to be short term gains today." [OPEC *Bulletin*, 4-87:5] A "loose" or "tight" market was something inherent, unconnected with monopoly. A Brookings symposium agreed that with a "surplus," the effects of a supply interruption or reduction would be negligible; with "tight supply," it would be important. [Fried and Blandin 1988]

Prices That Held

During the summer, OPEC overproduction cast "a shadow over the group's hopes of raising official prices later this year." Cooperation between Iran and Saudi Arabia was unaffected by the riot by Iranian pilgrims in Mecca. "OPEC's two largest oil producers still need their rapprochement on production curbs to support the benchmark price." [PIW 8-10-87:1] Saudi Arabia lodged a formal complaint against overproduction by other OPEC members. [NYT 8-4-87:A1]

Yet prices were firm for the usual reasons: "Most companies still seem happy to take more oil than they really need.... Many ... are still betting that OPEC ... [will] ... raise official prices in December, and want to build stocks ahead of an increase." [PIW 8-17-87:1]

Saudi Arabia now sounded a "warning": letting it be known that it was going slightly over quota. [PIW 8-24-87:1] When the price fell below $17, the Saudi monarchy denounced "criminal gangs" at the Mecca riots in "a bitterly anti-Iranian news conference." [NYT 8-27-87:D1] A very high-ranking official said the Saudis would "'demolish [Iran] politically and Islamically.' Western diplomats here are now exultant as they busily cable their home offices about the Saudis' 'sea change' and 'new activism.'" [NYT 8-29-87:A1] Some nondiplomats were unimpressed.[5] Iran made no reply but in early September said the Saudis should do more to hold down production by Kuwait, the UAE, and Iraq. [NYT 9-11-87:D3]

Minister Nazer ruled out any Saudi swing producer role and said his country would allow "field supervision of its production, providing that all the exporters agree to an on-site monitoring system." [WSJ 9-14-87:5]

As fears of a crash died away, the higher prices gave both confidence and disquiet. A consensus was building to increase from $18 to $20 in 1988, but higher prices were seen as "shattering OPEC's hard-won cohesion—which is already being stretched to the limit. While OPEC can probably hang together when there are hopes of volume gains for everyone, pressures on swing producers are likely to become intolerable if they must defend higher prices that continuously erode OPEC's shrinking market share." [PIW 9-7-87:5] Any set of fixed prices brought problems:

> OPEC producers and their customers are trying to preserve the term supply relationships they revived earlier this year, despite the price and supply pressures now evident in oil markets.... The common denominator for maintaining OPEC discipline is still likely *adherence to official prices, since it lets market forces determine who gets cut back.* [PIW 9-21-87:1, emphasis added]

But there was the rub. Holding to official prices left output allocation to the chances of the market. There were winners and losers. Some losers would not endure it and would cheat.

A September oil ministers' meeting was gloomy and frustrated over the lack of information. The Saudis had "strongly criticized the effectiveness of OPEC's efforts to track members' output and sales." [PIW 8-10-87:1] Now a special meeting of five ministers "admitted they don't really know how much more oil is being produced than [is] called for." [WSJ 9-11-87:2] Moreover, "proposals to set up permanent on-site production and export monitoring in all OPEC countries are seen by some ministers and officials as futile.... A previous $3-million exercise by Dutch auditors was thwarted by individual members' lack of cooperation, not by the auditors' lack of oil expertise." [PIW 9-21-87:5]

Saudi Arabia's downstream investment increasingly was "seen as the only way for [it] to jettison its unwelcome role as the world's main swing producer," as Kuwait had already done. [PIW 9-28-87:3] Not much has been done.

By late September, production was falling, but not by design. With overfull inventories, demand was lower. The gulf producers, especially Saudi Arabia, held to the price and lost market share. [WSJ 9-28-87:4] When Exxon indicated its preference for "flexibly-priced Iraqi crude," it was feared that Saudi Arabia, "OPEC's ever more reluctant linchpin," might be forced to produce below quota. [PIW 10-5-87:1] But the Saudis, underscoring their refusal to act as swing producer, let it be known that they had been discussing more discounts, in return for more volume from the Aramco companies (Exxon, Mobil, Texaco, Chevron). "The mere possibility ... sent shockwaves through OPEC and across the oil markets." [PIW 10-12-87:1] The report of a deal that would restore netbacking [PIW 10-19-87:1] was later denied. [PIW 11-23-87:1]

Then came a string of misperceptions: concern about overproduction was "fading fast" and companies were building, not depleting, stocks. [PIW 10-26-87:1] There was even "a drift in OPEC toward a $1 to $2 increase," since world oil markets had no difficulty in soaking up higher volumes [PIW 11-2-87:1, 2]. "The oil markets are looking healthier." [PMI 11-4-87:1]

Within a few days, spot prices were down sharply, markets were "unraveling," and "the central question is how vigorously Saudi Arabia will defend its 4.35 mbd output quota." [PIW 11-9-87:1] The Saudis "continue[d] to renounce the swing producer role." The gulf producers were all discounting. [PIW 11-16-87:1] Although Saudi Arabia tried to hang on to official prices [PIW 11-23-87:1], the production data gap grew worse. In the third quarter, the official ceiling was 16.6 mbd. The OPEC president estimated 17.8 mbd actual production, IEA made it a round 19.0 mbd and a consulting firm said 19.2 mbd. Aside from definitions and excuses, each country firmly believed that it deserved a higher quota. [WSJ 11-24-87:1] By the end of November 1987:

> Several oilmen who were in senior positions then are becoming increasingly alarmed at the uncanny parallels between events today and those of two years ago. [OPEC production is] far above official quotas ... and price discounting has become more widespread. Those were exactly the factors that both markets and OPEC producers resolutely ignored in 1985.... Saudi Arabia also knows that its readiness to bear the brunt of OPEC's volume earlier this year is at odds with public renunciation of any swing role. [PIW 11-30-87:3]

As previously, when the Saudis talked tough and acted soft, they got no respect. Doubtless hoping he would be believed this time, Minister Nazer repeated that "we would never be the swing producer, neither now or any other time." [PIW 12-21-87:SS3]

OPEC member statements were free of panic, but "the previous apparent consensus for a $2 a barrel price hike has evaporated, with only Iran now pushing for a $20 benchmark price." [PIW 12-7-87:1] Iran "accused the Arab members of OPEC ... of trying to keep down the price of oil to hurt Iran.... The Gulf producers had deliberately glutted world oil markets to pressure Iran's economy." [WSJ 12-2-87:2] In fact, the losses of the gulf producers were many times the damage to Iran.

At the December meeting, Iran threatened to double its production if its demand for a higher price was not met, but this threat was so empty that it upset nobody. [NYT 12-10-87:D1; WSJ 12-10-87:2] More bothersome was the persisting lack of basic information. "National political positions ... fill the vacuum created by the lack of hard data." Some believed inventories to be higher than a year ago; some thought they were lower; some thought them the same. Moreover, the "amount of oil at sea and where it is headed are issues of increasing interest" but not knowledge. [PIW 12-14-87:1] Minister Al Khalifa Al-Sabah of Kuwait said "the auditors—to put it mildly—got the runaround from various countries." [PIW 12-21-87:SS3]

The ministers "acknowledged that if they could not reach a credible agreement to limit output, world prices would crumble." [NYT 12-10-87:D1] They expected to agree within two days. [WSJ 12-10-87:2] In four days, they had agreed only to rebuff Iran, despised for its hypocrisy in discounting while calling for higher prices. Saudi Arabia again rejected a swing role. [NYT 12-13-87:1] The Iranian minister departed rather than accept what was offered. [NYT 12-14-87:D1] The meeting then reached what was called a "flimsy" agreement: keep the nominal $18 basket price and the 16.6 mbd limit—except that there was no limit on Iraq, whose minister "pledged to produce all the oil it can." [WSJ 12-15-87:3] No action was taken on quota monitoring. [NYT 12-15-87:D1] Prices were expected to drop, and they did, even as "ministers ... headed home vowing to cheat no more." [WSJ 12-16-87:16]

Arab-Iranian enmity was blamed but cannot have been important. Iranian price discounts would have been a minor irritation if the others had not matched them. An output cut would have raised or at least maintained prices. OPEC could not do it. The Saudis repeated "emphatically ... their resolve 'never again' to reduce their market share to save oil prices. ... 'The ball is in everyone's court, not just ours.'" They had stayed inside quota, while Kuwait, the UAE, and Iraq had "exceeded their quotas by absurdly large amounts." [NYT 12-17-87:D1] The Saudis had to keep repeating their refusal to cut back because the others had nothing else to suggest.

But lower prices would mean fewer discoveries and more consumption. "Eventually, *and certainly by the mid-1990s*, OPEC's excess capacity will have drained, oil analysts generally agree. That means that with only slight adjustments in its production levels, *OPEC will be able to manipulate world oil markets at will*." [WSJ 12-21-87:6, emphasis added] A year later, the consensus forecast was for a 41 percent price rise by 1993, and a decline in non-OPEC output after 1990. [OGJ 8-29-88:18]

The Year 1988

When the price dropped by January to the lowest point since October 1986, there was no panic. At the end of January, there was said to be "a new OPEC strategy ... a subtle but important shift in favor of volume restraint coupled with a more permissive attitude toward flexible pricing." [PIW 1-25-88:1] This was not new, but the emphasis was important. It amounted to this: sell no more than your quota, any way you can. Prices may fluctuate, but the average market level will not change much.

Cost Reduction and Capacity Expansion

There was increasing recognition of cost and supply. A Mobil executive said companies could live with a $15 price "for years to come." [PIW

2-1-88:5] Shell's worldwide coordinator for exploration and production said that "oil companies are accepting that they can't count on higher prices to stimulate the currently low global drilling pace." [PIW 4-18-88:3]

There was much discussion of reduced costs, mostly the meaningless "finding cost per barrel of oil equivalent." But there was also solid evidence, despite the lack of any general measure. The cost of drilling a well at a given place at a given depth in the United States decreased by about one-fourth from 1984 to 1992.[6] This holds for the rest of the world, which uses the same technology.

And while drilling wells became cheaper, fewer wells were needed per unit of new reserves and capacity because of other advances in knowledge. One was 3D—three-dimensional computer models of oil field structures, once limited chiefly to large offshore fields but now available for even small fields onshore or offshore. The other was horizontal drilling, the perfection of a technique—"deviated drilling"—first practiced mostly offshore. [Lohrenz 1991] Even as late as 1992, only 4 percent of U.S. wells were horizontal. There is a long way to go, and the trend will continue for years all over the world.

I noted in chapter 2 one dramatic example of North Sea cost reduction. A more general indicator was that the managing director of BP exploration estimated that in the right conditions a 10-million-barrel field was profitable. Twenty years earlier, the minimum had been several hundred million. He expected that more cost reductions would keep lowering the threshold. [OGJ 9-13-93:39] The CEO of Conoco estimated North Sea reserves in known fields at 22 billion barrels rather than the reported 13 billion. Instead of "overengineering our facilities to achieve high but short-lived production levels," they aimed to match "reservoir characteristics and optimize production levels." [OGJ 11-8-93:47,48]

Many new North Sea projects were announced in 1991 for 1992 through 1998. As we show in appendix 8B, a regression analysis estimates the average investment at about $12,000 per daily barrel. At 1992 prices, this seemed marginal or worse. Prices have actually declined since, yet there is no report of any project's being canceled or deferred. Moreover, I reckon the U.K. North Sea decline rate at an unusually high 13.5 percent. In only three years, lack of fresh investment would reduce output by one-third ($(1.1347)^{-3} = 0.68$). But U.K. North Sea output actually increased by about one-third from 1991 to early 1994. Lower and better-designed taxes also contributed. Moreover, a group of projects announced in 1994 were at substantially lower costs.

The IEA began to raise its supply estimates, heeding "criticism within the industry that the IEA has been consistently understating the amount of oil actually available, particularly missing large volumes of world supply from non-OPEC sources." [PIW 2-15-88:3]

Quota Impasse and Sagging Prices

In early 1988, the big OPEC producers mostly stayed within quota. The Saudis had made their point: "The unwillingness of any member or group of members to act as a swing supplier is giving rise to a search for alternative mechanisms for correcting sharp price movements." [PIW 3-7-88:3] The price increased from the January 1988 low and was stable through the first half.

There was a new effort to bring in non-OPEC producers and much publicity surrounding a meeting in May 1988, but the output reductions offered, totaling 183 tbd, only 1 percent of OPEC output, were too small even to acknowledge. The non-OPEC producers felt aggrieved by "the ungracious manner of rejection." [PIW 5-9-88:3] Months later, it still rankled. [PIW 10-17-88:2]

Rising Iraqi production was an increasing threat. Its "spot-linked pricing system [was] geared to specific markets and quite flexible." Iraq flouted OPEC constraints, "and output [is] near physical limits." [PIW 5-30-88:3, 18]

OPEC's June 1988 meeting faced the "basic disagreement over a credible and equitable production sharing system [which] is at the heart of the current OPEC impasse." [PIW 6-6-88:1] OPEC could do nothing but wait for consumption to turn up. [PIW 6-13-88:1] The June meeting simply extended the quotas. [PIW 6-20-88:1] But prices began to sag, not helped by Abu Dhabi's "no longer making any pretense of abiding by its quota". Kuwait also increased production. [PIW 7-4-88:3] So did Saudi Arabia. [PIW 7-18-88:1]

The End of the Iran-Iraq War

At the end of July 1988, the eight-year Iran-Iraq war finally ended; prices actually revived on word of peace. [PIW 8-1-88:3] Both countries were expected to try to increase production rapidly, but Iranian oil production facilities had deteriorated badly. [PIW 8-8-88:1] Iraq had hired French and Italian firms to develop new fields; the developers would be paid in oil. Producing capacity was slated to rise to 4.5 mbd "over the next few years.... Sixteen years after oil nationalization, Baghdad is actively seeking participation of US firms in oil exploration and development." It would not offer equity terms, but one way or another it would triple the number of operating rigs, from the current twenty-five or thirty to ninety by 1991. [PIW 5-30-88:3, 18] The oil minister who released this news visited the United States shortly after [PIW 6-6-88:1], but without known results.

Precise numbers aside, Iraq producing capacity was set for a strong increase. In 1980, Iraqi companies had drilled 40 oil wells and 67 wells of all types. Numbers for the next seven years were suppressed and probably near zero. But in 1988 they drilled 88 oil wells and 102 in total, and in 1989 137 oilers and 178 total. Since there were only 378 operating wells at the end of 1988, it is clear that a very large addition to capacity was underway. But Iraq was still in financial straits; this expansion would obviously take more effort. [PIW 8-22-88:7]

A Summing Up by Jennings and Ali Jaidah

J. S. Jennings of Shell expected that demand would grow more slowly than expected, and non-OPEC output would not decline, hence the rise in OPEC exports would be modest: "In such a world, OPEC, in order to defend an $18 price, would have to continue to live with a quota system for much longer than many of its members expected, and, with the passage of time, internal stresses within the organization could well intensify." [PIW 8-29-88:2]

Ali M. Jaidah, former OPEC secretary-general, repeated the myth that "because of the large discrepancy between investment costs and extraction costs in oil upstream and the very long time lags in energy investment, some sort of price regulation is required." The once-powerful oil companies had become "greedy and improvident. They could not resist the temptation to sell oil to newcomers outside the cartel for the sake of the odd buck." Hence OPEC had to replace them as price controller. But OPEC too became "greedy and improvident" in 1979–1980. For the present:

> We hear senior managers of oil companies haranguing OPEC, preaching to the organization that the state of the oil world, however depressed, will undoubtedly improve in a few years. They seem to say: Please remain strong and confident, we are going through a difficult period just now, but the wheels of fortune are bound to turn in your favor soon. Please hold tight until then and you will be in control once again....

> I just cannot understand how this low price can sustain investment in high-cost oil areas.... Somebody, somewhere, must be losing his shirt. [Annual Oxford Energy Seminar, reprinted in PIW 9-12-88:S2][7]

Price Decline

Prices slid from June through September as Saudi Arabia defended its market share. "The policy that it won't be swing producer has almost approached religion." [PIW 9-12-88:1] As in the previous year, the Saudis announced publicly that they were exceeding their 4.35 mbd quota as a warning to violators. [PIW 9-19-88:1] BP managing director Robert Horton made a plea for OPEC production discipline and an $18 price. [Ibid.] Venezuela, long a pillar of OPEC, let it be known that it was trying to obtain "special status in the US." [PIW 9-26-88:1] It was not clear what this meant.

The members could not agree to do what was feasible and profitable. OPEC could boost oil prices and increase short-term revenue by cutting output, but "the current standoff among members is almost a case of who will blink first.... Ironically, if market anticipation of ... a compromise arrests or reverses the price decline over the next few weeks, such a rebound may actually lessen the odds and the urgency of concrete action." [PIW 10-10-88:4]

But the urgency was manifest in a special meeting in late October, which prepared the agenda and probably a tentative agreement for a full meeting for November 21. Iran seemed ready to concede parity to Iraq, since its refusal of parity simply let Iraq produce at will, and Abu Dhabi seemed ready to cut back also. [PIW 10-31-88:1] Iran planned to expand capacity to 3.1 mbd in 1989 and to 3.6 in 1993, mainly through the long-delayed gas injection projects. [PIW 11-14-88:1] Iraq aimed at 5 to 6 million in the 1990s. [PIW 11-21-88:1]

In the meantime, reviving demand raised both prices and output, with fourth-quarter OPEC output looking to be around 21 mbd, "about 2 mbd more than forecasters were expecting just a couple of months ago." [PIW 11-7-88:1] Saudi Arabia produced 6 mbd, the highest in years. It was said to be only 0.5 mbd short of its actual capacity, far down from the 10 mbd of years ago. [PIW 11-14-88:1]

At the November 1988 meeting, "the big underlying worry ... was the threat of soaring production from Iraq by late next year." [PIW 11-28-88:1] Yet the meeting was a success. A new ceiling was set at 18.5 mbd, which, with expected leakage, would equal 19.0. This was 2.4 mbd over the previous program and allowed everyone to get more than previously. Iran gained more than most, as compensation for giving equality to Iraq, whose new quota of 2.64 mbd was far above its old quota but hardly more than the 2.60 it actually produced in 1988. [PIW 12-5-88:1] Nothing was said about the price. But spot and spot-linked prices (which meant practically all prices) promptly rose.

Expansion Brings Conflict, 1989–1990

The output expansion brought investment to the fore.[8] Table 8.2 shows a great decline in measured OPEC capacity, from 39.2 mbd (crude oil only) in 1979 to 27.4 at the end of 1989. [PIW 3-12-90:6] All reservoirs are always subject to two opposing forces: additions by drilling and connecting new oil wells, and enhanced recovery facilities, versus natural decline, a basic variable difficult to measure.

The Decline Rate, Four Gulf Producers

Line 1 of table 8.3 shows 1955 output, which equaled capacity, since there was then no restriction. Over the next twenty-five years, to the end of 1980, we credit each newly completed oil well with the national average output per well during the year of its completion. The aggregate for all wells is the total gross addition to capacity over twenty-five years (line 2). With no decline, capacity would have been as shown in line 3. But capacity was estimated by *Petroleum Intelligence Weekly* at a lower number (line 4). The loss is shown in line 5. Its percentage of the aggregate output, shown in line 6, is our best estimate of the average decline rate. In Kuwait, for example, for every hundred barrels produced in a year, capacity declined by 1.58 barrels.

In theory, the decline rate approaches the ratio of output to reserves.[9] This ratio, shown in line 7, is consistently lower than the estimated decline rate in line 6, suggesting that the published reserves of these countries, and probably of nearly all countries outside North America and the North Sea, include reservoirs or strata that are not being depleted. The bottom line indicates the possible overstatement of reserves; it ranges from 15 percent to 102 percent.

If my earlier suggestion is correct, that newly drilled wells tend to be bigger producers than average old wells, it follows that the true decline rate is higher than as estimated here and capacity is really cheaper to in-

Table 8.2
OPEC capacity, selected years, 1979–1994
(crude oil, mbd)

								Center for Global Energy Studies	
	1979	1983	1989	1990	1992	1993	1994	1993	1994
Saudi Arabia	10.84	11.30	7.75	8.50	8.70	9.00	10.20	9.30	9.90
Iran	6.99	3.00	3.10	3.25	3.60	3.90	3.80	3.90	4.10
Iraq	4.00	1.50	3.10	—	—	—	—	—	—
Kuwait	3.34	2.80	2.40	0.10	1.15	2.20	2.40	2.30	2.60
UAE	2.50	2.89	2.20	2.40	2.40	2.40	2.40	2.60	2.60
Qatar	0.65	0.65	0.40	0.40	0.60	0.45	0.43	0.40	0.44
N Zone	[a]	[a]	[a]	0.30	[a]	[a]	[a]	[a]	[a]
Total gulf	28.32	22.14	18.95	14.95	16.45	17.95	19.23	18.50	19.64
Venezuela	2.40	2.50	2.60	2.50	2.50	2.50	2.50	2.70	3.00
Nigeria	2.50	2.40	1.80	1.90	1.95	2.00	1.95	2.10	2.20
Indonesia	1.80	1.60	1.25	1.50	1.45	1.45	1.40	1.40	1.40
Libya	2.50	2.00	1.50	1.50	1.50	1.50	1.43	1.60	1.50
Algeria	1.23	1.10	0.75	0.80	0.80	0.80	0.78	0.80	0.80
Ecuador	0.23	0.25	0.31	0.30	0.33	—	—	—	—
Gabon	0.25	0.20	0.27	0.30	0.33	0.33	0.33	0.30	0.32
Total nongulf	10.91	10.05	8.48	8.80	8.85	8.58	8.39	8.90	9.22
Total OPEC	39.23	32.19	27.43	23.75	25.30	26.53	27.62	27.40	28.86
Natural gas, liquids, condensate	NA	NA	NA	1.98	2.16	2.29	2.45	NA	NA

Sources: PIW 3-12-90:6, 7, 1-7-91:7, 11-2-92:9, 7-5-93:7, 1-24-94:7; alternative estimates from Centre for Global Energy Studies (*Global Oil Report*) or International Energy Agency (*Monthly Oil Market Report*).

Notes: Natural gas liquids and condensate are not included in crude production or capacity. Iraq 1994 capacity is 600 according to PIW and 400 according to IEA.

The 1979 capacity in this table refers to preevolution Iran.

The previous table showed a loss of 4 mbd early in 1979.

[a] Equally divided, between Saudi and Kuwait capacity.

Table 8.3
Estimated decline rates, four Persian Gulf producers, 1955–1980 (output and capacity in tbd)

	Kuwait	Saudi Arabia	Iraq	Abu Dhabi (1962–1980)
1. Output, 1955	1,092	965	675	16
2. Capacity additions, 1956–1980	2,525	13,379	5,064	2,474
3. Total, end 1980[a]	3,617	14,344	5,739	2,490
4. Actual, end 1980	2,800	11,300	4,000	2,100
5. Loss, 1956–1980[b]	817	3,044	1,739	390
6. Percent of aggregate output lost (decline rate) per year	1.58	2.89	4.41	2.34
7. Average production/reserve	1.20	1.65	2.18	2.04
8. Line 6/line 7	1.31	1.75	2.02	1.15

Sources: Adelman and Shahi 1988; PIW 2-23-81:9.
Notes: Line 2 could understate additions because the average new well produces more than the average old well; this would increase line 6. Also, line 7 could be understated if published reserves exceeded reserves in actual contact with well bore and were being depleted. The correction would increase line 7.
[a] Line 1 + line 2.
[b] Line 3 − line 4.

stall than I have calculated in this book. These questions must be left for later research.

After 1980, output was severely restricted, lessening pressure on the reservoirs. Production per well might be reduced in fact or only in appearance, with some wells shut part or most of the year. Few will believe that the average production per newly drilled Saudi well really fell from 14,000 in 1980–1981 to 5,800 in 1991. I think it is a statistical artifact, but data that once permitted a check are no longer published. The days of open discussion (operation by multinational oil companies) are long gone. Some readers can substitute better numbers.

The decline after expulsion of the companies was greater than the natural decline. We cannot tell how much was due to less productive investment and how much to current underspending (neglect and undermaintenance). Above ground, equipment deteriorates. Wells need downhole cleaning by acidizing, fracturing, and workovers (partial redrilling). Maintenance and workovers are classified as direct operating expenditures but were probably neglected along with capital expenditures.

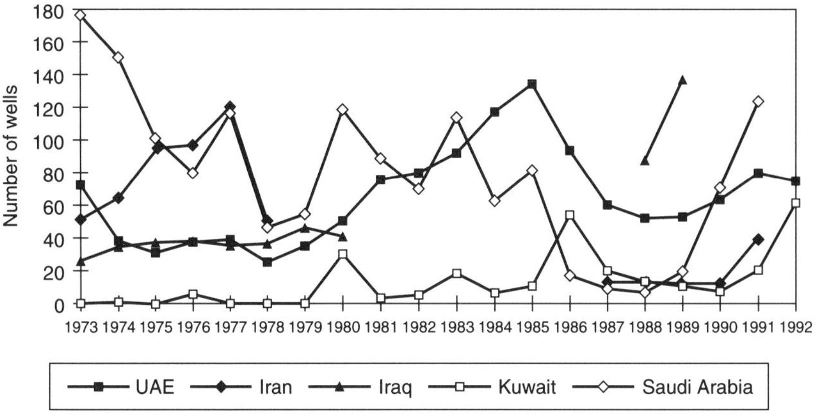

Figure 8.2
OPEC oil wells drilled in the Persian Gulf, 1973–1992. Source: WO:IOO.

Collapse and Revival of OPEC Drilling

Figures 8.2 and 8.3 show that drilling in Venezuela, nearly all by private companies, had dropped very far by 1976 because of the policy of no new concessions after 1958 and because of impending nationalization. But after nationalization the new national company PDVSA (Petroleos de Venezuela S.A.) invested and greatly increased reserves. Just as had happened earlier in the United States, there were few major discoveries but a great expansion of old fields.

Capacity grew substantially in Venezuela despite an unusually high decline rate of about 22 percent per year. But after 1982, long before the 1986 price crash, drilling collapsed for lack of money. It nearly ceased in 1986–1987. It was an achievement for PDVSA even to have maintained capacity. It found new high-quality fields, which could have been highly profitable but for a lack of capital and expertise.

In Africa, drilling declined through the 1970s, then severely after 1981. It was worse in Algeria than in Libya and Nigeria, which had partially private oil production.

At the Persian Gulf, Kuwait from 1973 to 1980 had only 4.5 completions per year, but drilling revived in the 1980s. Occupation and war stopped drilling from August 1990 through mid-1991, which makes the annual total all the more impressive.

Iranian drilling rose swiftly in the mid-1970s, then dropped to near-zero after the revolution. Iraq slowly increased to the eve of war with Iran

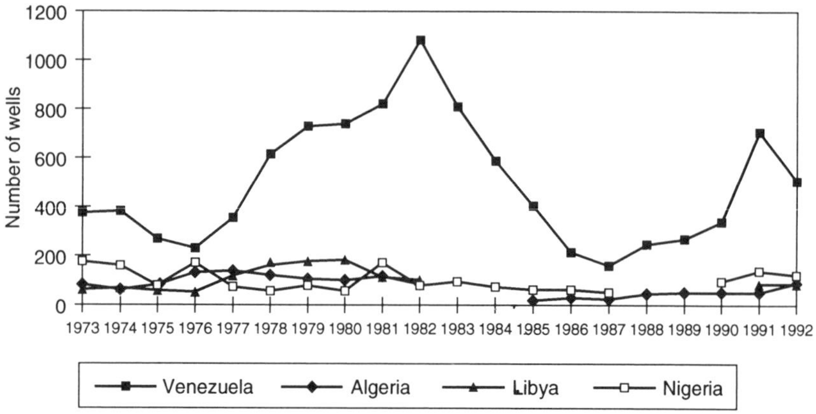

Figure 8.3
OPEC oil wells drilled outside the Persian Gulf, 1973–1992. Source: WO:IOO.

in September 1980. For the next eight years, the belligerents drilled few or no wells. After the war, it took three years for a revival in Iran. Iraq drilled an impressive 130 oil wells in 1989 and added perhaps 1 mbd capacity up to the attack on Kuwait.

The UAE profited by keeping private companies.[10] Alone in OPEC, these countries' drilling expanded, and their 1989 production exceeded that of 1979.

Instead of expanding as planned after 1973, Saudi Arabian oil well completions fell by half through 1985. From 1986 to 1989 there were only 13 oil wells completed per year; not until 1991 did oil completions reach 124. In 1992, Saudi Arabia stopped publishing the number, part of the general disappearance of information.

Barriers to Capacity Expansion

In 1987, I suggested that OPEC members would always provide some excess capacity for bargaining among themselves. [Adelman 1987] Beginning in 1989, members' announcements of programs became too numerous to mention, but it was soon apparent that the programs were going slowly or were reduced or postponed.

There have been many complaints of insufficient funds. In January 1990, it began to be repeated that $60 billion was needed over five years to provide an additional 5 mbd, or about $12,000 per daily barrel. Without the $60 billion, capacity would fall below demand by 1993! [PE 1-

90:6; NYT 1-26-90:D6; OGJ 2-5-90:NL1; PIW 2-5-90:7; WSJ 2-7-90:A7] Yet an OPEC average of $12,000 per daily barrel of new capacity exaggerates by a factor varying between 300 and 2,000 percent.[11] And even if the $60 billion were true, it would not be important.

Table 8.4 estimates the OPEC nations' development cost for 1990. It is upward biased because it includes exploration drilling and gas drilling, and it assumes the average new well is only as productive as the average old well. Of course, there are many year-to-year changes and sources of error in every item. Some readers will have better data. The point is that all components can be replicated and checked.

There is also an independent check, on the aggregate. For 1985–1987, the last three years of the Chase Manhattan Bank compilations, total Middle East–African oil production capital expenditures were $9,904 million. Estimated in table 8.5, they were $7,533 million, or 24 percent lower. The difference is easily accounted for. Our data exclude gas gathering systems, loading facilities, and, perhaps more important, local waste and payoffs.

Table 8.6 gathers a few publicly reported gulf projects for 1989–1990. Some of them are higher than as estimated in table 8.4 and some lower.

The $12,000 OPEC average and similar ridiculous numbers should be preserved as an example of the delusions ruling informed opinion. But even if true, $60 billion over five years would have been only 9 percent of 1991 OPEC revenues.

Line 12 of table 8.4 shows how trifling were the sums needed. But table 8.7 again uses Chase Manhattan as an independent check. Between 1976 and 1987, OPEC Middle East–Africa capital spending on oil and gas production was an average 1.7 percent of oil revenues, and in only three years did it exceed 2 percent of revenues.

It was cause and effect: a few years after capital spending ceased to be recorded, flights of fancy could begin.

The Flabby National Dinosaurs

The complaint about shortage of money is a clue to the real problem. Ibn Khaldun stated it 600 years ago: "Commercial activity by a ruler is harmful to his subjects and ruinous to the tax revenue." [Ibn Khaldun 1377, p. 232] Today, even after the collapse of socialism, most of the world's oil continues to be produced by governments, and most of these socialized enterprises are sloppy and corrupt. What they all have in common is irrational investment.

Table 8.4
OPEC nations: Development cost, 1990

	Abu Dhabi	Algeria	Indonesia[a]	Iran	Iraq	Kuwait	Libya[b]	Nigeria	Qatar	Saudi Arabia	OPEC Total or mean
1. Wells drilled, all types	51	80	738	24	178	13	98	80	19	98	1,379
2. Average depth of well	9,050	8,363	4,299	9,563	5,871	10,308	8,520	6,795	7,684	7,071	7,752
3. Drill and equipment cost: United States ($000s)	696	547	171	889	245	810	665	424	479	441	537
4. Investment/well ($000s)[c]	2,863	2,250	703	3,658	1,008	3,332	2,736	1,744	1,971	1,814	2,208
5. Producing oil wells	980	916	6,604	645	378	365	1,124	1,855	194	858	13,919
6. Production/oil well, db	1,758	1,289	221	4,961	7,275	4,389	1,290	987	2,108	7,293	3,157
7. Oil wells drilled	47	32	442	12	137	10	NA	53	12	71	816
8. Gross capacity added (tbd)	83	41	98	60	997	44	NA	52	25	518	1,917
9. Investment/db ($)	1,767	4,365	5,315	1,475	180	987	NA	2,668	1,480	343	928
10. Oil investment ($ millions)	146	180	519	88	179	43	268	140	37	178	1,778
11. 1989 oil revenues ($ millions)	10,020	7,000	5,716	12,500	14,500	10,863	7,500	8,700	2,000	24,000	102,799
12. Oil investment percentage	1.46	2.57	9.08	0.70	1.23	0.40	3.57	1.60	1.87	0.74	1.73

Sources: WO; API, Joint Association Survey; API, Survey of Oil and Gas Expenditures; OPEC-SB.

[a] The large number of producing oil wells makes it likely that the average new well is much more productive than the average old well. Hence line 9 is probably too high.

[b] Number of oil wells drilled unavailable.

[c] Drilling cost multiplied by 1.1×2.25 to allow for nonlinearity in depth-cost relation and for lack of supply/services network outside United States, as shown by wells/rigtime regression study. Drilling cost multiplied by 1.67 to allow for non-drilling cost.

Note 1: In the "OPEC Total" at far right, lines 1, 5, 7, and 8 are aggregates of the whole group. Lines 2, 3, 4, 6, and 9 are averages for the group. Hence one cannot derive a line for the whole group as one would do for an individual country.

Note 2: Calculations for Venezuela according to the above scheme are subject to unusually wide error because of the large number of small wells. However, data are available for a much better estimate, which follows:

VENEZUELA (from Annual Report, Petroleos de Venezuela) (Note: 47.7 bolivares per $US)

Capital expenditures ($ millions)	
Oil	804
Gas	420
Infrastructure	102
Oil, inc. infra.	871
Gross added capacity, tbd*	476
Investment/db, $	1,830
1989 revenues	13,313
Oil investment percent	6.54

*Net increase + 24.2 percent of production (average decline rate in 1987–1989)

Table 8.5
Middle East–Africa production capital expenditures

Year	Area	Wells drilled	Average depth	U.S. drilling cost per well ($T)	Total investment ($M)[a]
1985	Africa	658	8,295	584	1,660
	Middle East	868	6,703	393	1,474
1986	Africa	500	8,826	657	1,419
	Middle East	861	5,983	272	1,012
1987	Africa	475	8,368	450	924
	Middle East	667	7,277	362	1,043
Three-year total					7,533

Sources: WO-I00; Joint Association Survey.
[a] Drilling cost per well multiplied by 1.15 × 2.25 × 1.67 × (wells drilled).

Private investment is created and limited by expected profit. If there is money to be made, money will be found—if not by one company, then by another. Expenditures will be limited to projects expected to earn an acceptable rate. If an activity loses money, it will not last.

But a national company does not have its own assets subject to its own control. It cannot draw up a rational investment plan to maximize asset value. It invests to create capacity, not for maximum return but for jobs, contracts, and payoffs, as well as local amenities like housing.

In any government, one must build a political coalition to get a share of expenditures. In an oil-producing country, oil investment must get in line along with all other claimants to the cash stream of oil revenues. Each claimant tries to be a free rider, getting the benefit of oil investment while others cut back their demands. Spending for consumption, subsidies, and armaments tends to expand without limit. Oil maintenance and capital investment can always be deferred another year.

Between 1974 and 1982 Arab oil producers spent 35 percent of their oil revenues on their militaries, about seventeen times what they invested in oil production. During the next eight years the proportion increased. [Sadowski 1992, p. 8] The comprehensive study of Saudi Arabia by Askari [1990] confirms this. Much OPEC history is summed up in this example: "Algeria's revenue was used to keep the new privileged class comfortable. Investments in schools, housing, medical care, agriculture, and even the vital oil industry have stopped." [Yussuf Ibrahim, in NYT 1-19-92:E3]

Some kinds of overinvestment aggravate the waste. Fixing domestic oil product prices artificially low stimulates demand and "requires" refining-

Table 8.6
Comparison: Estimated and reported Persian Gulf investment per daily barrel, 1989–1990

	Abu Dhabi		Iraq	Kuwait	Qatar		Oman	Saudi Arabia
	PIW 5-14-90:3	OGJ 6-25-90:1	PIW 8-6-90:8	PIW 5-22-89:1	PIW 4-23-90:8	PIW 1-8-90:8	PIW 11-5-90:124	OGJ 4-9-90:21
Capital expenditures ($ millions)	500	500	70	1,000	300	95	500	3,000
Incremental output (tbd)	680	360	135	1,200	400	50	196	3,000
Reported investment: dollars/bd	735	1,389	519	833	750	1,900	2,551	1,000
Estimated investment: dollars/bd	1,768	1,768	180	987	1,480	1,480	2,269	343
Ratio, estimated: reported	2.4	1.3	0.3	1.2	2.0	0.8	0.9	0.3

Note: "Estimated" from table 8.4.

Table 8.7
OPEC oil production capital expenditures and revenues, Middle East and Africa, 1976–1987

Year	1976	1977	1978	1979	1980	1981	1982	1983	1984	1985	1986	1987	Totals
Middle East													
Total rigs	190	192	189	143	175	170	200	206	197	201	178	139	2,180
OPEC rigs	125	132	131	89	102	91	114	112	98	94	70	51	1,209
Percent	66	69	69	62	58	54	57	54	50	47	39	37	55
Africa													
Total rigs	138	173	175	200	219	212	209	150	105	118	97	90	1,886
OPEC rigs	122	153	149	166	180	160	151	102	70	82	74	66	1,475
Percent	88	88	85	83	82	75	72	68	67	69	76	73	78
M.E. + Afr. cap. exp.	2,375	2,290	2,250	3,120	4,100	4,820	6,730	5,970	4,530	4,010	3,160	2,770	46,125
OPEC rig percent	75	78	77	74	72	66	65	60	56	55	52	51	65
OPEC M.E. + Afr. cap. exp.	1,788	1,788	1,731	2,320	2,935	3,167	4,361	3,589	2,520	2,212	1,655	1,415	29,480
OPEC M.E. + Afr. revenues	118,565	129,627	123,916	187,193	254,952	236,581	182,784	140,668	129,189	117,019	74,882	92,138	1,787,514
Cap. exp./ revenues percent	1.51	1.38	1.40	1.24	1.15	1.34	2.39	2.55	1.95	1.89	2.21	1.54	1.65

Sources: IPE; OPEC-SB; Chase Manhattan Bank, *Capital Expenditures of the World Petroleum Industry*.

marketing investment. Petrochemical investment feeds the policymakers' ego and drains the economy, leaving even less money for production. A study of "resource-based industrialization" in four OPEC and four non-OPEC countries found that in all cases wealth was consumed not created. [Auty 1990]

Finally, the lack of engineering and management expertise aggravates the lack of money. True, during the international companies' tenure, host country nationals had filled the lower ranks and also had many higher-ranking jobs for years. But running a field is only the first step in making investment decisions to renew or expand. In exploration, development, and reservoir management the basic thinking had been done, and the data stored, in far-away memories and central corporate offices. Replacing them was bound to be slow.

To make things worse, foreign individuals and corporations were expelled from Iran and Iraq, and native-born technicians might fare worse. In Kuwait, advanced know-how seemed superfluous since vast low-cost oil fields remained in steady operation. Saudi Arabia was unique in keeping Aramco as a hired management team and only slowly easing out the expatriates. Its good sense has paid off. In Venezuela, there was constant tension between the PDVSA management, respected for its competence and honesty, and government plans and expenditures.

Bring Foreigners Back?

The problems of both money and expertise could be solved by bringing in foreign companies. Today, there are many newcomers, in addition to the former multinational majors. Entry into world oil has become far easier because markets are much wider and largely nonintegrated. Any group capable of exploration and development can enter because there is no worry about "finding a home for the oil."

In 1986, Algeria was first in OPEC to change its legal code to permit foreign participation, and it later improved the law. By late 1989, it had signed five contracts, although the results remained closely guarded. [PIW 10-23-89:7] There are more recent reports of discoveries, but after 1989, neither drilling rigs nor output increased.

By 1989, most OPEC countries were openly trying to get foreign oil companies back, particularly to renew and expand old oil fields. This stretch of old areas' reserves and production was the great achievement of American oilmen after 1930. [Adelman et al. 1983, chaps. 4, 6, 7] In a larger area it promised even more. But the OPEC effort was grudging when it needed to be strenuous. By the end of 1993, new foreign investment in the OPEC countries was still negligible. The OPEC governments could not bring themselves to call back foreigners on terms the latter would accept. Even when no equity participation was needed, Kuwait, one of the least inhibited, moved on feet of lead.[12]

Companies wanted to invest for the chance of an adequate return. They did not need to own the reserves. In fact, they had not owned them even before 1970. All they needed was the right to develop reserves and sell the oil as produced.

In time ingenious wording will get around the taboo on foreign ownership. The lawyer in Mozart's *Figaro* knew that with a synonym here and an equivocation there, some suitable mess of words could be concocted. But even when OPEC members agree in principle, they must agree on terms.

Investment in the OPEC countries requires a high rate of return after tax because risk is great. Companies were once expropriated, with meager compensation, and it could happen again. In some countries, some of the population is ferociously hostile, and murder is a form of political speech. OPEC quotas limit output and make it fluctuate more. By late 1992, in fact, Venezuela, Iran, Iraq, and Abu Dhabi were offering to exempt some foreign-owned operations from quota. [PIW 9-21-92:3] Particularly in Venezuela but also in Indonesia, some officials said "OPEC membership is actually a liability, particularly since the potential for quota obligations can scare away upstream investors." [PIW 10-26-92:1]

As for risk, one hears in OPEC, non-OPEC, and the former Soviet Union (FSU), "There is no risk; you know the oil is there!" But if the investment needed is too great, or the flow too small or uncertain, or tax too high or uncertain, the oil is not there.

Non-OPEC Countries

Non-OPEC countries are similar (and see above, pages 197–201). The legend is still powerful—that oil is a one-time resource, becoming more scarce and valuable, and better kept at home if possible to insure that our people will never do without. Private, especially foreign, capital aims only at profit and to steal our family jewels. Fetishism and xenophobia help local interests with political clout to preempt the good prospects. Safe from competition, they profit at the expense of the national income. Recently the Indian Petroleum Ministry decided "*in principle* to *consider* proposals for developing discovered oil fields on a joint-venture basis" with the state company. [NYT 5-4-92:D11, emphasis added] In other words, nothing will be done.

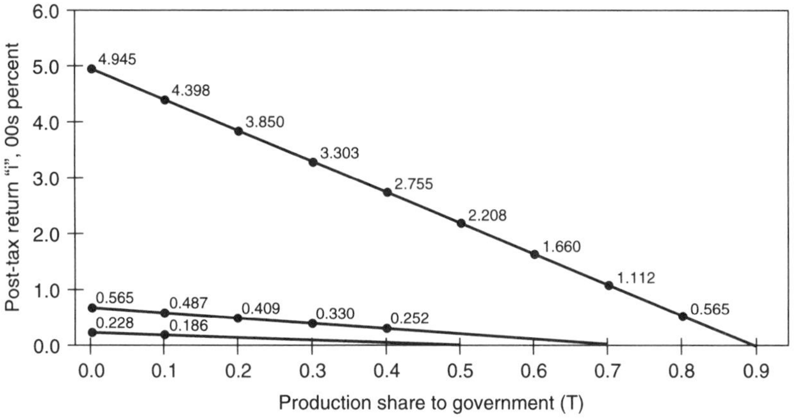

Figure 8.4
Rate of return, pretax and posttax, related to investment required (price $15/barrel).

As nations slowly shed these fantasies, and look at oil as a mere source of wealth, they must decide how to tax it. Figure 8.4 shows three hypothetical fields, with respective investment of $1,000, $7,000, and $13,000 per daily barrel. All are profitable. If they were taxed on income, all would be developed. But with royalties (production shares) even as low as 30 percent, the search in new areas is too risky, for the odds are that the small high cost fields are found first, and will be a deadweight loss. Even in a known area, at a 50 percent production share, the middle field is doubtful because the risk may be too high to justify investment for 17 percent.

The lowest-cost field looks like a sure thing, but it is not. In the relevant range, from 80 to 90 percent government share, the return is violently sensitive to small changes in the share. A small reinterpretation of the agreement, raising the share only slightly, rules out a highly profitable project. Wide margins are needed as insurance against repeating some of the history in this book.

The slow stirring of reason is seen in the failed projection of table 8.8. The National Petroleum Council can draw upon an incomparable richness of experience among its constituent oil companies. Its 1986 study assumed alternatively a $12 price increasing at 4 percent real annually and an $18 price rising at 5 percent real. It obviously (and I think, correctly) regarded the price drop of 1986 as being such a cataclysm or divide that one should not compute smooth functions across it. But at either $12 or $18 per barrel, it could see nothing but declining non-OPEC output. For the United States, its forecast trajectory was too low, but the implicit supply elasticity (.25) came close to the actual. But the other non-OPEC

Table 8.8
National Petroleum Council projections of noncartel output under lower prices

	1985 (actual)	1990 (projected)		1993 (actual)	1995 (projected)		2000 (projected)	
Oil price, real	25.67	14.04	21.88	11.08	17.08	27.92	20.78	35.64
Crude oil production (mbd)								
United States	8.9	7.1	8.0	6.8	5.7	7.0	4.5	6.4
W. Europe	3.8	3.3	3.5	4.8	2.6	3.1	2.0	2.6
Other[a]	9.7	6.5	7.6	11.6	6.6	7.9	6.4	7.5
NGL[b]	2.7	2.5	2.6	3.3	2.4	2.6	2.2	2.4
Total	25.1	19.5	21.7	26.5	17.2	20.6	15.2	18.9
Supply elasticities[c]								
United States			0.25	0.32		0.43		0.62
W. Europe			0.11			0.37		0.47
Other[a]			0.36			0.36		0.29
NGL[b]			0.11			0.18		0.19

Sources: PIW, 10-6-86:3; OGJ 10-13-86:40, 3-14-94-87; EIA, *International Petroleum Statistics Report* (July 1984): 8.
Note: Prices are for "total OPEC," published in MER. Because the NPC takes "$12 oil" as the basis for an increase at 4 percent, we use the price for "total OPEC," which was $12.21 in 1986 and $14.20 in 1993. Deflation by US:GDP:IPD, which increased by 28.2 percent from 1986 to 1993.
[a] Excludes communist blocs.
[b] Natural gas liquids.
[c] Elasticities computed by writer: $E = \ln(Q1/Q2)/\ln(P1/P2)$.

sources all produced substantially more even at prices half the 1985 levels.

The Cartel in 1989

In January and February, various non-OPEC producers resolved to help OPEC raise prices. [WSJ 1-27-89:C12, 2-22-89:A3] The skeptics advised: "Don't blink or you'll miss the non-OPEC cutbacks." [PIW 2-27-89:1] They were proved right. (For some far-fetched optimism see [PIW 3-27-89:SS].)

Pricing formulas became more precise. Among regions, the differentials over and above transport cost were small and transitory, and prices were highly correlated. [PIW 2-6-89:1] A special PIW survey found that both buyers and sellers preferred stability, but "buyers are no longer willing to give up the benefits of spot purchases. Hence ... term contracts ... mimicking the flexibility of spot markets ... now account for over half of world crude oil trade." [PIW 2-13-89:1] "Mideast exporters, using completely different formulas, [may] delicately set prices to undercut competitors by as little as 5c-10c a barrel." [PIW 7-24-89:1]

By late 1989, exporters were reverting to simpler schemes, based only on spot crude indicators, a *Petroleum Intelligence Weekly* survey indicated. [PIW 10-16-89:1] Venezuela long tried to avoid the spot market but was turning to it by late 1989, "thwarted by its cumbersome official posted price system." [PIW 9-11-89:1]

Rebuilding Capacity for Higher Quotas

Iran aimed at 4 mbd in 1989. [PIW 2-20-89:3] Its "highly visible ... ambitious reconstruction drive may be a first step ... to seek a higher oil production quota later this spring." [PIW 3-6-89:1] It achieved 3 mbd in 1989. In early 1990, Iran expected to be at 4 mbd capacity within a year "via foreign contracts." [PIW 3-26-90:1] It did not obtain the foreign firm contracts, and even to approach 4 mbd took not one but three years. Later, the Iranian oil minister thought term contract deals with large up-front payments would provide both markets and money. [PIW 4-10-89:5] It did not happen.

Beyond the Horizon

"For the long run, there is scarcely any way for prices to go but up.... In a few years, we should see a break-out of oil prices.... American imports should reach nine million barrels a day by 1991," and 12 to 13 mbd by the mid-1990s. [James R. Schlesinger, NYT 1-4-89 OpEd] (Imports had not reached 9 mbd by 1994.) The consensus view was that "global oil production capacity has been shrinking steadily, and oil demand is now rising faster than expected—[indicating] ... a much tighter international oil balance in the 1990s." [PIW 4-10-89:4]

PE [9-89:270] called a report by Michael Lynch [1989] "somewhat heretical." It was indeed unusual in five respects. (1) It was based on cost; (2) it recognized reserves as inventories and (3) the OPEC nations as revenue maximizers; (4) it forecast declining prices; and (5) it has been borne out. (I have had no occasion to change the view cited in chapter 7, note 26.)

In April 1989, both Kuwait and Iraq demanded higher quotas and backed up their words with deeds. Iraq restored the war-damaged Fao terminal, and Kuwait produced over quota. The GCC (Gulf Cooperation Council, of Kuwait, Saudi Arabia, the UAE, and Qatar) would meet in

May 1989 to discuss quotas. [PIW 4-24-89:1] The Saudis had given up some quota in November 1988 to get Iran and Iraq to sign on to the new agreement. Both would be asking for more by November 1989. Would the Saudis retreat again? [PIW 4-24-89:5]

Early in May 1989, King Fahd predicted the price would reach $26 before the end of the year. In Minister Nazer's language, he did not want the price to be "forcibly raised to $26 by reducing production," but adherence to production quotas "will allow the natural balance of supply and demand to work" to raise prices. [NYT 5-4-89:D1] Some may be able to distinguish these two scenarios. Saudi Arabia was openly exceeding its production quota, but the excess was only stored, not sold. [WSJ 5-5-89:A2] It might be called a first-phase warning. "Countries such as Kuwait, Iran and Abu Dhabi are clearly staking their claims for higher quotas by turning up the taps now." Iraq, surprised and discomfited by Iran's success in reaching 3 mbd, was also trying to raise its capacity. [PIW 5-8-89:1]

At the end of May 1989, Saudi Arabia proposed that quotas be increased but actual production decreased, with the cutbacks structured to favor the higher-capacity producers in the Gulf [WSJ 5-30-89:A6] It repudiated the $18 or any other target price, at least for the moment, and this was a shock. "If OPEC does not give the price signal, who will?" asked one industry analyst. [NYT 5-31-89:D1]

Kuwait's Policy Reversal

But Kuwait had made a historic reversal of policy.[13] Its "willingness to use its production capacity to press for its long-term goals could be a thorn in OPEC's side for the years to come.... It will rebuild its potential to 3.5 mbd over the next five years, with a $1 billion investment program. This reverses its early 1970s decision to reduce capacity in order to conserve reserves." [PIW 5-22-89:1]

Kuwait's recent capacity had been reckoned by the CIA at 2.2 mbd. Spending $1 billion to raise capacity by 1.3 mbd net meant $711 per additional daily barrel of gross capacity.[14] This is close to table 8.4. After the war, a consultant estimated the new investment needed to attain 2 mbd from ground zero at $1.25 billion, hence $625 per daily barrel. [PIW 7-15-91:7] The return on investment was about 980 percent.[15]

Because of discord over market share claims, despite rising sales, "a wide divergence of views" on prices, output, and quotas was expected at the June 1989 meeting. Kuwait and the UAE seemed "determined to overproduce if they are unsatisfied with new allocations." Saudi Arabia was "adamant" in refusing to give up its share of OPEC output. Its November sacrifice of 1.7 percent was not to be repeated. "The current rift between Saudi Arabia and Kuwait over price and market share goals seems to be fostering closer ties between Riyadh and Iraq." [PIW 5-29-89:1] This was to be proved a year later.

June 1989: Rising Demand and Market Share Struggle

The discord at the June 1989 meeting was even worse than expected. No matter how approached, "the problem is how to distribute any [production] increase among OPEC nations." Some countries wanted share to be held constant and any increase prorated. Kuwait demanded a larger share. [WSJ 6-2-89:B2] Saudi Arabia insisted on keeping its near-25 percent [NYT 6-5-89:D2] and supported prorating the increase. "In a show of brinkmanship ... the Kuwaiti oil minister has demanded that the quotas ... for Kuwait and the UAE be increased by some 30%." [WSJ 6-5-89:A3]

Iran supported Saudi Arabia against Kuwait. [WSJ 6-6-89:A2] The meeting ended in disarray after six days, after approving a production increase that Kuwait and the UAE refused to accept. Although both signed, the UAE minister laughed and said, "I always sign." [NYT 6-8-89:D1] Prices, which had risen since January, now turned down. Kuwaiti minister Ali al-Khalifa Al-Sabah called it "a gentleman's agreement" whereby his country would reduce output somewhat though not as much as the written agreement required. [WSJ 6-9-89:A2] He and Minister Nazer made a brave show of harmony, but Al-Sabah was considered the winner. The problem was permanent:

> OPEC has thought of overhauling its quotas for years. [But] oil is money—the stuff that pays the bills and buys the food.... And inherent in the very idea of an OPEC ceiling is that for one share to grow, another must shrink. Otherwise, everyone would produce at will and drive prices down for all of OPEC.... Kuwait seems to be usurping Saudi Arabia's role as the advocate of [price] moderation.... Indeed Saudi Arabia now wants higher prices.... Some ministers say, if OPEC members would just wait maybe five years, many would see their excess supply soaked up naturally. [WSJ 6-12-89:A1]

Minister Al-Sabah was cautious about the rosy future: "A consistent underestimation of potential supply and a consistent underestimation of the consumers' ability to adjust their demand led OPEC (and usually leads every other cartel) to overestimate its strength." [*OPEC Bulletin* 1–89:5] Kuwait was aggrieved because in March Saudi Arabia had signed a non-aggression pact with Iraq, without consulting any other GCC members. Iraq refused to sign a similar agreement with Kuwait.

June to September 1989

The June accord set the ceiling at 19.5 mbd, "below perceived demand to leave room for expected quota violations." Kuwait signed on only when promised a review of quotas and ceilings in three months. (It did the same thing in February 1993. [NYT 2-17-93:D1]) Rising demand and hoped-for decline in non-OPEC output built tension over market share. The paradox of higher demand's threatening lower prices was now familiar: "It is precisely the lure of rising oil demand that has opened the Pandora's box of market share for Mideast producers." [PIW 6-12-89:1] In September, "Iraq's rising export capacity ... will create new pressure on quotas. [Indeed, that was one reason for scheduling a meeting early, in September.] No point waiting until December, when Iraq would have staked out its position with overproduction.... Market share remains the [Saudi] top OPEC priority ... [but it] is no longer content with attaining only $18." [PIW 6-19-89:5]

But output did seem to decline [WSJ 6-27-89:A4], and Iraq was increasing export capacity more slowly than expected. [PIW 7-3-89:1] So was Iran; it was behind "but not irretrievably." Iraq could comfort itself with a goal of "6 mbd by the late 1990s." [PIW 7-10-89:3] Early in 1990, Iraq's oil minister claimed 4.5 mbd capacity, but no one believed him. He said the country had the potential for 6 to 8 mbd but had no plans to install it. Despite "financial constraints," Iraq was "spending $2 billion a year on the industry." As in Venezuela, the old state producing company had been dissolved and replaced by several regional companies. Proved reserve estimates were said to be based on 10 to 12 percent recovery of oil in place, hence too conservative. [WSJ 1-8-90:A1]

In mid-1989, OPEC capacity was expected to grow by 3.3 mbd by mid-1990, much more than the increase in consumption. This signaled greater excess capacity and more rivalry in getting larger quotas.

Forecasters continued to project deficient supply and higher prices. They pointed to OPEC members' use of excess capacity as a bargaining tool. "Individual country capacity is becoming increasingly important as OPEC members jockey for higher quotas and look for objective standards [reserves or capacity] for determining them.... Iraq, and to a lesser extent, Kuwait and Iran will have the potential to become swing producers over the next year." [PIW 7-31-89:1]

We will see that Saudi Arabia was very successful in restoring capacity, yet it was in financial straits and borrowed for the first time in thirty years. "Reducing government spending has been ruled out, lest serious social and political tensions emerge." Foreign financial assets were down from $140 billion in 1982 to about $40 billion, most of it not "readily available," and might be under $30 billion by the end of 1989. [PIW 8-7-89:1] In fact, it was probably smaller.[16] The ten-year pledge of a $500 million annual subsidy to Syria had just expired, and its renewal was in doubt. [WSJ 8-21-89:A6] In contrast, Kuwait continued to run current surpluses and add to foreign assets, estimated at over $85 billion. [PIW 8-21-89:5]

Saudi Arabia preferred to hold its quota but not exceed it and "lose ground to other OPEC members rather than stake a claim now for its traditional share of OPEC output, mainly because it fears undermining prices." [PIW 8-14-89:1] Indeed, the price had not weakened since the June 1989 meeting. [WSJ 8-17-89:A2] Iraq now aimed at higher output, with or without a higher quota. Higher export capacity was expected soon. [PIW 8-21-89:1]

By late August, OPEC production was up, but worldwide consumption was up even more, strengthening prices. "There are buyers for the oil and very little of it seems to be going into producer-owned storage." [WSJ 8-30-89:A2] In early September the Saudis were marketing more aggressively and producing 0.2 mbd above their 4.8 mbd quota. [PIW 9-4-89:1]

Venezuela announced it was aiming at 3 mbd capacity soon. [PIW 9-11-89:8] But OPEC worries were more immediate. "If the September meeting goes badly, Iraq may be tempted to use its rising capacity to stake out market share." [PMI 9-1-89:1] Kuwait, Libya, and Venezuela had some excess capacity "but difficulties in marketing their crude." In other words, they were overpricing, a barrier that could be quickly overcome. But for the moment, Iraq and Saudi Arabia, both with excess capacity, seemed not about to raise difficult and disruptive questions. [PIW 9-18-89:3]

As the September 1989 meeting approached, cheating had become "so widespread that it ... pushed OPEC's output to the highest level [of the] year—22 mbd." [WSJ 9-12-89:A2] The OPEC secretariat prepared studies of various "systematic" methods of quota setting. If based on capacity, it would benefit chiefly the Middle East producers. Members were already starting to lay the groundwork for claims of rising capacity to get larger quotas, but "verification [of capacity] would be an even thornier problem than it is in monitoring output levels." [PIW 9-18-89:3]

Still, the atmosphere was good and the time horizon short. "Few industry experts think the squabbling [over cheating on production quotas] will escalate to a danger point, at least not through the end of the year." And there was the usual comfort that "these new oil-producing areas already have begun to decline." [NYT 9-24-89:3-1]

The September Meeting

The squabbling nevertheless began immediately. Nobody's position had changed. [WSJ 9-25-89:A2] Kuwait would yield none of its claim for the additional output. Others opposed any increase in the output ceiling as long as the OPEC "basket" stayed below $18. Iraq waged a war of nerves, but no "quick surge of production" was feared. [PIW 9-25-89:3, 10-2-89:1] Three days of talks brought no agreement. Rather,

> Saudi Arabia has adamantly defended its 25 percent share of any ceiling the organization chooses to fix. Iraq wants a larger percentage than its 14.7 percent, and Kuwait and the Emirates want more than their present level of about 5 percent." [But all were daunted by] the difficult and complicated issue of reassigning new production levels to its members ... an exercise ... that the group had not been able to address with any success for the last four years. [NYT 9-25-89:D2]

A compromise fell apart at the last minute. [WSJ 9-26-89:A2] After "five days of sometimes acrimonious talk," they raised the production ceiling, thereby reducing the amount of cheating. The nations producing at capacity wanted an effective limit on output, which would raise prices; Kuwait and the UAE wanted more output for themselves. The effect would be to keep prices stable.

The meeting had been expected to begin work on a permanent allocation system, but "well before it began the ministers were backing off from that task because of its complexities." Iran offered a new scheme of division: it, Iraq, and Saudi Arabia would maintain current shares, while Kuwait, the UAE, and two others would receive larger shares and five others would receive smaller shares. "The hope was that the new system would help restore OPEC's credibility ... and thus firm up prices." The proposal was deferred. [WSJ 9-28-89:A2] Nevertheless, prices kept rising because the trade perceived "that demand for oil worldwide is increasing swiftly enough, and that OPEC's differences are not significant." [NYT 9-28-89:D1]

Minister al-Khalifa Al-Sabah of Kuwait summed up the year. "Everybody who could [overproduce], did; everyone who couldn't, complained about it." [PIW 10-2-89:SS]

From September to November

The future was still bright. Robert Mabro calculated that without a large expansion at the Gulf, oil would be in short supply by 1992, since non-OPEC production would decline and demand was growing. "There is no other way," agreed an American manager. [PIW 10-23-89:5] But for the next two quarters, demand was seasonally lower. "It's unlikely any member will want to reduce output during the struggle for market share." [WSJ 10-4-89:A14]

By late October, output had risen to nearly 23 mbd, some 3.5 mbd above the ceiling. Only a few members had excess capacity, which might in total be from 3 mbd to 7 mbd. The wide margin for error shows how loose the capacity estimates are. It was "enough ... to glut the market and cause an oil price collapse a few months from now if OPEC doesn't soon adopt a new quota system." The Saudis insisted on 24.5 percent, refusing to cut even one-half of 1 percent. The Iranian proposal was still being studied. [WSJ 10-23-89:A2]

"Saudi Arabia remains adamant that its quota share cannot be reduced, even symbolically." Some delegates believed that higher oil prices were weakening OPEC resolution. "If the oil market is sagging and looks likely to weaken early next year as demand slows seasonally, OPEC could find the political will to forge a compromise that has eluded it this autumn." [PIW 10-30-89:1]

Kuwait and Abu Dhabi were the principal gainers. [PIW 11-13-89:1] Obviously prices would have risen if they had not expanded output or if others had been willing to curtail. "The basic problem is finding a way to cut the current 23 mbd production level. But unless there's a solution to the quota claims of Kuwait and other countries, there is unlikely to be a viable agreement on total OPEC output for next year. And without that agreement, most ministers fear, prices could drop sharply."

Iran proposed raising the Kuwait and UAE quotas, but Saudi Arabia refused to accept any curtailment and remained "adamant that it will not waver from its claim to 24.5%.... The UAE remains the single biggest obstacle to a workable quota distribution." [PIW 11-20-89:1] The conflict persisted. Those without excess capacity wanted higher prices; those with excess wanted more volume. "The Gulf countries are unwilling to sacrifice output to provide higher prices for those members that now have little scope to raise production." [Ibid.; NYT 11-21-89:D5]

Rising OPEC capacity began to be more generally noticed. "One force driving OPEC nations to increase oil-production capacity undoubtedly is the notion that any rejiggering of quotas several years hence is apt to be based on production abilities." [WSJ 11-22-89:A1] In fact, the "rejiggering" had been a contested issue for at least three years.

Early in October, the Iraqi oil minister revealed that Iraq had no foreign companies producing oil, only some "engineering, design, and construction projects." [PIW 10-2-89:3] A spokesman boasted that Iraq could export more than 4 mbd (as against a 2.92 mbd quota), which "will help to make people think twice before violating the agreement." [NYT 10-10-89:D5] No one believed him. But with the addition of five fields currently under development, capacity might be over 4 mbd daily by 1991. [PIW 11-6-89:1]

"Breakthrough" at the November 1989 Meeting

The year-end meeting, brought forward to late November, was preceded by two months of private meetings among Saudi Arabia, Kuwait, Iran, and Iraq. Output well in excess of quota had been sold, yet prices had held. [WSJ 11-22-89:A5] Nevertheless, they feared slack demand in 1990. [WSJ 11-24-89:A2] Saudi Arabia would not retreat from 24.46 percent; Kuwait and the UAE wanted higher quotas to validate their higher production. [NYT 11-25-89:32] The trade was calm, except for worries over the lower spring demand. Again the meeting stretched out for a week. [NYT 11-27-89:D1; WSJ 11-27-89:A3]

The result was justly hailed as a breakthrough: "It bases production quotas on the ability of members to produce rather than on longstanding fixed percentages." Iran expressed satisfaction. [WSJ 11-28-89:A2] Kuwait received 6.8 percent instead of the previous 5.6 percent; Saudi Arabia accepted a cut from 24.46 percent to 23.90 percent and supported the $18 basket price. But Abu Dhabi still refused its 1.1 mbd quota, and its expected overproduction was "factored into the new ceiling calculations." [WSJ 11-29-89:A3]

There seemed to be a truce among the "Gulf countries, whose battles for market share and lack of unity have impeded any effective agreement for two years or longer." However, it pointed to weaker prices in the first quarter of 1990. The Iranian minister thought that quotas would be abandoned by 1991 because most members would be at full capacity then. More to the point, he claimed that Iran could produce about 3.65 mbd but would get to 4 mbd within two years and then to 4.3. [PIW 12-4-89:1, 4] This did not happen. "News of the agreement was interpreted on Wall Street as a sign that the price of oil would rise, increasing the value of the inventories of major oil companies." And Minister al-Khalifa Al-Sabah, "echoing a widespread judgment by a number of oil analysts ... said ... 'the quota issue will lose its relevance in the 90s.'" [NYT 11-29-89:D1] It is unclear what he meant; the minister was usually wiser than the experts.

But irrepressible Abu Dhabi soon announced that it wanted to go from its current capacity of 1.6 mbd to 2.1 or 2.2 mbd in the next few years. It had the advantage of foreign companies with equity production. Possibly it might need to offer the equity producers more than the current $1 per barrel margin. [PIW 12-18-89:5]

The First Half of 1990

Capacity Expansion

American contractors were said to have found a way to work in Iran, which would reach 4.5 mbd within two years. [PIW 1-1-90:1] Plans to raise capacity to 3.57 mbd by March 1991 were said to be included in the fiscal 1990–1991 budget [PIW 3-5-90:9], but capacity went barely above 3 mbd that year.

Saudi Arabia announced a capacity expansion "to over 10 mbd." Its current capacity was reckoned at 7.75 mbd (PIW), 7.0 (CIA), or 6.5 (the Japanese Institute of Energy Economics). One company estimated expenditures for expansion at $6 billion. "Potential contractors totally dismiss estimates by some Saudi sources that the expansion could cost up to $30 billion." [PIW 1-29-90:1] In March, Saudi Aramco announced it would raise capacity by about 2.5 mbd, at a cost of about $6 billion. [PIW 3-19-90:1] This would imply $2,400 per daily barrel, implausibly high, though very cheap. As usual, it cannot be checked as to amount or for nonproduction projects like the gas network.

Abu Dhabi increased its planned expansion. "Projects already on the drawing board should nudge short-term peak capacity to well over 3 mbd by 1994" [PIW 2-5-90:1], as compared with about 1.3 mbd at the time.

There was great frustration in Iraq: "Following a year of unsuccessful informal approaches to several majors, oil minister Al-Chalabi says that he is seeking non-production sharing deals with foreign partners.... However, as one oil [company] executive put it, 'We are not in the business of renting people for the cost of money.... We are looking for a much higher rate of return than a service fee can provide.'" [PIW 2-12-90:3]

Iraq offered the development of the Majnoon field, at the southeast border with Iran. It was said to hold 6 to 8 billion barrels and could produce 1 mbd. Petrobras had discovered the field in the 1970s but was not permitted to develop it. This history would raise the risk and minimum return. Lack of funds was chronic. This year only $3.2 billion was allocated to "oil, education, and health," and oil's share was uncertain. [Ibid.] Of course, only a fraction of the oil investment could go into production.

Other producers were also trying to attract foreign investment. Venezuela sought to keep "pace with output growth by major Mideast producers." [PIW 2-19-90:4] It failed. The North African states were also trying. The Algerian oil minister said crude oil capacity could be increased "quickly from current levels of about 700 tbd to 850 tbd." [PIW 2-26-90:4] It was not.

Some OPEC nations were doing better at expansion than others, and there was "a split in the group between the 'haves' and 'have-nots.'" [PIW 3-12-90:3] By March, prospects for a rapid jump in OPEC capacity were fading. In fact, up to half of the 6 mbd new capacity expected by the mid-1990s seemed in question. Budget constraints were blamed, as well as "the snail's pace at which oil companies and some producing-country governments are inching toward consensus on the kind of terms needed for the companies to help finance capacity development." [PIW 3-26-90:1]

In March and April 1990, Iraq sent teams to various countries to persuade investors to finance the development of some existing fields. They were presented as low-risk operations. But there was "a chilly response during oil minister Al-Chalabi's 4-day visit to Tokyo last week ... designed in part to hear proposals from potential investors." Ties were strained by some $10 billion in unpaid debts to Japanese companies, on which Iraq had suspended interest payment in 1986. [PIW 4-9-90:7]

Iraq also reported a deal with a consortium led by Occidental to develop Rumaila North (discovered before 1960 by IPC), to be paid for through a long-term crude oil supply agreement, as Iran was doing. Other international oil companies had been issued invitations in January 1990, but "most oil companies maintained they were not interested in service-type contracts, while the government was unwilling to offer exploration or concession arrangements." [PIW 9-10-90:7]

At this time, the Iranian oil minister said he was "near agreement with companies ... to develop established fields, with payment in crude.... The decision to seek foreign help in developing onshore oil fields is a clear policy reversal.... As recently as October [1989], Minister Aghazadeh said, 'Iran does not need foreign investment in oil.' ... His aim is to raise capacity from 3.3 mbd to 4 million next year [1991]." The aim is "to convince OPEC colleagues that Iran ... can turn up the taps unless it gets the price and quota that it desires." [PIW 4-16-90:3] Iran intended to spend $5 billion in hard currency on "its oil industry," including both oil production and natural gas. [NYT 5-28-90:29] This increase ended up taking about thirty months, not twelve, because it was apparently done without agreements with foreign firms, though discussions continued. [WSJ 10-19-92:A2; PIW 10-19-92:1]

In May 1990, Libya expressed confidence that it could raise capacity to 2.5 mbd "by the end of the year." Nine agreements with foreign companies were said to be signed or about to be signed. Industry observers thought it would take at least a year or two longer. [PIW 5-14-90:3] Even the skeptics were too hopeful. More than two years later, the goal had shrunk to 2 mbd by 1994. [PE 9-92:40] There was no progress. Abu Dhabi expected capacity to be just under 2.5 mbd by middecade. [PIW 5-21-90:3] It seems about on track.

In summary, by 1990 and afterward, the recorded excess capacity was so small that it was within the error of measurement. In and out of OPEC, the outlook was for supply growth:

> PIW's first worldwide survey of exploration hot spots shows that oil companies find themselves in an unexpected yet happy position: they are spoiled for choice [for the first time in almost 20 years].... Newly discovered, high-potential regions are competing for attention with large, established producing countries that are reopening to outside explorers after decades of dominance by national oil companies.... Even the Middle East ... is making tentative gestures toward Western companies.... Cash-strapped ... Iran and Iraq have publicly joined the competition [but] only as service contractors for known but undeveloped oil fields. [PIW 7-16-90:1, SS]

Expectations in the First Quarter of 1990

The trade did not fear price weakness. Internal projections surveyed by *Petroleum Intelligence Weekly* showed "a solid consensus ... on this basic outlook. Market power is seen swinging back to Mideast OPEC producers." [PIW 2-19-90:5]

The prime minister of Japan scheduled a ten-day Middle East tour, "aimed at strengthening Japan's relations with [Saudi Arabia] in view of the tightening of oil supply anticipated by the mid-1990s." It was scheduled to begin August 16. [*Japan Petroleum and Energy Trends*, 1-15-90:18]

In 1990, two dozen oil experts expected prices to rise by 50 percent by the start of 1995; the consensus price forecast was $30 in 1995. Supply could not expand enough to match demand. "The question about OPEC is ... whether it can expand its ability to pump oil enough to match rising demand, experts say." [NYT 1-24-90:1] The chairman of BP had predicted oil supply shortages by 1993 or 1994 "because of OPEC's inability to expand production," which could cost $50 billion to $60 billion, "an amount they cannot afford." [NYT 1-26-90:D6] They needed "some combination of loans, investments and technical help" from consuming nations. [NYT 1-26-90:D6]

The growing plight of the Soviet oil industry also made the trade more hopeful, although the production decline had not been fully reflected in Soviet exports. [WSJ 2-26-90:A2]

Another roundup in early March showed "a widespread consensus among petroleum experts that unless investments on the order of $60 billion are made during the next five years to increase production capacity in key petroleum producing countries, oil prices will surge." [NYT 3-6-90:D2] "Widespread consensus" continued to be gospel, above any profane demand for evidence. (For more of the rosy future, see [WSJ 2-20-90:A4, 2-22-90:C15; NYT 2-24-90:A1; OGJ 3-19-90:20, 3-26-90:43].) Minister Nazer of Saudi Arabia was more cautious: "The market will slowly but steadily turn into a sellers' market." [NYT 3-10-90:37] So was Shell: "Oil prices won't rise on the average much faster than the rate of inflation." [WSJ 4-3-90:C14]

In early January 1990, OPEC members began decreasing output [WSJ 1-19-90:A2], but firm prices persuaded them to stop reductions. [WSJ 1-29-90:C15] Prices then fell in February and March. In early March, there was some fear of a growing supply overhang, because futures prices had shifted into *contango*—that is, prompt supply was priced less than that further off. [PMI 3-8-90:1] Contango is an inducement to accumulate current inventories. The higher the degree of contango is, the greater is the perceived surplus.[17]

A Brief, Friendly Meeting

Saudi Arabia "appeared to be siding with Iraq against Kuwait in seeking higher prices." [NYT 3-5-90:D8] At the March meeting there was a general satisfaction, tempered by concern over quota violations, especially by Kuwait, which "with scant support, is advocating a rise in the production ceiling to stop prices from mounting above the $18 a barrel reference." [PIW 3-12-90:3]

Prices had been heading down since December and January, but the talk was all of whether to raise them. [NYT 3-16-90:D4] The meeting took only a weekend and "was most notable for its friendly and gentle tone." A shortage crisis was coming, although it was not imminent. [NYT 3-19-90:D1] If all countries respected their quotas, "we will have a stabilized market," said Minister Aghazadeh of Iran. [WSJ 3-19-90:A2] The price monitoring committee meeting was concerned with "a shift away from the problems of the 1980s—how to curtail production equitably and keep prices from collapsing—to a wholly new forward-looking agenda premised on tighter production capacity." [PIW 3-26-90:3]

Markets Awaken in April

But a week later, rising production seemed "to reflect a priority on securing markets and satisfying clients in disregard of production quotas and the slide in crude prices.... Both [Iran and Iraq] have had trouble selling crude and are offering some steep discounts." [PIW 4-2-90:1] The monthly IEA report suggested an inventory buildup, confirming the switch into contango a month earlier. [WSJ 4-4-90:A2] As competition intensified, there was some switching from term to spot contracts, particularly by Iraq and Kuwait. [PIW 4-9-90:1] A week later, "markets wake up knee-deep in oil, and prices tumble.... The basic problem is high OPEC production, which reflects efforts by several Mideast producers to stake out larger market shares.... This has led to intensifying price competition and points to market weakness through the spring." [PIW 4-16-90:1]

OPEC members were now reported "increasingly nervous," and an editorial in a Saudi newspaper was interpreted by some as an official threat of "another pricing war unless quota cheaters curb their output.... Kuwait is the chief target of the finger pointing in OPEC." [WSJ 4-11-90:C12] Minister Nazer called on members to adhere to quotas, and the ministers were "stepping up their contacts with each other." [WSJ 4-17-90:3]

Saudi Arabia, Kuwait, and the UAE, which had "been accused of contributing to the oil glut by producing more than their OPEC quotas," now issued a joint statement affirming their support for quotas and expressing grave concern over prices. [WSJ 4-18-90:C14] The next day the price fell by 70 cents. [WSJ 4-19-90:C14]

Kuwait and Abu Dhabi were expected to keep demanding larger quotas. "Iraq, Iran, and Saudi Arabia have previously used similar tactics of raising production to secure or preserve their large share of OPEC's production." [NYT 4-19-90:D1]

The Emergency Meeting

Saudi Arabia, Kuwait, and the UAE now demanded an emergency meeting. They sought

> an agreement about quotas for the second half of 1990 now ... rather than late May.... They tell PIW that a reallocation of quotas to more accurately reflect production capabilities is necessary for a workable agreement.... The $6 plunge in oil prices since January is putting untenable financial pressure on [other members]. Their only way out seems to be accommodating the main quota offenders. [PIW 4-23-90:1]

Iran and Iraq, producing at capacity, could not retaliate with increased production. [PIW 4-23-90:1] The hope was of a brief meeting, two days at most, that could give a firm assurance of output reductions. [WSJ 4-27-90:A2] It was held in Geneva rather than Vienna. An Austrian official had criticized Saddam Hussein for his threat to "burn half of Israel." The change was to rebuke Austria and show solidarity with Iraq.

The emergency meeting decided on a three-month cut to 22.1 mbd, the November 1989 ceiling. But Minister Mana Otaiba of the UAE revealed, after the agreement was struck, that his country had been producing not 1.9 mbd as generally supposed but 2.1 mbd. "In just one sentence, Otaiba torpedoed the whole meeting." [WSJ 5-4-90:C14] This conclusion seems exaggerated, but it suggests the hair-trigger sensitivity of the oil market. Prices certainly did fall on the day of the meeting. It was not made clear how much each member was allowed.

Still Good Times Ahead

Further ahead, there were blue skies. [NYT 5-6-90:F15] "Some oil ministers said it was only a matter of time before national production quotas—necessary in recent years of excess capacity—disappear just because most members will be physically unable to pump more oil." None dissented. [WSJ 5-8-90:A2]

A curious note was that "the Gulf Arabs maintain that sooner or later, all members will have to recognize that an end to the 8-year-old quota system is inevitable. They see the system as unmanageable with most countries at their production limits." [PIW 5-7-90:1] This statement is puzzling. Excess capacity concentrated into fewer hands seems to make output restriction easier. The few can restrict and simply disregard the others. Why even bother to hold an emergency meeting? The only explanation is that they wanted to force the members with little or no excess to accept some—that is, to cut back. Saudi Arabia had made good its claim to a market share, by continuing to produce, and forcing others to cut back and make some capacity idle, to make room for them. Kuwait and Abu Dhabi were imitating the Saudis.

Unfulfilled Pledged Reductions

OPEC members did cut rapidly but not to the full extent promised. Saudi Arabia cut but was not returning to the role of swing producer. "The Saudi move is contingent on output discipline from other members and will be reassessed in June." Iran and Iraq did have some excess capacity, as proved by their "continuing marketing problems." [PIW 5-14-90:1] Dubai Fateh, which (following *Petroleum Intelligence Weekly* and *Petroleum Economist*) I use as the successor to Arab Light since 1986, fell to an eighteen-month low.

On June 5, Saudi Arabia issued a statement in Riyadh and Washington reaffirming its commitment to abide by its quota. The statement stemmed from a sharp price drop following the confirmation of newly discounted pricing formulas, ranging from 25 to 70 cents per barrel. [WSJ 6-6-90:C6] Only a small fraction of the losses were regained the next day. "Saudis are known to be impatient with Kuwait and the UAE, chronic overproducers." [WSJ 6-7-90:C14]

By early June, the actual May production cuts were seen as only a fraction of pledged reductions, Saudi Arabia alone honoring its agreement to cut 430 tbd. "Both Iran and Iraq ... pumped at maximum." [PIW 6-4-90:1] In fact, "the main reason that OPEC's May cutbacks have been much smaller than expected is that Iran and Iraq pushed their production a combined 350 tbd higher last month." [PMI 6-7-90:1] Iraq's overproduction would soon be conveniently forgotten.

Formula prices were revised to new lows, "and it may soon force OPEC members to consider whether they are ready to relive the price wars of 1986 and 1988.... OPEC production remain[s] stubbornly high, particularly with flows rebounding from Iran and Iraq ... leaving Mideast producers such as Saudi Arabia with the unhappy choice of serving as a de facto swing producer or cutting prices to keep oil flowing." Both Kuwait and Saudi Arabia fell "under pressure to reduce prices." [PIW 6-11-90:1]

Late in June, the Kuwait oil minister, Sheikh Ali al-Khalifa Al Sabah, resigned to become finance minister. This was not believed to mean any policy change, but the historian loses an acute and well-spoken observer.

At this time, the "early promise of June OPEC cuts is evaporating." [PIW 6-25-90:1] Crude oil inventories touched their highest level in eight years, and despite the hope of relief from Iraqi threats and a Norwegian strike, "the hard fact [is] that there is a huge surplus of crude in the market, experts say.... OPEC strategists have wondered whether they are about to see a rerun of the 1986 price collapse." [WSJ 6-29-90:A3] June spot prices were at the lowest since September 1988.

Heads of State Take Over in July

Iraq ignored its own overproduction but warned Kuwait and the UAE to curb theirs. The deputy prime minister delivered the message in person and stated publicly that oil prices, which averaged about $14 a barrel, should rise to $25 a barrel. He was counting on the Saudis: "Saudi Arabia is no longer opposing a growing Iraqi role in the Gulf region's oil policies.... Last year, in a move that infuriated Kuwait, Saudi Arabia signed a non-aggression treaty with Iraq, leaving its Kuwaiti allies in the Gulf Cooperation Council to fend for themselves in their border dispute with Iraq." [NYT 6-28-90:D1] Kuwait had asked Iraq for a similar pact, which was brusquely refused. There is no evidence that the Saudis discussed the treaty with the United States, despite the supposed "special relationship."

Kuwait and Abu Dhabi said "they deserve higher quotas in light of their output capacity [and] the sacrifices they made for OPEC earlier in the decade." [PIW 7-2-90:1] But Kuwait "could be flexible on its demands." [NYT 7-6-90:D11] Early in the year, Iraq had gained sales from Kuwait by price cuts. Now both granted the buyer a special 70-cent allowance. [PIW 7-9-90:6] Obviously if both did so, neither was better off. Eight confidential company surveys showed oil markets threatening "to collapse from the sheer weight of excess oil supplies." [PIW 7-16-90:5]

By the second week in July, heads of state were taking over from their oil ministers to try to stop the price decline. Saddam Hussein of Iraq and the presidents of Venezuela and Indonesia were active, "but the King of Saudi Arabia ... seems to be emerging in the leading role." [WSJ 7-10-90:A3]

All five Arab Gulf producers met in Jiddah, Saudi Arabia, to agree to adhere to the November 1989 accord "until prices rise to an acceptable level," but they said nothing specific about Kuwait and the UAE. [WSJ 7-12-90:A4] Then prices "exploded" on a Saudi announcement that it would temporarily cut back its 24.46 percent share of OPEC output to 22 percent. Kuwait pledged to stay on quota, which one Saudi source said "took care of 50% of the problem. The other 50% was the UAE," which also agreed to adhere to its quota. The crude oil futures contract for August delivery went from $16.00 to $18.50 in a day [WSJ 7-13-90:A3] and held its gains over the next few days.

But disagreement was only deferred. Kuwait and Abu Dhabi wanted a 1 mbd quota increase in the fall, to let them use "the bulk of their spare production capacity." Iran and Iraq wanted stronger prices sooner. Saudi Arabia repeated that its yielding of market share was temporary. [PIW 7-16-90:1]

A week later, prospects looked good for "a short and sweet OPEC ministerial meeting" to ratify the Jiddah plan. But a stable fourth quarter

> depends on the willingness of Kuwait, the UAE and Saudi Arabia to forgo the hope of quota increases in the fourth quarter, which seems to have been a key factor motivating the original Jiddah compromise.... The escalation last week of Iraq's verbal assault on its Gulf Arab neighbors may be a symptom that the proposal is not as solid as it seems ... Iraq's threat ... was all the more striking in that it followed market indications that the UAE, Kuwait, and Saudi Arabia were already making [production] cuts.... "We all want prices back at $18 and are willing to take the necessary steps," observes one OPEC delegate, "but this spirit of cooperation could vanish if arm-twisting is used in an effort to achieve more ambitious goals." [PIW 7-23-90:1]

By this time, Iraq had become the chief player in the game.

Iraq: Savior into Pirate

After the 1972 expulsion of Iraq Petroleum Co. (IPC), Iraq profited, first by ignoring the 1973 production cuts ("embargo") and then by discounting prices. While Middle East output was flat, Iraq rose to a 3.5 mbd production record in 1979. These numbers speak well for their marketing and production. But it was one thing to expand existing underdeveloped fields. Iraq lacked the capability for large-scale investment in development and exploration. It let teams of French, Brazilians, and Russians discover fields, then pushed them out without development rights. Such actions eventually reduced Iraq's access to capital and management when it would need it most.

In 1980, with production and revenues at the peak, the Baath regime reached for a much bigger gain. The venture failed. Aside from dead and wounded, both countries lost heavily in oil revenue. But loans to Iraq from neighbors, Western suppliers, and the Soviet Union covered the costs of the war and left enough to develop impressive weapons and amass overseas cash to withstand over four years of embargo.

In 1990, the seizure of Kuwait offered a producing capacity and cash flow equal to Iraq and capable of great expansion. Even with Kuwait's foreign assets out of reach, the horizon would roll away.[18]

From Enforcer to Hijacker in 1990

By mid-July, Saddam Hussein was the OPEC enforcer. He was badly needed. The lowest spot price seems to have been as late as July 11. [WSJ 7-12-90:C12] Inventory buildup had ruined the prospect for the normal autumn demand pickup. "OPEC excesses pre-empt autumn stock build," headlined one trade weekly. [PIW:PMI 7-5-90:1] Exporters were "playing games" to hide discounts. [PIW 7-9-90:6] As late as July 16, after the accord of the five Gulf nations, markets were "teetering on edge of disaster again." [PIW 7-16-90:5]

Then Iraq began threatening Kuwait and Abu Dhabi, accusing them of the ultimate crime: being under American influence. "'Cutting necks is better than cutting means of living.' Saudi Arabia, Iran and Iraq have joined hands to bring about a greater sense of discipline to OPEC." The Saudi government warned that its "protection would not be extended to Kuwait in the face of Iraqi anger." [NYT 7-18-90:D1]

Oil experts called the cooperation of Saudi Arabia, Iran, and Iraq ("the Saddam factor") the most important event since 1979. [WSJ 7-24-90:A3] "The general accord of these three ... has set the stage for a credible policy by OPEC to push oil prices upward in the 1990s, experts say." [NYT 7-25-90:A8] One expert called this "a historic turning point"; another called Iraq "the OPEC policeman"; another called it "a landmark in the history of OPEC" [NYT 7-26-90:A1] or "a whole new ballgame," for which he thanked Saudi Arabia and Iraq. [WSJ 7-27-90:A2] This commentary is overblown, as usual, but an effective policeman would have been very important. In terms of table 7.1: if Saudi Arabia were the only one cutting back output, a price increase would cost it money, and it would veto the increase. If Iraq forced the totality of OPEC members to act together, the increase would profit them all.

The July OPEC meeting was short, limiting production and raising the target price of the OPEC basket from $18 to $21. "Discipline is guaranteed by a principal player which carries a loaded gun." [NYT 7-28-90:A1] Bloated inventories and stagnant consumption were overborne because "Iraq as the enforcer has become the key." [WSJ 7-30-90:A4] "The Saudis purred. They pretended not to see the gun Iraq was pointing at Kuwait's head." [*Economist* 8-4-90]

Yet the markets were skeptical. The *Oil and Gas Journal* noted a need for output reduction in addition to the Kuwait and UAE cuts. "Saudi Arabia, with token help from others, will trim output.... This is no small gesture. As they demonstrated in 1985–86, the Saudis take market share seriously. They'll want the ground they have yielded back when market conditions" permit. [OGJ 7-23-90:13, editorial] Others agreed: "OPEC, facing bloated inventories, faces long wait for higher prices." Excess supply was expected to bring prices down again. [WSJ 7-30-90:A6]

Up to this point, the "crisis" was on track. The challenge of oversupply and weak prices had brought the response of forceful action and rising prices.

The United States and Iraq

Recall that in 1986 Vice President Bush had urged the Saudis to cut back output to support prices. In 1990 his administration wanted "a rapprochement" with Persian Gulf countries, especially Saudi Arabia and Iraq, which "will be crucial to the USA's economic viability in the 1990s." Why the United States needed a rapprochement or anything else "to ensure that the oil will continue to flow" [PE 2-90:60] was, as usual, not explained.

Despite mounting threats and revelations about Iraqi weapons, the United States could see, hear, and speak no evil. Stroking a tiger would turn him into a tabby cat, it appeared to them. On July 25, a week before the Kuwaiti invasion, the American ambassador told Saddam Hussein that the United States sympathized with his "need" for more money (whatever it might be for), that some in the United States wanted an even higher oil price than the $25 Iraq demanded, and that the United States had no opinion on Iraq's border demands on Kuwait.[19] The ambassador was shocked only when Iraq occupied all of Kuwait.[20] [NYT 9-19-90: A29]

The message from Bush to Saddam Hussein parallels a message sent on the eve of World War II. President Roosevelt's April 1939 "message left Hitler with a feeling of oneupmanship over a man he should have regarded as potentially his most dangerous enemy. It does not bode well for the peace of the world when the President of the United States allows himself to be maneuvered into appearing as an inept and ignorant fool." [Watt 1989, p. 264]

U.S. help for Iraq was propelled by the self-flattering myth that the U.S. government would influence Saddam Hussein and change his behavior. The same U.S. ambassador to Iraq noted in April 1990 that "Iraq has modified its behavior and policies in large part because of our diplomatic efforts." [*Boston Globe* 6-15-92:1, 10]

Invasion, Blockade, and Dilatory OPEC

On August 2, Iraq seized Kuwait. The *Oil and Gas Journal* explained it:

> An OPEC accord that effectively lifts the group's second half quota by 400 tbd with demand soft and the world awash in oil stocks seemed more likely to collapse rather than support a $3 hike in the marker.... [Saddam Hussein] must also have recognized that it wouldn't work. So he invaded Kuwait. [OGJ 8-6-90:NL1, 19]

On August 5 came the UN blockade of Iraq and Kuwait. In January and February, military action drove Iraqi forces from Kuwait and forced it to agree to destroy chemical, biological, and nuclear weapons. This was the second time in three years that the United States protected Kuwait. (The first was in "reflagging" Kuwaiti tankers against Iranian attack in 1987.) Kuwait rightly gave us nothing for this protection, because our interest required us to give it. The consuming world could not tolerate any country having control of most Persian Gulf supply. Hence our supposed "special relation" with Saudi Arabia, trading protection for oil, was as superfluous as it was unreal.

Following the August 5 blockade, the price rose faster than in previous crises, to a monthly high of $31.55 in October. Much of the increase was due to deliberate OPEC inaction.

During the first half of 1990, Iraq and Kuwait together had produced 5.0 mbd, consumed 0.425 mbd, and exported 4.575 mbd.[21] The loss of their output almost precisely offset OPEC excess capacity, excluding Kuwait and Iraq. [PIW 3-12-90:6] But true excess capacity was probably 1.5 mbd larger than measured. *Petroleum Intelligence Weekly* had reckoned it for the gulf big five at 3.45 mbd. But just before the invasion, Sadad al-Husseini, an Aramco senior exploration and production vice president, had said the five had "more than 6 mbd" excess and were increasing their output. [OGJ 7-23-90:NL3] Less than 1 mbd of that 6 mbd excess can be ascribed to Kuwait and Iraq. Kuwait was credited with 2.4 mbd capacity and produced 1.7 in July. Iraq's capacity was reckoned at 3.1 mbd in February but produced 3.4 mbd in July. If al-Husseini is right, there was more than 5 mbd of excess capacity at the gulf, excluding Kuwait and Iraq. Thus the excess capacity exceeded the shortfall, by about 1.5 mbd.

Of course, the additional capacity could not simply be switched on overnight, but the market could take an interim strain because inventories were excessive. On January 1, 1990, oil stocks in the market economies had been 97.9 days of the previous quarter's consumption. They stood at 100.8 on April 1, 107.6 on July 1, 103.7 on October 1, and 104.4 on January 1. [DOE:IPSR, Tables 2.3, 2.4] Over and above this increase was the unrecorded gain in consumer stocks.

The inventory buildup shows that production exceeded consumption. Thus, the 1990 oil crisis was like the others: there was no shortage, but the *threat* of shortage generated precautionary demand for more inventories, which raised prices, which brought additional speculative demand. Expectation of a higher price is a self-fulfilling prophecy.

OPEC excess capacity was considered "ample to replace loss *if producers move.*" [PIW 8-13-90:1, emphasis added] OPEC could have prevented the jump in precautionary demand by publicly stating that it would take immediate action to produce to the limit as soon as possible. It did not do so. Saudi Arabia and Venezuela carefully dithered, seeking OPEC unanimity, which they did not need. By August 14, neither had even stated an intent to increase production. [WSJ 8-15-90:A3] Venezuela made "contradictory start-and-stop announcements on providing extra oil." [PIW 8-20-90:4] OPEC nations were "uncomfortable with the high levels of commercial stockpiles around the world." [NYT 8-16-90:A1]

Oil prices were helped up by "a Saudi request for an emergency OPEC meeting, without a date or time being established, and the report of *lower* Saudi production next month." [NYT 8-17-90:D1, emphasis added] But Saudi Arabia (and Venezuela) said they would increase output "to replace *much* of the 4 mbd lost ... [which] would slow the recent runup.... The Saudis also rescinded *indicated supply cuts* for next month." [WSJ 8-17-90:A3, emphasis added] The next day, a Saudi official who insisted on anonymity said the country "could increase its production by two mbd 'by tomorrow morning,' and could add an additional 0.5 mbd quickly thereafter." Venezuela promised a decision soon. [NYT 8-18-90:33] The next day, the Saudis said that they would increase by 2 mbd but not how soon or for how long. [NYT 8-19-90:1] A few days later, this became "as much as 2 mbd" [WSJ 8-23-90:A3] or "nearly" 2 mbd. [NYT 8-23-90:A14] It still was not clear a week later. [OGJ 9-3-90:31]

"Oil markets [had] remained jittery because neither the Saudis nor the Venezuelans had publicly declared their plans." Minister Nazer urged other OPEC countries to produce more. [WSJ 8-20-90:A3] President Perez "strongly suggested that Venezuela and other OPEC countries will soon increase production" but not quotas. [WSJ 8-20-90:A4] On August 23, Venezuela "appear[ed] to end three weeks of indecision" by announcing that September production would be up 325 tbd and December by 500 tbd. [NYT 8-24-90:D4]

A week after the invasion, the IEA estimated the September shortfall at 0.6 to 2.5 mbd, and on August 20, it raised its estimated gap to 1.4 to 3.2 mbd because of "little evidence that any additional oil has actually been pumped to replace the lost output of Iraq and Kuwait." [NYT 8-21-90:D1] There were varying reports on how much OPEC would produce above quotas—perhaps 3 mbd [WSJ 8-28-90:A3] or 3.6 mbd. [NYT 8-28-90:D6]

The UAE, "one of the most frequent quota breakers in the past, called on all OPEC states to strictly adhere to [stay within] quotas." The president of Iran said higher output would be treachery, but Iran raised output immediately, reaching capacity in two weeks. [OGJ 8-20-90:24] So did others, even while arguing it was not their mission to save the West from an economic crisis. Not "special relationships" but sellers' self-interest helped consumers. The head of the Purvin and Gertz consultancy attributed the price rise to delays by Saudi Arabia and Venezuela: "The market thinks the additional production just isn't coming on schedule, and the clock keeps ticking away." [NYT 8-23-90:A1]

Some senior OPEC ministers had an interesting excuse for keeping capacity unused: "The only thing keeping a lid on prices is the promise that OPEC can produce more when it becomes necessary. Once that capacity is spent, any added shortfall will create chaos." [PIW 8-27-90:9] Thus, producing *less* meant *lower* prices. The message of the sophistry was that they would produce less.

On August 29, OPEC finally met and lifted restrictions [WSJ 8-30-90:A3], but this agreement did not guarantee capacity output. At the end of the month, the IEA predicted a substantial oil shortage for later in the year. [NYT 8-31-90:D1] Such a perception probably underlay the resumed price increase. [WSJ 8-31-90:A3]

By early September, it was known that Saudi production for that month would be up by over 2 mbd, with smaller increases in Abu Dhabi and Venezuela. But "the Saudis offer no indication of likely production levels for the fourth quarter." [OGJ 9-10-90:24] Supplies were ample, but the mere prospect of fourth-quarter shortages raised prices [WSJ 9-6-90:A3], and there were reports of hoarding by distributors and retail consumers. [NYT 9-6-90:D4] Private economists were more pessimistic: "usable commercial crude oil stocks will be depleted by the end of October." [NYT 9-6-90:D1] In hindsight, of course, they were far wrong, but Saddam Hussein was still in Kuwait. When he issued new threats, there was a strong interaction of precautionary and speculative demand and a new price peak at the end of September.[22]

"Ridiculous Surplus"

In the first half of October, there were predictions of oil prices' reaching $45 before winter ended; Herman Franssen predicted $60 if war came. [WSJ 10-19-90:A2] James R. Schlesinger thought "a simple war," with no damage to Saudi fields, could send the price to $60. [WSJ 10-22-90:B3A] With more insight, the Venezuelan oil minister wanted an agreement by "some of the world's biggest oil consuming and producing countries" to fix the price somewhere between $25 and $30. [WSJ 10-19-90:B9] (There was such a meeting in Geneva a month later, but it took no action. [NYT 11-7-90:D8])

But prices dropped on October 19, and then it was all downhill. Soon Iran softened "price terms in an effort to increase flagging Western sales." [PIW 10-29-90:1] That week, Saudi production surpassed 8.2 mbd, and it was estimated that the lost output had been made up. [NYT 11-4-90:1] A week later, OPEC production exceeded the preinvasion ceiling, demand was "sluggish," and an oil expert said "OPEC could end up in the ridiculous situation of a production surplus even without Iraq and Kuwait." [WSJ 11-12-90:A3] And so it was: "Excess capacity is back. In November, for the first time since Iraq's invasion of Kuwait, OPEC members produced less than they could have, and sold even less than they produced." [PIW 12-3-90:1] IEA noted the same phenomenon. [WSJ 12-5-90:A6]

At this point, all that kept the price up was the uncertainty surrounding the expected war. Even so, contract prices, influenced by declining futures prices, were between June and July levels. The December OPEC meeting was openly worried about price collapse, even down to $10. [NYT 12-11-90:D1; WSJ 12-12-90:A2] Both Iraq and Kuwait attended as usual. It was agreed without dissent that quotas would be restored once the crisis was over. [WSJ 12-13-90:A2] By early January, "oil glut seems likely as demand slows, even if there is a short-lived war." [WSJ 1-8-91:A5]

There is no evidence that consumption had slowed, except as caused by recession, but there was an end to precautionary and speculative demand. The IEA had projected a fourth-quarter stock drawdown of 500 tbd; there had actually been a 300 tbd buildup. [WSJ 1-10-91:A16] No one should reproach IEA; its basic data were too delayed and fragmentary.

The war in January and February ended the uncertainty. The spot price had declined to $27.40 by January 16, the first day of the bombing of Iraq. Two days later, it was down to $16.40 and stayed in the $16 to $19 range. Essentially prices were back to where they had been in early 1990, before the trade had been scared by excess inventories. The speed of reversal was unprecedented.

Appraisal of the 1990 Crisis

This price upheaval started, like the other two, with excess supply and weak prices. This situation brought drastic action to restrict supply. Overnight, two major suppliers vanished. The upward price spiral was quicker than in 1973 and 1979. The supply loss was soon made up, but so it had been in the earlier crises.

But something was new. The price doubled from July ($14.03) to October ($30.86) and then declined. Several factors explain this. First, the Saudis acted differently. After a month's silence let the price rise, they increased output and let it be known they would keep it high. That was a far cry from 1979–1980, when their prolonged refusal to ensure more supply kept driving up the price for over a year.

Second, there were no price controls in the consuming countries. There was no incentive to buy crude oil or products at low controlled prices to hold for inevitable and guaranteed higher prices. Hence there was no additional kick to speculative demand. Third, the use of futures markets helped. Buyers bought futures contracts instead of bidding for physical barrels. Of course, this raised futures prices and made it profitable to buy now for sale later, but the effect was damped by postponement and hedging and stopped altogether as soon as the news ceased to be threatening. This appraisal is supported by the work of Weiner [1992].

Fourth, strategic petroleum reserves in the United States, Japan, and Germany were also important although not used. There were token sales after the war began and prices collapsed, when they were no longer needed. The reserves were saved for a "real physical shortage." This was nonsense. Shortage can happen to an individual but not to the whole market. Unless the price is fixed below the market-clearing level, it rises to equate the amount offered with the amount demanded. The price rise does the economic damage. If the SPR had offered large or unlimited amounts for sale (or options for future sale), it would have prevented the price upheaval.[23]

But knowing that the reserve might be used moderated the surge of precautionary-speculative demand. Perhaps even more important, the knowledge quelled the panic in governments. There was no feeling of a gun pointed at the decision makers' heads. The reserves bought time to take diplomatic and military action. Once war began, the price dropped to earlier levels so quickly that the OPEC nations had no chance even to discuss how to peg it at a higher level.

The war changed nothing. "Mideast nations are eager to purchase some of the powerful high-tech gadgetry that won the Persian Gulf war.... American defense contractors are salivating over the prospect of big Mideast sales." [WSJ 3-4-91:A1] The weapons trade has flourished.

Official Washington believed, with Daniel Yergin, "What we had before was a special relationship [with Saudi Arabia]. Now we have a more special relationship." [NYT 6-19-91:D1] They expected the United States would have "more influence in OPEC than any industrial nation has ever exercised.... They are now just beginning to discuss how they might use their new franchise.... If crude oil price plunged.... Washington might lean on a reluctant Saudi Arabia to cut production and push prices back up to a range of $18 to $22." [NYT 3-5-91:D1] Thus raising prices would demonstrate U.S. influence. Others thought that U.S. influence was shown by OPEC's not raising prices. [Hogan 1992]

In the real world, "reluctant" Saudi Arabia had months earlier cut output by 15 percent. When prices recovered it rescinded "most" of the cuts. [WSJ 1-22-91:A2] OPEC output for January was substantially down, for lack of demand. At the end of February:

> Opinions differ even in the kingdom on how much the Saudis should reduce output on OPEC's return to quotas. But there are signals from the kingdom's oil-policy makers that they will be willing to cut output as much as necessary to support a $20 or $21 price *so long as others in OPEC cooperate....* "We hope that we get more commitments from the others—and a better understanding of our sacrifice in not using all our capacity," [said a Saudi official]. [WSJ 2-25-91:A1, emphasis added]

In fact, the Saudis did not get the cooperation and were steadfast in refusing to be swing suppliers. The lesson had been learned. The supposed American influence is nowhere to be seen. But a former assistant secretary of state thought that "Saudi Arabia now almost assumes a Japan-like importance in the economic field."[24] [WSJ 10-26-92:A1]

The Postwar Market, 1991–1993

Figure 7.1 showed that prices in 1991–1992 look like those of 1986–June 1990. Figure 8.1 showed that measured excess capacity has stayed very low, yet by November 1990 members felt burdened by it and the feeling continued.

Pricing Methods Show Chronic Oversupply

The lines between spot and term contracts remained unclear. [PIW 7-29-91:3] But a *Petroleum Intelligence Weekly* survey found that term prices actually averaged below spot: "For the past four years or so, since formula pricing was introduced, the long-term nature of supply contracts appears to have been more valuable to producers than to buyers. The producers are effectively willing to pay for the insurance of having steady customers by discounting their crudes relative to spot markets." [PIW 1-6-92:1] This is evidence of an oversupplied market over the whole period, with the amount offered chronically in excess of the amount demanded. Moreover, an increasing number of suppliers signed frame contracts, which bound sellers, but not buyers, who had "no obligation to take the oil if they don't want it" and "without canceling the contract." Prices were set cargo by cargo, and the exporter benefited by keeping a wider customer list. [PIW 5-11-92:3, and SS] There is an obvious saving in transaction costs, but the buyer gained more.

There were price differentials according to cargo destination. In spot markets, premiums tended to be rapidly arbitraged away. But for term contracts on major Persian Gulf crudes, the advantage would last "anywhere from two to nine months at a time." In rising markets, Western buyers overpaid; in declining markets, Asian buyers did. Apparently buyers expected the premiums to balance the discounts. [PIW 4-13-92:1] This does not look like a permanent systematic difference (price discrimination). The prices of many crudes sold under term contracts were linked to one or another actively traded benchmark grade. For at least some important crude oils, the particular benchmark did not much matter; the result was nearly identical. [PIW 8-10-92:1]

Cooperation, Dialogue, and Interdependence

Japan briefly considered funding its contribution to the Persian Gulf war by an import tax. "Some Saudi and Kuwaiti oilmen are furious.... The energy tax proposal is seen as actively hostile." [PIW 2-4-91:3] The issue was kept alive at OPEC meetings. In March, some members said that "the group should publicly distance itself from *old policies of large volume cutbacks aimed at creating price rallies* in favor of an identity that emphasizes the need to prop up prices in time of surplus to ensure that capacity will be available to meet demands." [PIW 3-18-91:1, emphasis added] This is a useful reminder that (1) "old policies" had been "large volume cutbacks" to boost prices, (2) there was a "surplus" even with Kuwait and Iraq shut down, and (3) one could still play on consumer fears of not enough oil. In a May 1991 speech at Harvard, Minister Nazer called for "reciprocal energy security." In return for "even modest demonstrations of goodwill toward [Saudi Arabia] ... the US and other consumer nations would gain guaranteed access to a fairly priced ocean of oil." The modest goodwill would consist of cash to finance oil investment. [PIW 5-27-91:16] The terms of that investment, what "access" meant, or how it was to be "guaranteed," or what was "fairly priced," were all left unsaid.[25]

Consumer country viewpoints were in harmony. U.S. experts called low oil prices "bad." [NYT 3-12-91:D6] At a Paris meeting of twenty-one consumer and producer nations held in July 1991, the convening governments, France and Venezuela, said the conference would avoid pricing issues, "focusing instead on finance and investment." [PIW 5-27-91:10] At the meeting,

> Rising European and Japanese oil product taxes were the target of the most intense attack by Saudi Arabia, Iran, and other producer countries.... The producers shared their nightmare vision of billions of dollars in new investments—made partly at the behest of consumers—lying fallow because said consumers won't consume enough.... The softer message being whispered in the corridors was that the clear battle lines from the oil wars of the 1970s and 1980s are becoming increasingly blurred. [PIW 7-8-91:1]

Money invested "at the behest of consumers" is an artful suggestion that producing countries have invested to help consumers. "Clear battle lines" imply that there was organized resistance to the oil producers during 1970 to 1990. This could not be more untrue.

The IEA called a meeting of experts from producer and consumer countries. First scheduled for autumn 1991, it was held in Paris in March 1992. It also ignored prices and production but addressed "industrial cooperation" and funding for upstream and downstream investment. [PIW 3-2-92:6] And a meeting of ministers was called in Bergen in July 1992 to pursue similar topics. Meanwhile the Asia-Pacific Economic Cooperation Conference set up an energy task force, which held at least four meetings to discuss Asian energy shortages. [PIW 5-25-2:4] Oil was everywhere in surplus, but consumer representatives could see only shortage.

In September 1992 the Directorate for Energy of the European Communities issued a special report [EC 1992] and summarized the outlook for oil:

> Lessons of the 1990s: ... We shall be leaving the period of surplus and entering a period of sufficiency. The marginal barrel will matter, and the marginal investment necessary to produce that barrel. It is a decade of investments! In total, over the next decade, some $250 billion will be needed to bring supply into line with demand. Outside OPEC, some $80–100 billion will be required to sustain output of over 40 mbd. And in the OPEC countries, investment of over $150 billion will be needed in order to maintain output at current levels and to bring on stream the 9 mbd of extra production required.... Is the oil sector attractive to investors? [EC 1992, p. 31]

If non-OPEC output is maintained at 40 mbd, then assuming a low production decline rate of 6 percent (reserves fifteen times output), 2.4 mbd new capacity will be provided each year. If it costs $9 billion each year, that amounts to $3,750 per daily barrel. (See chapter 2, and appendix 8B to this chapter.) The cost, including 20 percent return on investment, is about $3.46 per barrel. Conversely, at a price of $15.00 per barrel, the rate of return is 132 percent. Contrary to its message, the directorate's numbers would be comforting if it offered any support for them. They are evidence only of a state of mind, to put it kindly.

Foreign Investment in OPEC

Reports of foreign investment in the OPEC countries became even more frequent after 1990.[26] But through 1993 there was mostly frustration.

Venezuela invited bids to reactivate old fields under service contracts, which did not require congressional approval. [PIW 2-4-91:3, 4-8-91:3] Nearly ninety bids were received, but only five bidders stayed because of the restrictions in the contracts. Negotiations dragged on over fees. [PIW 3-23-92:5; NYT 6-23-92:D4] But some contracts were signed in June. Foreign firms were guaranteed that production from the restored marginal fields would not be reduced by OPEC quotas.[27] [PIW 9-21-92:3] One deal fell through because Shell's insistence on international arbitration of disputes would have required congressional approval. [PIW 10-26-92:1] Another, the Cristobal Colon liquefied natural gas (LNG) project, seems to be only marginal at prevailing natural gas prices in the United States, the designated market, but perhaps viable for Europe.

There seemed to be continuing slow progress in 1993. "Strategic associations" were joint ventures with foreign companies that would put up capital to upgrade Orinoco crude. The presidential elections in late 1993 seemed to confirm the progress. But foreign investment is still peripheral. The state company PDVSA still receives insufficient funds for drilling and makes large, uneconomic investment in oil refining and petrochemicals. It has not reached its end-of-1991 target of 3 mbd capacity. However, in August 1994, Petroleos de Venezuela proposed to allow foreign investment in new oil fields. The taxes will be based on profits, not production. A summary account makes tax rates look unacceptably high [WSJ 8-23-94:A5A], but terms can be changed. If the Congress approves and negotiations begin, Venezuela may have taken an upward fork in the road.

In Kuwait, diplomats estimated the cost of rebuilding "the oil industry" at $5 billion to $10 billion—one-tenth the sums being bandied about two months earlier. [WSJ 5-16-91:A1] It was a small part of their foreign assets.

Putting out the fires and capping the wells exploded by Iraqi forces before they withdrew was expected to take as long as two years. [WSJ 4-5-91:A11] Although experts said they did not expect "significant" oil revenue for a year or more [WSJ 5-30-91:A11], by the end of June 1991, the Kuwaiti oil minister expected that more than half the fires would be out by the end of March 1992. [NYT 6-25-91:C4] In fact, all fires were extinguished and wells capped by November 6, in "less time than almost anyone expected, at less expense in lives and money, and with very few technical innovations." The first half of the work took six months, the second half less than two, with the cost variously estimated at from $1.5 to $2.2 billion. [NYT 11-7-91:A3] The incentive was strong: "The longer it takes Kuwait to get its industry up and running ... the less clout it carries in OPEC power structure, where muscle is measured by oil output." [WSJ 2-24-92:B3C]

But Kuwait had managerial problems. Had Kuwait Oil Company (KOC) installed downhole blowout preventers in wells, there might have been fewer fires. [PIW 5-6-91:1] It needed expert help for reservoir damage assessment and correction but lacked this experience itself. [PIW 9-30-91:3] In 1991, KOC was offered several equity deals but refused them on principle. BP, a former operator, had a natural inside track and asked for equity or production sharing but was refused. The earlier-than-expected capping of wells improved KOC confidence [PIW 11-18-91:3], but "Kuwait's lack of expertise in reservoir management" persisted, and an agreement was reached in March 1992, with BP reimbursement apparently in both cash and crude oil. [PIW 3-23-92:1] Some expected future costs were said to be three times prewar because of damage and the need for pumping and secondary recovery.[28] [WSJ 4-26-91:A1]

Kuwait recovery in production exceeded expectations. It hoped to increase light crude oil output, which was believed to require the help not of engineering firms but of oil companies, and there was no sign of lessening resistance. [PIW 7-13-92:1] Yet it remained hopeful of some kind of arrangement involving margins per barrel produced. [PIW 10-26-92:3]

As noted earlier, Algeria had been the first OPEC nation to begin trying to win back international companies. The prime minister's trial balloon, a proposal to sell off part of the oil fields, was quelled by "vocal domestic opposition." [PIW 7-29-91:7] No more was heard of it. By December 1991, Algeria claimed "expressions of interest from at least 19 firms in its plan to attract some $6 to $7 billion in cash bonuses for entry into enhanced oil-recovery programs," to increase oil production to 1.0 and then 1.2 mbd. [PIW 12-23-91:3] But nothing more was heard of this or of the ingenuous proposal that foreign companies take up to 49 percent of production of existing fields by reimbursing expenditures made by the national company Sonatrach. [NYT 12-2-91:D5]

In 1992, the returning oil minister, Nordine Ait-Laouissine, successfully proposed to privatize natural gas. "I had, 20 years earlier ... advocated the very measures that I was now proposing be overturned. But the national company was without money and the technical and human resources." [PIW 3-9-92:6] The minister was soon forced out. Privatization remained as an option but not a fact. However, there was no evidence by the end of 1993 of an actual increase in Algerian investment. Drilling rigs active in 1993 were one-fourth down from 1986. In recent years, Islamic militants have ordered foreigners to leave and killed some to encourage the others.

Iraq remained cut off from foreign investment after the war. But during August 1991, the United States suggested allowing the Kurds to use some of the Iraqi assets and funds held by Western governments "and to begin to exploit and export the huge oil resources in the north of Iraq." [NYT 8-20-91:A8] This would have added a formidable competitor, with large resources, anxious to attract investment and sell oil. It would have endangered OPEC. Nothing more was heard of it.

Iraq actively negotiated with foreign companies. In contrast to the

> flurry of talks before Iraq's takeover of Kuwait last August, this round is likely to bring equity positions for foreign companies and an active role for these firms well into the next century.... Officials want to move quickly and seem likely to accept

what was rejected out of hand [in April 1990].... Baghdad would probably accept an agreement ... which would look a lot like a standard production-sharing contract but would be called something else." [PIW 10-28-91:1]

Nine months later, a senior official said, "We are prepared to discuss all conditions, including production-sharing and profit oil, which will meet our requirement to increase production capacity to 5 to 6 mbd within four years." They were offering a production share of "5% and lower." [PIW 7-13-92:1, SS] A 5 percent share is around 80 cents per barrel. It must provide a high rate of return for the unusual risk. That is a hint of very low Iraqi costs.[29]

Petroleum Intelligence Weekly reported after an on-the-ground tour that much of the Iraqi oil production industry had been restored. This paralleled other reports of "reconstruction efforts, financed by multibillion dollar slush funds still held by Iraq largely in Switzerland." [NYT 7-27-92:A7] The cash reserve was estimated at nearly $30 billion before the war [NYT 4-25-93:14] and about $20 billion in early 1993. [Salomon Brothers *International Oil* 4-20-93] Iraqi borrowing from foreign governments (over and above that from private banks and corporations) was estimated at from $86 billion to $90 billion. [WSJ 2-25-91:A5; NYT 3-1-91:A8] It took skill to divert such a large proportion to safe haven.

One cannot tell how soon Iraq will resume large-scale production. Saudi Arabia wishes to delay this as long as possible, and the Clinton administration is sympathetic, as its predecessors were.[30] The new Iraqi willingness to accept large-scale foreign equity investment is unique among OPEC governments. If it follows through, it will probably force other OPEC members to do the same. But one cannot be sure anything will happen. In 1972, Iraq set a goal of 6 mbd for 1980. [NYT 3-22-75:41] In 1982, it set the same goal for 1990. [OGJ 1-4-82:73] In 1993, it set it for 2000 A.D. [PIW 8-2-93:6] The goal was as feasible then as now.

In May 1991, Iran set out a goal of 5 mbd by 1993. [NYT 5-27-91:35] It sought "long term crude sales contracts on favorable terms to companies that invest in exploration and development in Iran. However, [the minister] ruled out any equity or production sharing agreements.... Talks mainly involve existing discoveries that they can't afford to develop on their own." [WSJ 5-30-91:A6] There was even an NIOC-Total letter of intent on offshore oil field development. [WSJ 5-31-91:A10] Late in 1991, three offshore projects seemed "all remarkably close to equity deals." [PIW 10-14-91:1] Oil minister Aghazadeh called them service contracts with a payout in crude oil.

What actually happened was a series of relatively short-term loans. Iran had no debt at the end of the war in 1988 and as late as 1991 was debt free. [Hunter 1992] But by the end of 1993, it had incurred $30 billion of debt by borrowing and using suppliers' credits, for restoration of all kinds, including oil. Financing of this type tends to be expensive. By the end of 1992 some $10 billion of interest and principal payment was in default. [Indyk 1994, 6; PE 2-93:3]

No development contracts are known to have been signed. Iran and the companies were still far apart on terms. Operators wanted to recover costs by crude liftings within five to seven years; NIOC wanted up to 18. The fields would revert to NIOC after full recovery of all costs and a one-time return. [PIW 5-4-92:1] Two months later, Iran was said to be receptive to onshore exploration by foreign firms. Again, equity could not be considered, but some risk-reward element might be. [PIW 7-27-92:1] By the end of November 1992, Iran expected to "conclude negotiations soon on some" projects. "Wartime neglect, along with damage to surface facilities from Iraqi attacks, did greatly reduce pressure in oil reservoirs over the past decade." [WSJ 11-25-92:A1] Nothing had been done by the end of 1993.

By their own efforts, the Iranians have made substantial progress. In 1989 they aimed for 4 mbd in one year, and they almost made it in four years. In 1991, they planned to spend $2 billion (by March 1993) to increase onshore output by 650 tbd. [WO:IOI 8-92:98] In October 1992, Iran produced briefly at 4 mbd. This was not yet sustainable capacity. It aimed to "convince OPEC colleagues that Iran—like Saudi Arabia—can turn up the taps unless it gets the price and quota that it desires." [PIW 10-19-92:1; WSJ 11-5-92:A5] But even the forecast of 4.5 mbd soon [WSJ 10-19-92:A4] was treated with respect because Iran permitted a visit by foreign journalists and gave estimates by individual fields, onshore and offshore. [OGJ 11-9-92:37] Such detail had not been released since 1978, although it was not audited.

Moreover, NIOC aimed to install sixty drilling rigs. [OGJ 11-9-92:37] The previous (1978) high had been only 31 rigs. [AAPG-B 10-79:1894] After zero rigs in 1980–1981, there had been a gradual climb-back. By early 1993, forty-eight were reported operating. [OGJ 5-17-93:76] Forty-plus rigs would promise a steady growth in capacity, but the rig number collapsed to ten in early 1994, a clear signal of trouble.

An underlying 10 percent annual decline rate was mentioned. Minister Aghazadeh said average output per well had steeply declined since the peak of the 1970s but gave no data. He expected capacity to exceed

5 mbd by 1999. This now seems out of reach, along with "enormous" prospects to explore. [PIW 10-26-92:7]

OPEC: Market Share Contention in 1991–1993

Capacity as a bargaining tool was again a major theme from the start of 1991. "Additions to capacity should ensure that Iran is better placed to argue for larger OPEC quota allocation the next time the issue is discussed." [PIW 1-28-91:3]

The first postwar meeting was in March. Despite a 97 percent capacity utilization rate and "despite all that has happened since Iraq invaded Kuwait ... [OPEC] delegates are making many of the same old arguments about the same thorny controversies that plagued them through most of the past decade." Should they restrict output, or was the price perhaps too high? Should the low-population gulf producers make disproportionate cuts? Who was to blame for excess capacity? Should quotas be changed? But there was one constant: "Saudi Arabia is adamant in its refusal to take on the swing-producer role even after the war." [PIW 2-25-91:1]

As a glut loomed, OPEC recognized a need to cut output collectively for the second quarter, but was at a loss of how to do it. [PIW 3-4-91:1; NYT 3-11-91:D1] The Saudis would not consider going down from the current 8 mbd to 7 mbd. If others insisted, there would be no agreement. [WSJ 3-11-91:A3] As in the past, the key issues were "whether oil glut and price collapse threaten and, if they do, who should bear the burden of fending them off." [PIW 3-11-91:1] The final decision, which was to cut by 5 percent with the Saudis keeping 8 mbd, caused discontent among the others.

In form, quotas were suspended after the invasion. Although published, yet they were not considered quotas because they were "voluntary." [NYT 3-13-91:D1; WSJ 3-13-91:A3] The pretense continued through November 1992. [WSJ 11-30-92:A2] In fact, total output and its division continued to be an obsessive interest of the group, no matter what the formal agenda (or lack of it) at the meetings.

OPEC's "seemingly intractable conflict over how much each and all members should produce" remained. Any attempt to base quotas permanently on capacity would destroy OPEC, said two nongulf delegates. "Can the Saudis really afford to have us quit [OPEC]?" [PIW 3-18-91:1] Iran made "a stunning departure" from past policy to become a Saudi ally, a pricing "moderate," no longer calling for production cuts. [WSJ 3-18-91:A5] Nobody bothered to recall that only eight months earlier, "the general accord" of Saudi Arabia, Iran, and Iraq had been a new dawn for higher prices. [NYT 7-25-90:8]

After the March 1991 meeting set the temporary quotas, "the burden of excess supply facing spot crude markets" kept prices weak. [PMI 4-5-91:1] The estimate of excess capacity was near zero, but there was no way to test it. Saudi Arabia estimated its capacity was 8.5 mbd. Contractors thought sustainable capacity was below 7.5, but they were interested parties, "given the potential for lucrative contracts if more work is done." [PIW 5-20-91:1]

At the June 1991 meeting, "the conflict created by a shortage of customers now and a potential shortage of capacity later on could strain relations ... between Saudi Arabia and Iran." [PIW 6-3-91:1] The Saudis wanted cooperation and output restraint now; the Iranians said the Saudis could restrain now and wait for the approaching shortage. There was worry over phasing in Kuwait and Iraq production, but neither issue had to be faced just then. [NYT 6-6-91:D6] It was "the shortest conference in more than a decade" and simply extended the production ceiling and the nonquotas. But sluggish demand led to inventory buildup and worry over prices. [WSJ 7-25-91:A2] Kuwait produced a little, and the possibility of resumed Iraq exports depressed prices. [WSJ 7-30-91:A2]

In September, Saudi Arabia demanded a higher OPEC ceiling, which would raise its nonquota to over 8 mbd. [WSJ 9-19-91:C14] "Some insiders" worried over inadequate maintenance of wells. [PIW 9-23-91:1] The ministers disagreed on what level of demand to expect but were gratified by a price increase since early September. [WSJ 9-24-91:C18] At the meeting, "Securing agreement from the others to supply 23.65 mbd—even given that this level implies little, if any shut-in capacity—was no easy task.... Riyadh managed to hold most of its ground, insisting that it would produce 8.5 mbd [whatever others did]. As always, the smaller producers would be expected to continue pumping at capacity." [PIW 9-23-91:1] The last two sentences are not consistent. If smaller producers stay at capacity, the Saudi ultimatum is flouted. Yet the Saudis made an issue of saying they would not go below 8.5 mbd. [NYT 9-25-91:D10] This could only mean that others must cut. It explains why other ministers were hostile to Minister Nazer's refusal to budge. [NYT 9-26-91:D1; WSJ 9-27-91:A4] "All 13 countries now are living with a strong revenue imperative" [WSJ 9-30-91:A1]—as though they had ever lived with anything else. They could all hope for a higher demand to bail them out, but they could never count on its prompt arrival.

The OPEC nations asked consumers for oil development aid. [WSJ 4-9-91:A3] In August 1992, the OPEC need was raised from $60 billion to $80 billion to provide 5 mbd "in the next five years," and non-OPEC was said to need $170 billion. [OGJ 8-31-92:25; WO 9-92:13] These big, round numbers are invulnerable to any analysis because there is no hint of how they were arrived at.

In December 1991, there was the usual far-off glow of the day when demand would rise, perhaps beyond "technical capacity to produce." [NYT 11-25-91:D2] But there was fear of weak prices in spring 1992. [WSJ 11-25-91:A2] Accordingly, the end-of-November meeting agreed to continue at current production levels through the first quarter of 1992 but to cut in the second quarter. The nonquotas were expected to become quotas. [WSJ 11-27-91:C12] Saudi Arabia stated that its higher level of output since the invasion was permanent and that capacity would be raised to 10 mbd by 1994. [NYT 11-27-91:D1] There was "fear of a stormy spring.... All, including Saudi Arabia, agree on the probable need for production cuts. But the abnormally fuzzy outlook for both supply and demand reinforced the usual conflicts ... [and left] the group deeply divided on how best to cope with this prospect." [PIW 12-2-91:3]

IEA now reduced its estimates of expected demand: "The market consensus is that too much crude is being produced." [WSJ 12-5-91:A2] The Saudis refused to be the only ones to cut back, "holding oil taps wide open in soggy market.... Oil prices may have dipped by several dollars in the past few weeks, but don't expect [production] cutbacks from Saudi Arabia as a result." [PIW 12-16-91:1] Algeria called for an emergency meeting but was rebuffed. [NYT 12-24-91:D1] The Algerian oil minister suggested a 10 percent OPEC production cut. Otherwise, some experts thought, prices would fall by $4. [WSJ 12-26-91:C12] This implies a short-run demand elasticity of about .125.[31] These and similar estimates are a little higher than those assumed in table 7.2.

The Saudis were not bluffing; they increased December output. Iran cried "treason," but the Saudis repeated that they would not cut output unless all other members did so. [WSJ 1-7-92:A3] Venezuela cut by 2 percent and said it hoped others would. [NYT 1-11-92:33] Others did but very little. "Forecasts of a sharp drop in exports from the former Soviet Union ... have not materialized." [NYT 1-21-92:D5] But on the eve of an emergency meeting and after some more announcements of small cuts, the Saudis offered to cut 100 tbd, or 1.2 percent. [WSJ 1-22-92:A2] Called in haste, the December 1991 meeting was "the most important for OPEC since December 1985." One estimate was that a cut of 2 mbd (from 24 to 22) would raise the OPEC basket from $17 to $21 [WSJ 2-10-92:A2], which implies a surprisingly high short-run elasticity of .157.[32] "Saudi Arabia will likely threaten to withdraw from the agreement and [produce more] if there is significant cheating by other members." [PIW 2-10-92:1]

"Capacity is king in OPEC efforts to overhaul quotas." The ministers agreed unanimously to cut output to raise prices, but the Saudis immediately said they wanted to retain their 35 percent of OPEC output. They recalled a bitter past when they were the only ones to honor an agreement to cut. They professed to want to keep prices "at a modest level—not much higher than ... $18." [NYT 2-13-92:D6]

The Saudis wanted acceptance in principle of capacity or production as the basis for quotas. There was the rub. "Smaller members of OPEC such as Algeria ... want Saudi Arabia to account for most of the cutbacks because its production increased the most in the wake of the Persian Gulf crisis." The Saudis hinted they would be willing to cut to 8 mbd but only if others matched the percentage cuts. [WSJ 2-13-92:A2] After a few days' wrangling, OPEC agreed to reduce total output from 24.2 to 22.9 mbd, of which Saudi Arabia claimed over a third. The accord was viewed as "shaky." [NYT 2-16-92:19] Some wanted to come down to 22.5. [WSJ 2-18-92:A2]

It seems like a remarkably narrow range of disagreement, yet it had stretched out the meeting and was never resolved. The Saudis kept insisting that their capacity was 9 mbd and backed this up by deliberately producing that much in early February 1992. It was "a demonstration of capacity tied to the kingdom's demand during the Geneva OPEC meeting that new quotas and prorata cuts be based on production potential." [PIW 2-24-92:1]

Obviously governments were overstating both capacity and production to get higher quotas. [PIW 2-24-92:1; see also PE 10-92:54] Minister Aghazadeh said Iran would have 4 mbd by March 1992, "but it doesn't mean that we will need to produce at full capacity all the time. That depends on demand." [PIW 2-24-92:7] Like Saudi Arabia, Iran was using capacity for bargaining power.

There was general disbelief of the 4 mbd capacity, but later Iran came close to it. In March, Saudi Aramco said it was awarding contracts for development to bring total capacity near 10 mbd but did not name a date. Current sustainable capacity was estimated officially at 9 mbd, unofficially at 8.6 to 8.7 mbd. [PIW 3-30-92:1]

The cuts agreed upon in February amounted to only 740 tbd, of which Saudi Arabia accounted for 500, short of the 1 mbd "seen as necessary to stabilize markets." [PIW 4-6-92:1]

Prices were up in early April 1992, on reports of actual OPEC March cuts of 0.9 mbd. But production was still considered too high, and Saudi Arabia, it was believed, would rebuff any suggestions that it produce less than 8 mbd. It had cut enough; Iran had not. "Virtually all members agree that production restraint is still needed to preserve the fragile positive psychology found in oil markets." Moreover, Kuwait was recovering faster than expected, reporting 920 tbd for the third week of April, and others would have to make room. [NYT 4-20-92:D2] "Although the high output declared is probably motivated in part by a need to increase [Kuwait's] quota allocation, there is no doubt that it is making rapid gains." But output data were becoming less trustworthy and also less meaningful as a measure of export sales. Current Saudi oil production might be going into, or past output coming out of, its worldwide storage network. [PIW 4-20-92:1, SS]

At a meeting late in April and another meeting with ten non-OPEC producers, not including Norway, Mexico, and Colombia but including the oil-producing republics of the former Soviet Union, "a very useful exchange of views" took place, according to Minister Nazer. [WSJ 4-24-92:A4] An OPEC meeting the next day froze nonquotas—and, they hoped, output—at existing levels. [NYT 4-25-92:45] But they instructed the OPEC secretariat to report on "new ways to monitor effectively members' actual output." The market balance was considered "fragile." [WSJ 4-27-92:A2]

By late May 1992, prices had risen because of three months of static output despite rising demand. "Core Saudi policy goals remain intact, including ... at least 8 mbd." True, it had not been "legitimized" by the others, "but OPEC's poor track record in observing its ceiling undermines claims that Saudi Arabia made any major concessions." [PIW 6-1-92:1]

Their chief concern was the threat of taxes on oil products. The May agreement was a "'momentary signal' to Western governments that if higher oil prices are wanted to thwart demand, oil producers would be 'perfectly happy to oblige.' ... [Saudi oil minister] Nazer recently warned the EC that adoption of an environmental, tax-oriented posture 'introduces elements of uncertainty that would [negatively] affect investment to expand production capacities.'" [PIW 6-1-92:1]

At the end of May, Saudi Arabia abandoned its "moderate oil pricing policy" and said it favored a $3 increase. Some observers considered it a show of disapproval of oil product taxes. Others said that having gained market share, Saudi Arabia was willing to see higher prices. Some gulf oil officials cautioned that the Saudis were leaving much room for maneuver, and their policy was reversible. If other nations violated the "informal pledge" (the nonquota quota), they would raise their own output to retain their 35 percent share. Or if Japan and the United States dissuaded the EC from product taxes, "the Saudis might reward the industrialized countries with lower oil prices." [NYT 5-27-92:D1] "This is a shot across their [Europeans'] bows," said a Saudi official. A carbon tax "wouldn't discourage energy use but would simply siphon off revenue rightfully [*sic*] due to the oil exporters." Others paraphrased OPEC sentiment: if they want higher prices, we'll give it to them. [WSJ 5-26-92:A3]

OPEC opposed taxes on oil products because they would depress demand, OPEC's market share, and crude oil prices. Between 1990 and 1992, most consuming country governments made substantial increases in taxes on oil products: the twelve EEC members, by roughly $10 per barrel on average. They "scooped" the decreases in crude oil prices, diverting them from the producers and consumers into the national treasuries. [CGES 5–6-93:35–44] The *Economist* in 1985 had urged this on the United States, which continued impotent to tax gasoline.

OPEC threats to "retaliate" for higher taxes with higher prices were ridiculous, but they were made, because OPEC knew that many in the consuming countries would take them seriously. "The Bush Administration reacted with disappointment to the decision of its moderate Arab ally." [WSJ 5-27-92:A2] Daniel Yergin said that oil producers had sought higher prices "after" European countries had raised oil product taxes. [NYT 6-6-92:37] It was about fifty years "after." At Yergin's consultancy, Minister Nazer denounced "taxes aimed at reducing oil consumption and imports." (NYT 2-11-93:D2)

At any rate, OPEC's "surprise endorsement of an output ceiling below most estimates of the third-quarter call on its crude gave off a loud bang in oil markets last month. Now ... strong prices [are] the runaway favorite." [PMI 6-4-92:1] They continued strong through August, and at the end of the month "winter looks tight." [PIW 8-24-92:1] But only a week later, the market was having difficulty absorbing OPEC crude. [PIW 8-31-92:1] Prices began moving down. There was unusually good agreement on estimates of fourth-quarter demand, the range being only from 24.5 to 25.1 mbd. "The debate is more over what each country should produce to meet the demand." [WSJ 9-15-92:A4]

Iran warned that if others exceeded their quotas, they would produce an additional 800 tbd. It was a credible threat, *if* one believed they had capacity of 4 mbd and would be up to 4.5 mbd by March 1993. [PIW 9-27-92:1]

At the September meeting, all members favored higher prices but feared a "tailspin should Saudi Arabia and Iran lock horns over their respective production capabilities and related claims for market share." The safest way to avoid a row over quotas was to extend the existing agreement and let Kuwait increase to 1.5 mbd by the end of 1992. [PIW 9-14-92:1, 5] "With the passage of time, the group's repeated inability to compromise in periods of comfort bodes ill for its chances of reexerting discipline when times get tough." [PIW 9-21-92:1] Prices declined on Iran's assurance that they "won't flood the market, but pump as much as the market will bear." [WSJ 9-22-92:C14]

But Libyan plans to expand to 2 mbd by 1993 had slipped a year [PIW 9-28-92:4], and Nigeria had refused to improve terms for resident companies. This could delay or derail plans to reach 2.5 mbd by 1995. "A nagging source of uncertainty" was how far Saudi Arabia would go to defend its market share claim of 34.7 percent. [PIW 10-5-92:2] For the year 1992, however, it was just under 33 percent.

Iran's escalation was matched by its neighbors. Kuwait's claim for a 2 mbd quota "is seen as positioning in response to Iran's attempts to stake out a 4 mbd capacity figure.... Recent UAE claims that it plans to expand its current 2.4 mbd capacity are seen in the same light." [PIW 11-3-92:1]

Prices had been strong throughout 1992, but futures began to decline in early October, and the spot price fell more, going into contango by the beginning of December. Prompt oil was selling at a discount, the usual symptom of overly full inventories. OPEC output in October had gone over 25 mbd, a twelve-year high. At the November 25 meeting, OPEC returned from nonquota quotas to simple quotas. [NYT 11-4-92:D7] The Saudis called for lower output but were not willing to be the only one to cut back, despite Iran and others' urging that honor upon them. [WSJ 11-10-92:A2, 11-23-92:C12]

Suggestions for an output cut were of the order of 2 to 3 percent. [WSJ 11-24-92:A2] "But no one wants to make the first or deepest cut." [NYT 11-27-92:D10] Finally they decided that all must make "temporary allocations" [11-28-92:34]. It took them another two days to arrive at an output cut of only 0.55 percent. Essentially it was a return to the preinvasion schedules, except that Iran received 3.5 mbd instead of its former 3.2 mbd.

It took Iran nearly a day to accept this. [WSJ 11-30-92:A2] "Saudi Arabia [refused] to budge from its 8.4 mbd allocation, although the kingdom did accept a modest drop in its percentage." Iran, which had gained respect with the 4 mbd output demonstration in October, again promised 4.5 mbd for March 1993 and demanded that its quota be raised to 4 mbd. [PIW 12-7-92:1]

November output was not reduced, "underscoring the 'you first' problem that members have when it comes to ceding tangible market share in order to defend prices." [PIW 1-11-93:1] It was estimated that perhaps 2 mbd additional capacity would be in place by midsummer, mostly in Saudi Arabia and Iran. "Others seem to be running to stay in place." [PIW 1-4-93:1]

Two constants remained. One was "the standing position of Saudi Arabia that production cuts must be shared proportionately, leaving it a one-third share of OPEC's output." [WSJ 1-26-93:A3] The other was the chronic financial crisis. Only the UAE and Kuwait were still clearly creditor nations. All the others currently "seek extra funds to finance budget deficits and industrial imports ... or simply to subsidize layabout relatives. None of OPEC's main members is living within its means, nor is likely to any time soon." [WSJ 1-29-93:A11]

The Saudi foreign debt was estimated at a modest $10 billion. [WP 8-29-93:C2]

> The bloated public sector can no longer absorb every young Saudi, and the private sector prefers cheap, well-trained foreigners to Saudi graduates whose education still relies heavily on rote and on Islam. Private sector employment remains only 10% Saudi.... The infrastructure and welfare state built in the boom years of the late 1970s and early 1980s also are starting to creak. Some Jidda streets get water only two days a week, doctors often deliver babies in the emergency room because hospital beds are scarce.[33] [WSJ 1-13-93:A1]

Nigeria was the OPEC nations' plight at its worst:

> Plagued by debts of nearly $30 billion, fear of civil unrest if subsidies are cut and IMF wrath if they aren't ... the Nigerian government ... in hopes of reducing its deficit for 1993 ... is opening the crude taps every side and—despite earlier laments from partners that this would delay capacity expansion—cutting oil field budgets from 1992 levels. [PIW 1-25-93:1]

The private companies wanted to invest more, for larger production and revenues in the future, but if the government could not match the outlays, its share of ownership would have been reduced. Nigeria preferred to cut back on both capital and current spending. This "strange and self-abuse" continued into 1994. [OGJ 4-4-94:80]

From mid-October 1992 to the end of January 1993, the OPEC basket fell from $19.60 to $16.33; the Dubai Fateh marker, from $18.70 to $14.70. [PIW 10-19-92:8, 1-25-93:8] All now looked to the mid-February meeting.

The Saudis wanted the OPEC basket not to exceed $22; higher prices would curb demand. [WSJ 2-18-93:A2] At that time, the basket was $17.65. By year's end, it was at $12.88. OPEC said the North Sea was "a primary reason for lower oil prices." It was in the same mental rut as in 1983. The new chief executive of BP knew better: "Our job is to produce oil and make a profit. We're not playing the same game [as OPEC]." Horizontal drilling and increased recovery factors "have made existing fields more productive than expected." [NYT 12-17-93:D6]

The year 1993 was a sample of the whole period since 1986. The continued rise in exports in effect went all to Kuwait. The price rose at first, then weakened to its lowest since 1986. The first half of 1994 saw a rebound similar to 1993.

The basic problem for OPEC continued to be as stated in table 7.1. As before, there were informal calculations of output cuts needed to raise prices. "Industry experts say OPEC needs to cut production by at least one mbd ... in order to boost crude oil prices beyond $16 a barrel beyond recent $14 levels." [WSJ 3-23-94:A2] The implicit short-run elasticity of demand was .125 for the world industry, .370 for OPEC as a whole, and 1.08 for the Saudis.[34] A higher price would have been a huge gain for the industry and very profitable to OPEC but not to Saudi Arabia if most members cheated. Therefore Saudi Arabia insisted on market share. Until the other members agree to share the burden of cutting back output and give credible assurances that they will continue sharing, there is no basis for a bargain.

When Iraq returns to production, as it must eventually, OPEC's problem will be much more acute, and for that reason, it may be more quickly solved. I leave the last word to Sheik Yamani: "He said oil prices could plunge below $10 a barrel if OPEC did not cut output to make way for Iraqi exports. [But the end of OPEC would be] 'far, far worse.' ... OPEC should avoid the mistakes of its past by insuring that its largest members do not 'disproportionately share' the burden of stabilizing prices." [NYT 4-12-94:D9]

Appendix 8A: The Soviet Implosion

I first briefly review what has happened inside the FSU (former Soviet Union) oil industry and then its effect on the world oil market.[35]

FSU Oil Industry

Before the breakup, the consensus was rising marginal cost. Output could be expanded only at sharply increasing capital expenditures per unit of incremental capacity. This was long the conventional wisdom in Soviet research offices and institutes. Soviet experts openly deplored the Soviet Union's investing ever more heavily in order to maintain production and exports. They called for more coal, nuclear power, and conservation. (The advice sounds like that in the market economies.) Aleksandr Arbatov, probably the expert best known in the United States, attacked oil and gas exports as a "narcotic." [Gustafson 1989, pp. 135, 269, 271, 286] Gustafson shared and voiced the consensus: "Will Soviet planners begin to take account of the rapidly rising marginal cost of oil and re-examine their energy-trade policy?" [p. 287]

A necessary assumption was that aside from much incidental waste, the Soviet oil industry was not radically inefficient as compared with the capitalist world. Hence the "rapidly rising marginal cost" was a fact of nature. But this was not true.

Lack of accountability and rigid adherence to plans and rules were the external signs of a severe distortion of effort. An outsider might have suspected this.[36] But what proved it, and made the idea of imports grotesque, was the fact of numerous private oil companies' crowding into the FSU, trying to obtain production rights. They were willing to bet large amounts on the proposition that there could be great and profitable expansion. "The Soviets offer an oil and gas resource unmatched outside the Middle East." [OGJ Ed 4-30-90:21]

Because of state ownership, there was in the FSU no idea of the marginal efficiency of capital to guide investment. No producing unit owned assets, with a value it would try to expand. It brought factors together by using funds made available by the next higher command unit, carried out its plan as best it could, and transferred revenues back to the command unit. There was no scanning mechanism to evaluate courses of action and choose the most profitable in the long run. Managers were rewarded with a bonus for output, not return. The discussion of regressive taxes based on output, not profit (in chapter 7), fits here too. Damage to the environment was reckoned at zero and was heavy.

Many have noted how similar was Soviet agriculture to the harsh paper-stacking bureaucracy of ancient Egypt. The same was true with investment in industry. In antiquity there were trade, private property, and widespread laissez faire but no money capital. There was no such thing as the value-maximizing firm, spending a known amount of money to generate a future income stream with a present discounted value greater than the outlay, seeking the better investment and shedding the worse. There was no banking system, no means of pooling savings, no negotiable credit instruments. There were not even rudimentary notions of investment, return, and risk. Hence the lack of innovation and economic thought. [Finley 1981, pp. 179–190; Green 1990, pp. 362–375; Schumpeter 1954, pp. 54–78]

No such things were known in antiquity or in the Soviet Union. The FSU oil industry is a huge fossil. Looking at it helps us understand the OPEC and non-OPEC national oil companies.

Failure in the FSU

The worst mistake in the FSU was probably to continue price controls and then only slowly relax them. The same bad reason was given as in the United States decades earlier: decontrol would hurt individuals and promote inflation. But the effect was much worse in the FSU because of the strong inflation generated when the banking system extended unlimited credit and spending power to industries. The slowly decontrolled oil prices lagged the rise in other prices. Until the end of January 1993, the natural gas field price was 1.1 cents per mc; industrial customers paid 5.7 cents and European customers $2.40. [WSJ 4-9-93:A6] The drastic fall in the real price of oil, which at the end of 1992 was about 5 percent of the world price, kept producers from buying current supplies and services and it depressed workers' living standards further in locations often harsh and remote. Above all, it penalized the good and bad alike and postponed triage.

Failure to decontrol prices also meant failure to begin rationalizing consumption. Energy was grossly wasted in industry and households; for example, in overheated apartment buildings, the only way to lower the temperature was to open the windows (the "Soviet thermostat"). Energy costs were a small part of the total production or living cost, and increases could have been borne relatively easily. Hardship subsidies could have been provided at a small fraction of the cost. Private motorists were few and grass-roots grumbles over gasoline prices accordingly faint.

The second mistake was a sort of *nomenklatura* privatization, de facto takeovers by the old management of producing groups or associations. To offset their losses because of price controls, the associations were propped up by "loans" that kept all of them—the best and the worst—going. This was made worse by the "arrears crisis," of bad debts due from nonpaying customers, which became acute in 1992. The Central Bank of Russia saw its job as preventing collapse by providing credit to all enterprises. [Fischer 1993] Thereby it blocked the most important task, to sort out the assets and choose which wells and fields to shut in and which to expand. This process has barely started. Presumably the oil industry has not reached the delirium of coal, where output has been cut at all mines instead of at only the worst losers, the workforce has actually increased, and mounting subsidies now amount to nearly 2 percent of Russian GDP. [*Economist* 12-4-93:70]

Corruption and side deals profit managers at the expense of companies. This is harmful to company performance and has delayed the urgent task of restructuring and sorting out.

The gap between controlled and uncontrolled price levels generated enormous profits. Of course the "owners" tried to sell all they could in noncontrolled markets, particularly as exports, where some of the proceeds could be diverted into foreign bank accounts. The bizarre result was an FSU export of capital. These sales were often illegal, and huge sums were reportedly paid in bribes, provoking the FSU governments into heavy and haphazard taxation in order to keep some of that income flowing into the national treasuries.

Many Western oil companies tried to obtain concessions in the FSU, but the great activity had generated little investment by the end of 1993. There was no set of laws within which companies could work. To settle terms of investment is inherently difficult. Here it was made harder or impossible by unfamiliarity with concepts like risk bearing and return. The tax system seemed to combine most of the bad features of tax systems around the world. Of course, there was the usual populist suspicion of foreign capitalists and fear of giving away the family jewels. The extreme devaluation of the ruble fed these fears, yet it was irrelevant, because foreign companies seeking production rights were not acquiring tangible assets, only rights to invest.

In 1994 it was announced that the proposed Sakhalin Island project would be governed by a hybrid regime of profit-sharing along with revenue sharing. The government share of output would depend on the internal rate of return. [PIW 5-2-94:1] A progressive tax on profits would be the most hopeful change since the end of the Soviet Union.

Domestic "squatters" wanted to have preference to exploitation rights. They could dispose of patronage, perquisites, supply contracts, and products at below-market price. They could resell some rights to foreign companies. The usual protectionist slogan was jobs for the local boys. In the great Stokmanovskoye gas field, a Russian company said it could do the job more cheaply and provide "more jobs for Russians and ... a large part of its equipment orders with domestic enterprises." [OGJ 11-16-92:NL1] When the cheapness of the local company is seen to be illusory, it will be too late to avoid waste and delays. But doubtless some would rather lose the whole patrimony than let foreign capitalists earn some of it.

Within each FSU republic, each region tried to get a larger share of the investment spending and eventual income, a familiar pattern worldwide in any country with a federal or provincial system of government. Moreover, republics through which pipelines passed could demand high fees or lower prices for transit rights. In October 1992, Ukraine reduced the flow of Russian gas to make its threats credible, jolting foreign customers and injuring both Ukraine and Russia. With some justice, Ukrainians see Russians as bullies, and Russians see Ukrainians as deadbeats. The Russian threat to landlocked Kazakhstan is even more acute and threatens the factors of ventures like Chevron's Tengiz project (see appendix 8B). The worst effects of these power games are long run: increased risk makes foreign and domestic investors more reluctant, and they hold out for bigger returns.

Thus, although the consensus of high marginal cost was quickly proved wrong and the FSU oil industry was shown to be capable of great profitable expansion, by the end of 1993 its potential had not checked the steep decline in output. Natural gas had not declined far, but undermaintenance threatened a decline. Possibly Azerbaijan and Kazakhstan, free to make their own bargains with foreign companies, will show Russia the way.

Billions of barrels will be discovered and developed, but only if the FSU nations provide a framework of law and ownership for home and foreign investment. The state must refrain from taxes, threats, confiscations, and extortions. The principle hammered into the old *nomenklatura* was, "If you see something you want, grab it, and extort the maximum for its use."

That these maneuvers increase risk and make oil development much less likely is a new and alien idea, not soon learned.

World Market

The effects on the world market were already well advanced by the end of 1992. By dissolving the old Soviet enclave, the world market became larger, and the OPEC share decreased. FSU exports have been maintained. The decline in economic activity has lessened consumption. As the economy recovers, the use of prices will discourage wasteful use and limit the consumption increase. The FSU, which was so confidently expected to be a net importer by 1985, will be a large exporter twenty or more years thereafter.

The FSU integration into the world market is still far from complete, but shipments within an FSU republic in time will be paid for at world market prices, as exports are already. Urals crude has emerged as a marker or designator crude for Western Europe, analogous to West Texas Intermediate in the United States. It "already influences markets for dozens of crudes, including most Mideast barrels moving into Europe." [PIW 5-4-92:3] And in August 1992, deliveries by FSU republics had "risen ... despite falling production levels, *leading OPEC to complain of market disruption.*" [OGJ 8-17-92:NL3, emphasis added] Blaming "market disruption" on this or that individual seller or area—Kuwait, Russia, the North Sea, or any other—is wrong in itself and also symptomatic of a higher-than-competitive price.

Appendix 8B: Investment and Rates of Return on Oil Development

In chapter 2, I set up two alternative break-even equations:

$$\text{“}i\text{”} = P/(K/Q) - c - a$$

or

$$\text{“}P\text{”} = (K/Q)(i + c + a)$$

where P is the wellhead price; Q is peak annual output, assumed to be in the initial year, and to decline at a percent per year; K is capital expenditures, assumed all spent in the year before production starts; and c is the levelized operating cost per unit, which is 7.5 percent rather than the conventional 5 percent of capital expenditures.

We can enter a price and calculate the rate of discount "i" that would make the investment barely worthwhile; then we can compare "i" with the firm's actual cost of capital i. Alternatively, we can enter a rate of return i and reckon the cost or supply price, that is, the price "P" that would make the investment barely worthwhile.

As an OPEC example, take first (from table 8.4) the Persian Gulf member with the highest 1989/1990 investment requirements, Abu Dhabi: $K/Q = \$1,767/365 = \$4.84$. We take conventional values: $c = .075$, and an additional charge per barrel $v = \$1$. From table 8.3 we have the decline rate $a = 2.34$ percent, disregarding unstable years after 1980. If $P' = \$15$:

"i" $= \$14 - .075 - .0234 = 2.79$, or 279 percent/year.

For Venezuela, the usual estimation method cannot work. The very large number of wells, most very low output, makes it too far wrong to equate the capacity of the average new well to that of the average old well. However, we have better data: the *Annual Reports* of Petroleos de Venezuela. [PDVSA 1992, 9, 20, 62] Investment for "creation and maintenance of production capacity" of oil was \$1,289 million, after allocating "infrastructure" between oil and natural gas according to respective expenditures. There was no net gain in capacity. Hence, the gross gain of 617 tbd also equals the year's decline, which was 26.0 percent of the year's production. (The decline rate has long been very high in Venezuela. [Adelman 1972] [PIW 3-23-92:5] gives a "natural decline rate" of 22 percent.) Incremental investment was therefore \$2,090 per additional daily barrel, or \$5.73 per annual barrel. Using the same price assumption:

"i" $= 14/5.73 - .075 - .26 = 2.14$, or 214 percent/year.

For a non-OPEC example, we take the upper-quartile country from table 7.3, which is $K/Q = \$10,210/365 = \$28$. Using the conventional $Q/R = 1/15 = .067$, and $a = Q/R - (Q/R)^2$

"i" $= 14/28 - .075 - .062 = .363$.

This is pretax. Assume now fifty-fifty production sharing, which amounts to the government's taking half the wellhead price:

"i" $= 6.50/28 - .137 = .095$

The pretax return could be split to benefit both parties. The after-tax return is no more than marginal.

Example: West Siberian Waterflood Development Project

The basic data [Jack A. Krug and William Connelly, in OGJ 2-8-93:72] are: $R = 120$ mb, $K = \$295$ m, and peak output $Q = 32.6$ tbd. The price assumed in the paper is $20, but we use $15. We can calculate the decline rate from $Q_f/Q = e^{-aT}$. Of the four variables the paper gives us three, allowing us to calculate: $3.08/11.9 = e^{-20a}$, hence $a = .0675$.

$$\text{"}i\text{"} = 14/24.79 - .075 - .0675 = .422$$

Our basic assumption is too pessimistic: that capital expenditures are all made at the outset. In fact, the project capital expenditures extend over the first seven years, half of them in or after year 4. After year 2, revenues actually exceed costs. However, when the authors incorporate taxes, the posttax DCF return, even with the unrealistic $20 price, is only 13.3 percent. Unless the parties use a much lower tax and a much more flexible tax schedule, there is no hope of such projects' ever starting, despite their being very profitable before tax.

Example: Tengiz (Kazakhstan)

The Tengiz field in Kazakhstan is said to hold 25 billion barrels of oil in place, of which 6 to 9 billion are considered recoverable. (There are also large amounts of natural gas.)

From three news reports [NYT 4-6-93:D1; WSJ 4-8-93:C15, 5-13-93: A6] it is possible to approximate the first phase of Tengiz development. We are given: $K = \$1.65$ (± 0.15) billion spent "over three to five years," $Q = 260$ tbd, or 95 mby, by 1997, and $R = 1.1$ billion barrels. As calculated:

$$Q/R = .086$$

$$a = .086 - .086^2 = .0786$$

$$\text{"}i\text{"} = P/(K/Q) - c - a$$

$$\text{"}i\text{"} = \$14/\$17.37 - .075 - .079 = .652$$

Chevron's share is stated as 20 percent of profits, or .130 after tax. Since the value of Chevron stock rose after each of two announcements, the true return is probably higher.

I have made no allowance for natural gas produced with the oil. Initial reserves are 1.5 tcf, of which 8.6 percent would be 129 billion cubic feet per year (bcfy). At a wellhead value of \$1, net of operating cost, this would amount to \$129 million annually.

Example: U.K. North Sea and the United States

In Chapter 2, I showed that in 1992 it cost about \$3.32 to develop an additional reserve barrel in the United States. The consultancy County Natwest Woodmac have compiled expenditure and reserve data on fifty-seven new U.K. North Sea fields to be started between 1992 and 1998, to develop 2.3 billion barrels of oil and 14.3 trillion cubic feet of gas. [OGJ 8-17-92:50] Since 1991 prices were used, I convert to dollars at \$1.767. The reserves are given separately for each field, enabling a regression estimate for expenditures per barrel or mcf in the ground, as follows (t-statistics in parentheses):[37]

Expenditures per barrel:	\$5.46	(10.8)
Expenditures per mcf:	\$0.93	(15.3)

We translate expenditures per incremental reserve barrel into dollars per barrel of incremental producing capacity. Reserves R are cumulated output, starting with Q barrels per year, declining at a constant percentage rate, a, over T years. Integrating, we have $R = Q/a\ (1 - e^{-aT})$. The final year's output is Qe^{-aT}. We can write it as Q_f, so $R = Q/a\ (1 - Q_f/Q)$, and $a = Q/R\ (1 - Q_f/Q)$.

In Chapter 2, I explained why we can approximate $a = (Q/R) - (Q/R)^2$. Moreover, since $Q_f/Q = e^{-aT}$ we can write it $e^{-aT} = (Q/R)$, or $T = -\ln(Q/R)/a$. It is the same rationale: the more intensive the depletion, the less time the project takes before closing down. Thus if K = capital expenditures, $K/Q = K/R\ (1 - e^{-aT})/a$.

For the United States in 1992 and the U.K. North Sea after 1991:

Area	Q/R	a	T	K/R	$(1 - e^{-aT})$	K/Q	\$*IDB*
United States	.1010	.0908	25.2	\$3.32	.899	\$32.85	11,992
U.K./North Sea	.1605	.1347	13.6	\$5.46	.840	\$34.04	12,426

The cost per daily barrel is almost the same, but there is a very important difference. As pointed out in chapter 2, marginal cost in the United States was lowered by eliminating the poorer prospects. The United States is a shrinking province; the U.K. North Sea is expanding. The supply curve is moving leftward in the United States, rightward in the North Sea.

The Centre for Global Energy Studies has reviewed the same body of North Sea data but has not tried to summarize as I have. [*CGES* 1–2-92] It asks: "[W]hy are many oil companies contemplating investing in the North Sea when the returns they can reasonably expect are likely to be inadequate? ... [They] have probably factored higher oil prices into their calculations than it would be prudent to expect." [p. 17] Perhaps, but there is also contrary evidence. Adelman et al. [1991] calculate an expected price change that derives from the price paid for reserves as compared with the current wellhead price. It is flat for U.S. oil, and price expectations are uniform worldwide. Adelman and Watkins [1992] have shown the same flat expected price for Canada for recent years. Another piece of evidence is Philip K. Verleger's analysis of the BP royalty trust certificate, which he concludes shows an expected flat price. [Verleger 1989–1992, various times]

An alternative explanation is that reserves added are understated, or investment requirements and costs are overstated. There is important confirming evidence: no cancellations since 1991, despite deteriorating prices. But the question raised by CGES is always worth asking.

Notes

1. For a classic demonstration of the effect in the competitive oil tanker market, see Zannetos [1966, chap. 6, esp. p. 172]. There is a fairly abrupt bend in the curve at about 88 percent of capacity.

2. In 1986, "Renewed Relations with Iraq Fall Short of U.S. Expectations ... 'We hoped for it. We still hope for it,' says a State Department official." [WSJ 3-17-86:24] A year later: "U.S. is Tilting Toward Iraqis ... Some [in the State Department] say Iraq's support for terrorism is relatively minor, and that better relations with Washington have moderated Iraq's behavior considerably." [WSJ 3-31-87:33]

3. "Saudi Arabia has secretly contributed billions of dollars since the early 1970s to movements and governments in a dozen countries ... particularly in areas where the executive branch has been unwilling or unable to gain Congressional support.... The Saudi ability to finance foreign policy efforts promoted by Washington has declined recently with the slump in oil prices, but as the payment to the Nicaraguan Contras demonstrates, Riyadh remains willing to provide cash at key moments." [NYT 6-21-87:1]

4. Kuwait minister al-Khalifa Al Sabah showed again why he was the most unpopular of ministers. He was asked: "Some members believe this agreement will stabilize the market and make the world safe for term contracts of longer than three months. Do you agree? A. If spot is above official, no country is going to sell on term contracts. Any country ... is going to increase its spot volume. We know from experience." [PIW 7-6-87:SS1]

5. Karen Elliott House of the *Wall Street Journal* wrote: "They are busier than ever trying to buy insurance policies from every peddler.... The Saudis are putting up a brave front, but there's little evidence they've stiffened their spines.... While talking of standing up to Iran,

the Saudis secretly sent the Algerian foreign minister to plead for their safety in Tehran.... While American newspapers describe Saudi largesse on behalf of William Casey's covert operations in the Middle East, that, too, is only part of a larger Saudi strategy of making payoffs to every piper in the region.... There's probably little America can do that would give the Saudis the courage to stand up for their own self-interests." [WSJ 10-8-87]

6. Percentage decrease in drilling cost, holding depth, and location constant [JAS 1992]:

Well type	Onshore	Offshore
Oil	35.1	23.5
Gas	16.1	19.9
Dry	26.8	22.3

7. There were also two false statements: that "Professor Adelman wanted to see ... 10c a barrel of oil" and that a UN report on the price of oil "never saw the light of day because the ever-powerful oil companies succeeded in suppressing it." In fact, it had been published [UN 1955].

8. There have been many econometric studies of recent OPEC behavior. I found most useful Griffin [1985] and Dahl and Yucel [1991].

9. That is, reserves are defined as cumulative output starting with current output. In infinite time:

$$R = Q \int_0^\infty e^{-aT}\, dt$$

$$= Q(1 - e^{-aT})/a = Q/a$$

The correction for finite time is very small at low values of a. See chapter 2, appendix 1.

10. The United Arab Emirates is composed of Abu Dhabi (1.1 mbd in 1988), Dubai (0.4), and Sharjah (0.04). References are not always clear, but since Dubai and Sharjah grew only slowly, changes in UAE production and capacity are nearly always due to changes in Abu Dhabi.

11. One gropes for some way to reconcile this myth to the real world, like using archaeology on Homer's *Iliad*. Was the "average" weighted or unweighted? A net rather than a gross capacity increase? Those putting forth the numbers did not even mention such problems, let alone work them out. Robert Horton, chairman of BP, repeated the myth. Later he was ousted by the nonexecutive directors because "he had lost credibility with analysts and investors." [PIW 6-29-92:7] One wonders if the $60 billion story contributed. (See p. 284.)

12. "Because Kuwait oil wells flowed prolifically with little mechanical prompting before the damage by the Iraqis, the oil men there say they haven't the sophistication needed to do the best job of reconfiguring the fields. Negotiations have been dragging on for almost six months, with the oil company asking unusually high fees for its services, the Kuwaitis say. BP was very active there before the nationalization of the oil industry, and it has retained much of the geological data." [WSJ 2-24-92:B3C] Some kind of arrangement was made. [PIW 3-23-92:7]

13. Kuwait downstream integration into refining and marketing made expansion somewhat easier. It could always sell most or all or its quota as product in numerous markets. There was no need to bargain with a few large buyers over crude oil prices. But the advantage was minor. Kuwait's profit on refining and marketing was unknown and may have been negative but could not have been substantial because refining is a competitive industry in which it had no advantage. The supposed "stability" conferred by vertical integration is imperceptible.

14. Assuming the new production declines at the same rate as the old, one must install 1.41 mbd (gross) to have 1.30 mbd (net) after five years production. That is, 1.41 $(1.0158)^{-5} = 1.30$. Then \$1b/1.41 mbd = \$711/bd.

15. See chapter 2 and appendix 8B for the formula: $i = P/(K/Q) - a - c$, taking $P = \$17$, $(K/Q) = (\$625/365)$, $a = .016$, $c = .075$

16. Askari [1991, pp. 37–39] estimates end-of-1981 holdings at about \$160 billion and end-of-1985 holdings at \$50 billion, excluding so-called loans, chiefly to Iraq. See also Askari [1990]. During 1986–1989 inclusive, the cumulative Saudi Arabian current account deficit was \$38.1 billion. [IMF-IFS, September 1991, p. 517] Thus Saudi Arabian holdings were around \$12 billion at the end of 1989. In 1990, there was an additional \$4.1 billion deficit, and \$25.7 billion in 1991. [OPEC-SB 1992, p. 8] Hence official net assets are probably below zero. Of course, private parties have large holdings.

17. With statistics on inventories as bad as ever or worse, contango or its contrary, backwardation, served as a proxy perhaps superior to the original. It expressed the view of buyers and sellers over the adequacy of inventories relative to expected demand in the near future. Philip K. Verleger, Jr., was the first to perceive the importance of this inventory surrogate and analyzed it in detail. See [Verleger PEM]

18. Cf. Patrick Clawson, in the *Brookings Review* (Spring 1991):3: "The invasion of Kuwait was motivated not by dire need, but by greed on a gargantuan scale." An opposing view is in Taylor [1991]: "Saddam Hussein regarded a special connection with the PLO as vital to his image as the emerging hero of the Arab masses. *The cultivation of such an image was more important to him than the annexation of Kuwait.*" [p. 110, emphasis added] Taylor's book is written to vindicate "the efforts of the foreign policy bureaucracy to get the [U.S.] president to deal with the Middle East in terms of 'on the ground realities." [p. 44] Their grasp of those "realities" seems feeble.

19. The ambassador later said this press report [e.g., NYT 9-23-90:19] was "fabrication ... disinformation." [NYT 3-21-91:A1, A15] This is not credible. "An unidentified senior State Department official ... said it [the news report] was 'essentially correct.'" ([NYT 3-22-91:A1]) See also Elaine Sciolino, *The Outlaw State: Saddam Hussein's Quest for Power and the Gulf Crisis* (New York: John Wiley & Sons, 1991), pp. 179, 271. A year later, senators who had finally secured access to the original documents were indignant over the ambassador's evasive and inaccurate version of them. [NYT 7-13-91:A1]

20. Even after the attack, Zbigniew Brzezinski thought Iraq had "financial and territorial claims (not all of which were unfounded)." [NYT 10-7-90:OpEd] Peter Tarnoff, president of the Council on Foreign Relations (undersecretary of state in the Clinton administration), urged Kuwaiti "concessions to Iraq regarding oil pricing and production, territory and debt." [NYT 11-30-90 OpEd] But since the end of 1989, Kuwait had reduced output by 200 tbd, while Iraq had increased by 400 tbd. Plainly Iraq was the bigger influence for lower prices. George McGovern wrote in 1994 that "Iraq ... suffered grievously from the oil price squeeze engineered in part by Kuwait"—not, of course, from using billions of oil revenues for weapons. [*Middle East Policy* 3, no. 1 (1994): 163] Kuwait had no interest in driving prices down; it cost them money, like all other sellers. Nor had they the power to do so.

21. Production is from PE, which includes condensate and natural gas liquids. Consumption is from the CIA, inventories from DOE:IPSR.

22. Former Minister Yamani said on October 8 "that the price of crude oil should drop to \$15 to \$18 ... once the Persian Gulf crisis was defused." [NYT 10-9-90:D25] He was on target.

23. This is explained in my 1983 presidential address to the IAEE. [Adelman 1993, ch. 25]

24. The Saudi government has kept its sure instinct for American politics. In January 1994, it needed to "restructure" (stretch out) payments due to some large defense contractors. Prince Bandar bin Sultan, the ambassador to the United States, gave an interview that amounted to "a pointed message to the Clinton administration.... If oil prices sag further, cancellation of some arms orders would be considered. ... [I]t would deal a severe blow to the aerospace industry [and] ... hurt Bill Clinton's chances of carrying [California] in 1996, which is considered crucial to his re-election." [WSJ 1-10-94:A4] Therefore the embargo on Iraq exports should stay.

25. The thesis is elaborated but not further clarified in Mabro [1991]. But he suggests that for security, consumers should finance surplus capacity in OPEC countries instead of reserves accumulated at home.

26. Some OPEC countries have tried to privatize state-owned industries and free the private sector from price and other controls. Harik and Sullivan [1992] survey Algeria, Iraq, and Saudi Arabia. The authors are not economists. Their assumptions seem favorable to *dirigisme* under strain. But my impression of their findings is that reforms have been too little to help but enough to make greed, corruption, and vulgar new-rich display more visible, hence more easily blamed on capitalism.

27. Similar provisions were being discussed in Iran, Iraq, and especially Abu Dhabi: "Companies say that they would be prepared to invest in further boosting capacity if the increased flows are guaranteed once the outlays are made."

28. There are some rough cost indicators. "The total cost of rehabilitating Kuwait's *oil industry* is put at $8b to $10b." [NYT 5-4-92:A1, emphasis added.] According to CMB [1984], in the Middle East between 1980 and 1984 (the last five years available) expenditures on "crude oil and natural gas" were 23 percent of total oil industry capital expenditures. If so, crude oil production expenditures would have been $1.8 to $2.3 billion. If the newly established capacity is 2 mbd, the required investment would be $900 to $1150 per daily barrel. Table 8.4 estimates $987.

29. But it also shows how unstable and risky is a production-sharing contract in a low-cost area. Assume the price $P = \$15$, the investment needed is $750/bd or $K/Q = \$2.05$ per annual barrel, the decline rate $a = .05$, the operating cost allowance $c = .075$, and the investor's minimum acceptable return is 20 percent per year. Then a production share of 5 percent is acceptable:

$$i' = .05\ (\$15)/(\$2.05) - .125 = .241$$

But a slight massaging of the contract to bring the production share down to 4 percent brings the return well below the line:

$$i' = .04\ (\$15)/(\$2.05) - .125 = .167$$

30. Saudi Arabia and Kuwait sought and received assurance from Secretary of State Warren Christopher that the Clinton administration would continue sanctions against Iraq. [WP 3-21-93:A37]

31. Let the price ratio P equal $17.15/$13.15 and the sales ratio Q equal $59/61 = .967$. Then if $Q = P^x$, and $.967 = (17.15/13.15)^x$, $x = -.125$.

32. Let $P = \$21/\$17 = 1.235$, $Q = 59/61 = .967$ and if $Q = P^x$, then $x = -.157$.

33. "The Saudi establishment [has failed] to deliver ... a consumer driven, industrialized, developed world ... in which other people did the work.... Those whose brothers ten or fifteen

years ago could expect to receive government jobs, free loans, free cash for housing, all the perks that went with a successful rentier society, their expectations have been dashed." The only available forum for discontent is Islam, a threat to the establishment, but with no clear vision of what is to replace it. [Roberts 1993]

34. As usual, let $Q = P^x$, where Q is the ratio of new output to old output and P the ratio of new price to old. Here $16/$14 = 1.143. For the world, excluding 5.1 mbd OPEC, consumption taken in the last quarter of 1993, Q = 59.4 mbd/60.4 mbd. For OPEC exports Q = 19.7 mbd/20.7 mbd, and for Saudi Arabia (home consumption 750 tbd) Q = 6.4 mbd/7.4 mbd.

35. The following sketch is based on press reports and on Smith [1992, 1993]. It has benefited from the comments of Joseph S. Berliner.

36. I may be permitted a personal reminiscence. In 1971, I chaired a session at a conference on Arctic oil and gas of the American Association of Petroleum Geologists. There was much Soviet attendance. After my session was over, two Soviet petroleum engineers walked up to the podium, where I was the only remaining person. I wish I could have recorded what they said. They were quite apolitical but articulate and disgusted with the rigid rules on drilling, testing, producing, and so on, and the waste it entailed, in time and capital equipment and supplies. Why use trained engineers, one of them asked, to do stupid things that revolted their knowledge and intelligence, because somebody had written them into the rule book, for reasons that had long since ceased to make sense, if they ever had.

37. Note added in press: a report on 27 U.K. and Norway North Sea fields to begin production after 1995 (prepared by the consultancy Wood Mackenzie [OGJ 1-2-95:20]) shows a striking cost reduction:

Expenditures per barrel:	$3.58	(4.3)
Expenditures per mcf:	$0.33	(9.7)

9 Conclusion

The price of a barrel of crude oil at the Persian Gulf sums up the last twenty-five years:*

Year	Price in current dollars	Price in 1993 dollars
1970	1.21	4.19
1974	12.12	32.70
1981	33.50	51.92
1985	24.31	31.94
1993	14.52	14.52

The 1970 price could not last because it was in the unstable zone: well above the competitive level and far below the monopoly level. The price is higher today but still in the zone. Without OPEC as price guardian, erosion would go much faster than twenty-five years ago.

OPEC has been unable to raise the price since 1986. Table 7.1 sets out its plight. A higher price goes to all producers of oil, but the burden of lower sales is borne only by OPEC and then largely shifted to Saudi Arabia. Thus, a price boost highly profitable for the world industry may on balance be middling profitable for OPEC and marginal or worse for Saudi Arabia. (Above, p. 313.)

OPEC members cannot escape this trap by their efforts alone. But they could be rescued, if in time their market share rose high enough so they could again give up some of it for a higher price. In 1993, their share was 36 percent. On optimistic assumptions, it could reach 50 percent in a decade. The Saudi share will, if anything, decline from the current 12 percent

*1970, Middle East spot FOB. Others are the Saudi Arabian average revenue per export barrel. Inflation adjustment by the U.S. gross domestic product implicit price deflator.

when Iraq returns. Escape from the market share trap does not appear likely.

The consensus that we have often reviewed has always seen a "tighter oil balance" soon and higher prices. Years pass, demand grows, excess capacity dwindles, the former Soviet industry totters, Iraq is out of the market—but the real price is lower. Yet the glow is still on the horizon in 1995, lit by impressive numbers that are safe from analysis because they cannot be checked, or replicated, or even explained. Chapter 7 mentioned a consultancy that in 1986 made a forecast "based on reserves": prices to rise, non-OPEC output to drop. Prices dropped and non-OPEC output rose. Now its "worldwide database" shows only 1,100 billion barrels left in the earth; therefore, the king of Saudi Arabia should raise prices "to better meet future realities" [*Economist* 8-6-94:10]—or present fantasies. These numbers are not even wrong. They have no meaning.

The odds are good that non-OPEC output of both oil and natural gas will keep growing because at current "low" prices, development is highly profitable pretax. The obstacle is government. Regressive taxes abort development and increase exploration risk. But past mistakes are opportunities now. Non-OPEC countries have lowered rates and have moved, too slowly for their own good, toward taxes aimed at skimming rents, not aborting them. There is a long way to go in tax reform. Progress would guarantee rising non-OPEC output even if costs tend to increase. In fact, if anything, costs seem to be decreasing. Moreover, state companies are being made private, to aim for maximum asset value, to cultivate what provides a return on investment, and to weed out what does not. On both counts, taxes and state ownership, the greatest underachievers are the former Soviet republics.

I would expect the cartel to be under increasing strain. It may break down and revive, with perhaps a changed membership, but the monopoly gains will not be given up without a struggle. The latest violence, in 1990–1991, was probably not the last.

Competition in oil is as workable as elsewhere. There is no original stock or store of wealth to be doled out on any special criterion. Oil reserves are inventory. Reserves and capacity are created by investment. As in other industries, projects range from low risk to high risk, from long lead times to very short ones. Capital markets are equipped to handle the whole range in risk and time.

Markets are information networks. Errors and glitches abound, and they waste resources. Monopoly adds errors, distorts the adjustment, and magnifies the waste. The larger the industry is, the more the waste. Government monopolies last longer and waste more.

Worldwide, oil has been growing more plentiful, but for all we know it may some day become more scarce. The warning would be a persistent rise in development cost and in values in-ground. These can be monitored at a very low cost. Buyers and sellers of oil and natural gas producing properties are forced to estimate future prices to decide on present values. Money concentrates the minds of oilmen, bankers, engineers, and geologists. They may all be wrong at any given time, but a market rewards good guesses and penalizes bad ones.

Oil demand and supply are more stable than most other minerals. Oil is so widely used that changes in one or a few sectors of the economy cannot heavily affect its consumption. There are also more crude streams than sources of other minerals. (But many of those streams are clustered around the Persian Gulf.)

A competitive price for oil would not only be lower than current; it would also be more stable. The oil price explosions were unrelated to scarcity and entirely due to the cartel. The irregular decline after 1981 was its defeat. Along the way, buyers and sellers built up excess inventories and drew them down, according to hopes and fears of what the cartel might do. By making stocks volatile, the cartel has made prices volatile.

The political benefits of a lower oil price might be much greater than the economic. The flow of oil wealth makes some producing countries worth invading and gives others the means to invade or to threaten a shutdown around the Persian Gulf. The smaller the oil revenues are, the less is the chance of aggression and of the producing governments' buying nuclear and other weapons. They are undemocratic, and there is no change of regime except by violence or conspiracy. Many of their inhabitants are in a pious rage. Those who disagree with them serve the Great Satan and deserve only death and the Fire. The less hard currency there is in such hands, the better off is the rest of the world.

The weaker the cartel is, the better for others. But it has existed for years, and it is composed of sovereign states with which the United States wants orderly and correct relations. Anticartel action would have been a fitting response in the mid-1970s or the early 1980s; it no longer is.

The right public policy, in my view, is in the doctors' oath: do no harm. Do not make things worse. Rule out price and import controls. Make no agreement with the cartel or any of its members because one cannot hold a sovereign monopolist to its word. The record of broken agreements in this book is dismal and long. Preserve and increase the strategic reserves. Even when used unwisely or not at all, their mere existence helps calm the fear that has convulsed the market, even with no shortage of oil.

The United States must protect the Persian Gulf nations from invasion or aggression. We fended off Iran in the 1980s and Iraq in 1990. The countries there rightly give us nothing for what they will have anyway. Friendly or barely correct relations will have no effect on oil price and supply.

We should do nothing to help or protect the cartel from the subversive force of competition. If some useful measure has the incidental effect of damaging the cartel, so much the better. The best example is a tax on gasoline or other oil products.

If the cartel disappears, there will be a burst of investment by the member nations. Their imports of weapons will dry up, but producing and service companies will be newly welcome. The nations will try to save something from the wreck, by expanding output as partial compensation for lower prices. It would be imprudent to expect this any time soon, just as it was imprudent in 1989 to expect the collapse of the Soviet empire.

References

Abbreviations

AAPG	American Association of Petroleum Geologists, *Bulletin*, monthly
BP	*British Petroleum Statistical Review of World Energy*, annual
BW	*Business Week*
CIA:IESR	Central Intelligence Agency, *International Energy Statistical Review*, monthly
CMB	Chase Manhattan Bank, *Capital Expenditures of the World Petroleum Industry*, annual, 1955–1987
DOE:AER	U.S. Department of Energy, *Annual Energy Review*
DOE:IPSR	U.S. Department of Energy, *International Petroleum Statistics Report*, monthly
DOE:MER	U.S. Department of Energy, *Monthly Energy Review*
FNC Energy News	First National City Bank (New York), *Energy News*, monthly
IMF-IFS	International Monetary Fund, *International Financial Statistics*, annual
IPE	*Interational Petroleum Encyclopedia*, annual
JAS	American Petroleum Institute et al., *Joint Association Survey of Drilling Costs*, annual
LM	*Le Monde*, Paris
MEES	*Middle East Economic Survey*, weekly
MP	"Meet the Press," NBC television program
NYT	*New York Times*
OGJ	*Oil and Gas Journal*, weekly
OGJ:WWO	*Oil and Gas Journal* annual supplement, *Worldwide Oil*
OPEC-B	*OPEC Bulletin*, monthly, ten issues per year
OPEC-R	*OPEC Review*, quarterly
OPEC-SB	*OPEC Statistical Bulletin*, annual
PDVSA	Petroleos de Venezuela S.A
PE	*Petroleum Economist* (formerly *Petroleum Press Service*), monthly
Pemex ML	Petroleos Mexicanos, *Memoria de Labores*, annual

Pemex DG	Petroleos Mexicanos, *March 18 Discourse of the Director General*, annual
PIW	*Petroleum Intelligence Weekly*
PMI	*Petroleum Market Intelligence*, monthly
PODE	Venezuela, Ministry of Mines and Hydrocarbons, *Petroleo y Otros Datos Estadisticos*, annual
PONS	*Platt's Oilgram News Service*, daily
POPS	*Platt's Oilgram Price Service*, daily
PPS	*Petroleum Press Service* (name later changed to *Petroleum Economist*), monthly
SOGE	American Petroleum Institute, *Survey on Oil and Gas Expenditures*
WO	*World Oil*, monthly
WO-IO	*World Oil* annual: *International Outlook issue*
WP	*Washington Post*
WSJ	*Wall Street Journal*
WSJ:A	*Asian Wall Street Journal Weekly*

Publications

[Adelman 1962] M. A. Adelman, *The Supply and Price of Natural Gas* (Oxford: Blackwell).

[Adelman 1963, 1964] ———, "Les Prix petroliers a long terme," *Revue de l'institut français du petrole* (December 1963); trans. "Oil Prices in the Long Run," *Journal of Business of the University of Chicago* 43:143–161.

[Adelman 1964] ———, "Efficiency of Resource Use in Crude Petroleum," *Southern Economic Journal* 31:101–122.

[Adelman 1966] ———, "Oil Production Costs in Four Areas," *Proceedings of the Council on Economics* (American Institute of Mining Mechanical and Petroleum Engineers).

[Adelman 1972] ———, *The World Petroleum Market* (Baltimore: Johns Hopkins University Press).

[Adelman 1985] ———, "An Unstable World Oil Market," *Energy Journal* 6, no. 1.

[Adelman 1987] ———, "Are We Heading Towards Another Energy Crisis?" Remarks at the National Press Club, September 29.

[Adelman 1990] ———, "Mineral Depletion, with Special Reference to Petroleum," *Review of Economics and Statistics* 72:1–10.

[Adelman 1991] ———, Review of [Petroleum Finance 1988], *Energy Journal* 12, no. 4:145–147.

[Adelman 1992] ———, "Finding and Developing Costs in the United States, 1945–86," in J. R. Moroney, ed., *Advances in the Economics of Energy and Resources*, vol. 7 (Greenwich, Conn.: JAI Press).

[Adelman 1993] ———, *Economics of Petroleum Supply* (Cambridge, Mass.: MIT Press).

[Adelman and Shahi 1989] ———, and Manoj Shahi, "Oil Development-Operating Cost Estimates, 1955–1985," *Energy Economics* 11.

[Adelman and Ward 1980] ———, and Geoffrey L. Ward, "Worldwide Production Costs for Oil and Gas," *Advances in the Economics of Energy and Resources*, vol. 3 (Greenwich, Conn.: JAI Press).

[Adelman and Watkins 1995] M. A. Adelman and G. Campbell Watkins, "Reserve Asset Values and the Hotelling Valuation Principle: Further Evidence," forthcoming in *Southern Economic Journal.*

[Adelman et al. 1983] ———, John C. Houghton, Gordon M. Kaufman, and Martin B. Zimmerman, *Energy Resources in an Uncertain Future* (Cambridge, Mass.: Ballinger).

[Adelman et al. 1991] ———, Harindar DeSilva, and Michael F. Koehn, "User Cost in Oil Production," *Resources and Energy* 13:217–240.

[Agawa 1979] Hiroyuki Agawa, *The Reluctant Admiral* (Tokyo: Kodansha International).

[Ahmad et al. 1989] Yusuf J. Ahmad, Salah El Serafy, and Ernst Lutz, *Environmental Accounting for Sustainable Development* (Washington, D.C.: World Bank).

[Ajami 1981] Fouad Ajami, *The Arab Predicament* (Cambridge: At the press).

[Akins 1973] James E. Akins, "This Time the Wolf Is Here," *Foreign Affairs* 11, no. 1 (April).

[Akins 1981] ———, "The Influence of Politics on Oil Pricing and Production Policies," *Arab Oil and Gas* 10 (October 16):19–28.

[Akins 1982a] James E. Akins, Book review [Kissinger 1982], *American Arab Affairs*, no. 1 (Summer):147–151.

[Akins 1982b] ———, "Prospects of Supply Interruptions from OPEC in the Near Future," in George Horwich and Edward J. Mitchell, eds., *Policies for Coping with Oil Supply Disruptions* (Washington, D.C.: 1982) American Enterprise Institute.

[Akins 1983] ———, "Evolution of OPEC in the Next Ten Years" (paper presented to the Laval University symposium, September 7).

[Akins 1985] ———, interview in *Oil and Gas Journal*, April 15.

[Akins 1991] ———, "The New Arabia," *Foreign Affairs* 70, no. 3 (Summer): 36.

[API 1959] American Petroleum Institute, *Petroleum Facts and Figures: Centennial Edition* (Washington).

[API:JAS] American Petroleum Institute et al., *Joint Association Survey on Drilling Costs*, annual (Washington).

[API:SOGE] American Petroleum Institute, *Survey on Oil & Gas Expenditures*, annual 1955–1956, 1959–1973, 1983–1991 (Washington). (no longer published)

[Arshi 1992] Ali Asghar Arshi, "Energy Swaps as Profit Motive Instruments in Oil Markets," *OPEC Review* 16, no. 2:201–216.

[Askari 1990] Hossein Askari, *Saudi Arabia's Economy: Oil and the Search for Economic Development* (Greenwich, Conn.: JAI Press).

[Askari 1991] Hossein Askari, "Saudi Arabia's Oil Policy: Its Motivations and Impacts," in [Kohl, ed., 1991].

[Auty 1990] R. M. Auty, *Resource-Based Industrialization: Sowing the Oil in Eight Developing Countries* (Oxford: Clarendon Press).

[Balogh 1980] Thomas Balogh, "Oil Recycling—the Need for a New Lending Facility?" *Lloyds Bank Review*, no. 137 (July):16–25.

[Barran 1971] Sir David Barran, speech to Fuel Luncheon Club, London, February 16.

[Berndt and Wood 1987] Ernst R. Berndt and David O. Wood, "Energy Price Shocks and Productivity Growth: A Survey," in Richard L. Gordon, Henry D. Jacoby, and Martin B. Zimmerman, eds., *Markets and Regulation* (Cambridge, Mass.: MIT Press).

[Bill 1988] James A. Bill, *The Eagle and the Lion: The Tragedy of American-Iranian Relations* (New Haven: Yale University Press).

[Binks 1992] Adrian Binks, "Price Reporting and Price Formulas: Minimizing Price Risk," *OPEC Bulletin* 23, no. 10 (November–December):13–16.

[Blitzer Cavoulacos Lessard and Paddock 1985] Charles Blitzer, Panos Cavoulacos, Don Lessard, and James L. Paddock, "An Analysis of Financial and Fiscal Impediments to Oil and Gas Exploration in Developing Countries," *Energy Journal* 6, special tax issue: 59–72.

[Bloch 1961] Marc Bloch, *Apologie pour l'histoire, ou métier d'historien* [1942], Cahiers des Annales no. 3 (Paris: Armand Colin).

[Boskin et al. 1985] Michael J. Boskin, Marc S. Robinson, Terrance O'Reilly, and Praveen Kumar, "New Estimates of the Value of Federal Mineral Rights and Land," *American Economic Review* 75 (December), 923–936.

[Bosworth 1990] Barry Bosworth, "International Differences in Saving," *Papers and Proceedings of the American Economic Association* 80, no. 2:377–381.

[Bosworth and Lawrence 1982] Barry P. Bosworth and Robert Z. Lawrence, *Commodity Prices and the New Inflation* (Washington: Brookings Institution).

[BP 1970] British Petroleum Ltd., Annual Report for 1970.

[Bradley 1986] Paul G. Bradley, *Report of the Mineral Revenues Inquiry in Regard to the Study into Mineral (including Petroleum) Revenues in Western Australia* (Perth, Western Australia: Government Printer).

[Bradley and Watkins 1987] Paul G. Bradley and G. Campbell Watkins, "Net Value Royalties: Practical Tool or Economist's Illusion," *Resources Policy* 13, no. 4.

[Brandon 1988] Henry Brandon, *Special Relationships: A Foreign Correspondent's Memoirs from Roosevelt to Reagan* (New York: Atheneum).

[Brealey and Myers 1984] Richard Brealey and Stewart Myers, *Principles of Corporate Finance* (New York: McGraw-Hill Book Co.).

[Burns 1934] Arthur F. Burns, *Production Trends in the United States Since 1870* (New York, National Bureau of Economic Research).

[Burton 1982] Ellen Burton, "Private Investment in Petroleum Inventories" (Ph.D. diss. MIT).

[Canada 1980] Government of Canada, Department of Energy Mines and Resources, *The National Energy Program* (Ottawa).

[Canada 1982] ———, *The National Energy Program: Update 1982* (Ottawa).

[CEA] Council of Economic Advisers, *Economic Report of the President*, annual (Washington, D.C.: Government Printing Office).

[CGES 1992] Center for Global Energy Studies [London], *Global Oil Report.*

[Al Chalabi 1980] Fadhil J. Al-Chalabi, *OPEC and the International Oil Industry: A Changing Structure* (New York: Oxford University Press).

[Chandler 1973] Geoffrey Chandler, "Some Current Thoughts on the Oil Industry," *Petroleum Review* 27 (January).

[Church 1974] *Multinational Corporations and United States Foreign Policy. Hearings before the Subcommittee on Multinational Corporations of the Committee on Foreign Relations.* [Senator Frank P. Church, chairman] U.S. Senate. 93d Cong., 2d sess. 11 vols.

[Church Documents] *The International Petroleum Cartel, the Iranian Consortium and U.S. National Security.* Prepared for the use of Subcommittee on Multinational Corporations of the Committee on Foreign Relations. U.S. Senate, 93d Cong., 2d sess. (A collection of hitherto secret documents, covering the years 1952–1961, in addition to declassified documents and pages released in 8 Church).

[Church Report 1975] *Report of the Subcommittee on Multinational Oil Companies and U.S. Foreign Policy* (Washington, D.C.: Government Printing Office).

[CIA 1977] Central Intelligence Agency, *The International Energy Situation: Outlook to 1985* (Washington, D.C.: April).

[CIA 1979] ———, *The World Oil Market in the Years Ahead* (Washington, D.C.: August).

[CIA 1980] ———, "The Geopolitics of Energy," statement submitted to the Committee on Energy, U.S. Senate. Reprinted in full in *Petroleum Intelligence Weekly*, May 19, pp. S1–S4.

[CIA-IESR 1978] ———, *International Energy Statistical Review*, May 17, p. 19.

[CMB] Chase Manhattan Bank, *Capital Investments of the World Petroleum Industry.*

[Cooper 1986] Richard N. Cooper, *Economic Policy in an Interdependent World: Essays in World Economics* (Cambridge, Mass.: MIT Press).

[Craig and George 1983] Gordon A. Craig and Alexander L. George, *Force and Statecraft: Diplomatic Problems of Our Time* (New York: Oxford University Press).

[CRS 1977] Congressional Research Service, *Toward Project Interdependence* (Washington, D.C.: US Government Printing Office).

[Dahl and Sterner 1991] Carol Dahl and Thomas Sterner, "Analyzing Gasoline Demand Elasticities," *Energy Economics* 13, no. 3 (July).

[Dahl and Yucel 1991] Carol Dahl and Mine Yucel, "Testing Alternative Hypotheses of Oil Producer Behavior," *Energy Journal* 12, no. 4 (December): 117–138.

[Das Gupta and Heal 1979] Partha Das Gupta and Geoffrey M. Heal, *Economic Theory and Exhaustible Resources*, (Cambridge: Cambridge University Press, 1979).

[DeGolyer and MacNaughton 1993] *World Energy Statistics* (Dallas, annual compiled and pubished by DeGolyer and MacNaughton).

[Dirlam and Kahn 1958] Joel B. Dirlam, "Natural Gas: Cost, Conservation, and Pricing," in *Papers and Proceedings of the American Economic Association* 48 (May): 481–501, incorporating by reference the testimony of Alfred E. Kahn, Federal Power Commission, *Tidewater Oil Co.*, Docket G-9932 (1957).

[DOE 1977] Department of Energy, *Annual Report to Congress* (Washington, D.C.: Government Printing Office).

[DOE 1987] ———, *Energy Security: A Report to the President* (Washington, D.C.: Government Printing Office, March).

[DOE/EIA 1989] Department of Energy, Energy Information Administration, *International Energy Outlook: 1989* (Washington, D.C.: Government Printing Office).

[DOE/EIA 1991] ———, *Symposium on Short-Term Energy Forecasting* (November 21).

[DOE/EP 1993] Department of Energy, Office of Domestic and International Energy Policy, *Development Costs of Undeveloped Nonassociated Gas Reserves in Selected Countries* (DOE/EP-003P) (Washington, D.C.: Government Printing Office, January).

[Dominguez Strong and Weiner 1989] Kathryn M. Dominguez, John S. Strong, and Robert J. Weiner, *Oil and Money: Coping with Price Risk Through Financial Markets* (Cambridge, Mass.: Harvard University).

[Dowty 1984] Alan Dowty, *Middle East Crisis: U.S. Decision-Making in 1958, 1970, and 1973* (Berkeley: University of California Press).

[EC 1992] European Communities, Directorate for Energy. *Energy in Europe: a View to the Future* (Luxembourg).

[Eckbo 1976] Paul L. Eckbo, *The Future of World Oil* (Cambridge, Mass.: Ballinger).

[Eckbo 1987] ———, "Worldwide Petroleum Taxation: The Pressure for Revision," in [Gordon, Jacoby, and Zimmerman, eds. 1987], pp. 214–233.

[EMF 6, 1982] Energy Modeling Forum, *World Oil: Complete Report*, Energy Modeling Report 6 (Stanford, Calif., December).

[EMF11, 1992] ———, *International Oil Supplies and Demands*, EMF Report 11 (Stanford, Calif., April).

[Exxon 1976] Exxon Corporation, Public Affairs Department, *Middle East Oil.*

[Farmanfarmaian et al. 1975] Khadidad Farmanfarmaian, Armin Gutowski, Saburo Okita, Robert V. Koosa, and Carroll L. Wilson, "How Can the World Afford OPEC Oil?" *Foreign Affairs* 53 (January): 202–222.

[FEA/Church] Federal Energy Administration, Office of International Energy Affairs, *U.S. Oil Companies and the Arab Oil Embargo: The International Allocation of Constricted Supplies*, prepared for the Subcommittee on Multinational Corporations of the Committee on Foreign Relations, U.S. Senate, 94th Cong., 1st sess.

[Fesharaki 1980] Fereidoon Fesharaki, "Global Petroleum Supplies in the 1980s," *OPEC Review* (Summer).

[Finley 1981] M. I. Finley, *Economy and Society in Ancient Greece* (New York: Viking Press).

[Fischer 1988] Stanley Fischer, ed., "The Slowdown in Productivity Growth," *Journal of Economic Perspectives* 2, no. 4 (Fall): 3–98.

[Fischer 1993] ———, "Economic Performance in the FSU, mid-1993," MIT Department of Economics Working Paper 93-04.

[Fontaine 1971] André Fontaine, "La Tentation de la rupture," *Le Monde*, January 14.

[FPC 1972] U.S. Federal Power Commission, Bureau of Natural Gas, *National Gas Supply and Demand 1971–1990*, Staff Report 2 (Washington, D.C.: Government Printing Office).

[Fried and Blandin 1988] Edward R. Fried and Nanette M. Blandin, eds, *Oil and America's Security* (Washington, D.C.: The Brookings Institution).

[FTC 1952] *The International Petroleum Cartel*. Staff Report to the Federal Trade Commission submitted to the Subcommittee on Monopoly of the Select Committee on Small Business. U.S. Senate (Washington, D.C.: Government Printing Office).

[FTC-Church] Certain pages of [FTC 1952] were excised before publication but later declassified. They are reprinted in [8 Church 1974, pp. 529–532]. In addition, pp. 508–529 seem to be documentation for the previously excised pages.

[Fuller 1991] Graham E. Fuller, *"The Center of the Universe": The Geopolitics of Iran* (Boulder, Colo.: Westview Press).

[GAO 1982] General Accounting Office, *The Changing Structure of the International Oil Market* (Washington, D.C.: Government Printing Office).

[Garnaut and Clunies-Ross 1975] R. Garnaut and H. Clunies-Ross, "Uncertainty, Risk Aversion, and the Taxing of Natural Resource Projects," *Economic Journal* 85 (June): 272–287.

[Gately 1993] Dermot Gately, "The Imperfect Price-Reversibility of World Oil Demand," *Energy Journal* 14, no. 4:163–182.

[Gause 1994] F. Gregory Gause III, *Oil Monarchies: Domestic and Security Challenges in the Arab Gulf States* (New York: Council on Foreign Relations Press).

[Gelb 1988] Alan Gelb & Associates, *Oil Windfall: Blessing or Curse?* (Washington, D.C.: Oxford University Press for the World Bank).

[Gilbert 1953] Felix Gilbert, "Two British Ambassadors: Perth and Henderson," in: Gordon A. Craig and Felix Gilbert, eds., *The Diplomats, 1919–1939* (Princeton: Princeton University Press).

[Goodwin et al. 1981] Neil de Marchi, "Energy Policy under Nixon: Putting Out Fires," and "The Ford Administration: Energy as a Political Good," in Craufurd D. Goodwin, ed., *Energy Policy in Perspective: Today's Problems, Yesterday's Solutions* (Washington, D.C.: Brookings Institution).

[Gordon 1966] Richard L. Gordon, "Conservation and the Theory of Exhaustible Resources," *Canadian Journal of Economics and Political Science* 75 (3):19–26.

[Gordon 1970] Richard L. Gordon, *The Evolution of Energy Policy in Western Europe* (New York: Praeger).

[Gordon Jacoby and Zimmerman 1987] Richard L. Gordon, Henry D. Jacoby, and Martin B. Zimmerman, eds., *Energy: Markets and Regulation* (Cambridge: MIT Press).

[Gordon 1993] Robert J. Gordon, "The Jobless Recovery: Does It Signal a New Era of Productivity-Led Growth?" in *Brookings Papers on Economic Activity*, no. 1 (Washington, D.C.: Brookings Institution).

[Green 1990] Peter Green, *Alexander to Actium: The Historical Evolution of the Hellenistic Age* (Berkeley and Los Angeles: University of California Press).

[Griffin 1979] James M. Griffin, *Energy Conservation in the OECD: 1980 to 2000* (Cambridge, Mass.: Ballinger).

[Griffin 1985] James M. Griffin, "OPEC Behavior: A Test of 'Alternative Hypotheses,' *American Economic Review* 75, no. 5:954–963.

[Griffin Daley and Steele 1982] James M. Griffin, George Daly, and Henry B. Steele, "Recent Price Escalations: Implications for OPEC Stability," in [Griffin and Teece 1982], pp. 145–174.

[Griffin and Teece, eds. 1982] James M. Griffin and David J. Teece, eds., *OPEC Behavior and World Oil Prices* (London: George Allen & Unwin).

[Gustafson 1989] Thane Gustafson, *Crisis Amid Plenty: The Politics of Soviet Energy under Brezhnev and Gorbachev* (Princeton: Princeton University Press).

[Hackett 1978] John W. Hackett, *The Third World War* (London: Sidgwick and Jackson).

[Hammer 1987] Armand Hammer (with Neil Lyndon), *Hammer* (New York: G. P. Putnam's Sons).

[Harik and Sullivan 1992] Iliya Harik and Denis J. Sullivan, *Privatization and Liberalization in the Middle East* (Bloomington: Indiana University Press).

[Hartshorn 1967] Jack E. Hartshorn, *Oil Companies and Governments* (London: Faber & Faber). U.S. title: *Politics and World Oil Economics.*

[Hartwick 1991] John M. Hartwick, "The Non-renewable Resource Exploring-extracting Firm and the r% Rule," *Resources & Energy* 13:129–143.

[Heal and Chichilnisky 1991] Geoffrey Heal and Graciela Chichilnsky, *Oil and the International Economy* (New York: Oxford University Press).

[Helfat 1988] Constance E. Helfat, *Investment Choices in Industry* (Cambridge, Mass.: MIT Press).

[Hogan 1989] William W. Hogan, *World Oil Price Projections: A Sensitivity Analysis* (Cambridge, Mass.: Kennedy School of Government).

[Hogan 1992] ———, *Oil Market Adjustments and the New World Order* (Cambridge, Mass.: Kennedy School of Government, July).

[Hubbard and Weiner 1989] R. Glenn Hubbard and Robert J. Weiner, "Contracting and Price Adjustment in Commodity Markets: Evidence from Copper and Oil," *Review of Economics and Statistics* 71 (February).

[Hunter 1992] Shireen T. Hunter, *Iran After Khomeini* (New York: Praeger for Center for Strategic and International Studies).

[Ibn Khaldun, 1377, (1967)] Ibn Khaldun, *The Muqaddimah: An Introduction to History* (1377), Tran. Franz Rosenthal, ed. N. J. Dawood (Princeton: Princeton University Press).

[Indyk 1994] Martin Indyk (National Security Council), at a symposium February 24, 1994, in *Middle East Policy* 3, no. 1.

[IPA] U.S. Bureau of Mines, *International Petroleum Annual.*

[IRS 1991] United States Tax Court. Exxon Corporations and Affiliated Corporations et al., Petitioners, v. Commissioner of Internal Revenue, Respondent. Nos. 18618-89, 24855-89, 18432-90. (This lawsuit concerned the prices received by two Aramco companies during 1979–1981. The writer was a consultant and witness on behalf of the Internal Revenue Service.)

[Irving 17:30] Federal Court of Canada, Trial Division, T3100–79 Between Irving Oil Ltd. and Her Majesty the Queen, *Reasons for Judgment (Muldoon, J.) Exhibit 17: Partial Agreed Statement of Facts*, par. 30, Ottawa, Ontario, March 4, 1988 (The subject of this lawsuit was the set of prices charged by Chevron to a Canadian refiner between 1955 and 1975. The writer was a consultant and witness on behalf of Irving Oil Ltd.)

[Isaacson 1992] Walter Isaacson, *Kissinger: A Biography* (New York: Simon & Schuster).

[Ismail 1991] Ibrahim A. H. Ismail, "OPEC Production Capacity: the need for expansion," *OPEC Review* 6, no. 3.

[Kaldor 1983] Nicholas Kaldor, "The Role of Commodity Prices in Economic Recovery," *Lloyds Bank Review* (July):21–34.

[Kapstein 1990] Ethan B. Kapstein, *The Insecure Alliance: Energy Crises and Western Politics since 1944* (New York: Oxford University Press).

[Kemp 1987] Alexander Kemp, *Petroleum Rent Collection Around the World* (Halifax, Nova Scotia: Institute for Research on Public Policy).

[Kemp Reading and Macdonald 1992] Alexander Kemp, David Reading, and Bruce Macdonald, "The Effects of the Fiscal Terms Applied to Offshore Petroleum Exploitation of New Fields," North Sea Study Occasional Paper 37 (University of Aberdeen Department of Economics, April).

[Kimura 1986] Shuzo Kimura, "Japan's Middle East Policy," *American-Arab Affairs*, no. 17 (Summer).

[Kissinger 1982] Henry A. Kissinger, *Years of Upheaval* (Boston: Little, Brown).

[Kohl, ed. 1991] Wilfrid L. Kohl, ed., *After the Oil Price Collapse: OPEC, the United States, and the World Oil Market* (Baltimore: the Johns Hopkins University Press).

[Kosik 1975] Joseph T. Kosik, *Natural Gas Imports from the Soviet Union: Financing the North Star Project* (New York: Praeger Publishers).

[Kouris 1983] George Kouris, "Energy Demand Elasticities in Industrialized Countries: A Survey," *Energy Journal* 4 no. 3 (July).

[Krasner 1978] Stephen D. Krasner, *Defending the National Interest: Raw Materials Investments and U.S. Foreign Policy* (Princeton: Princeton University Press).

[Levy 1987] Frank Levy, *Dollars and Dreams: The Changing American Income Distribution* (New York: Russell Sage/Basic Books).

[Lewis 1954] Bernard Lewis, "Communism and Islam," *International Affairs* (January).

[Lohrenz 1991] John Lohrenz, "Horizontal Oil and Gas Wells: the Engineering and Economic Nexus," *The Energy Journal* 12:35–54.

[Lohrenz 1992] John Lohrenz, "Exploration: a Misunderstood Business," in Richard Steinmetz, ed., *The Business of Petroleum Exploration* (Tulsa: American Association of Petroleum Geologists 1992).

[Lohrenz et al. 1981] John Lohrenz, Bernard H. Burzlaff, and Elmer Dougherty, "How Policies Affect Rates of Recovery from Mineral Sources," *Society of Petroleum Engineers Journal* 271 (December): 645–657.

[Lund 1992] Diderik Lund, "Petroleum Taxation under Uncertainty: Contingent Claims Analysis with an Application to Norway," *Energy Economics* 14, No. 1 (January).

[Lynch 1987] Michael C. Lynch, "The Underprojection of Non-OPEC Third World Oil Production," *Proceedings of the Ninth International Conference, International Association of Energy Economists* (Calgary, July).

[Lynch 1989] ———, *Oil Prices to 2000: The Economics of the Oil Market*, Special Report 1160 (London: Economist Intelligence Unit, May).

[Lynch 1992] ———, "The Fog of Commerce: The Failure of Long-term Oil Market Forecasting" (MIT Center for International Studies, September).

[Mabro 1991] *A Dialogue between Oil Producers and Consumers* (New York: Oxford Institute for Energy Studies).

[Maddison 1991] Angus Maddison, *Dynamic Forces in Capitalist Development: A Long-Run Comparative View* (New York: Oxord University Press).

[Mancke 1979] Richard B. Mancke, *Mexican Oil and Natural Gas* (New York: Praeger).

[Al-Mani and Al-Shaikhly 1983] Saleh Al-Mani and Salah Al-Shaikhly, *The Euro-Arab Dialogue* (New York: St. Martin's Press).

[Masters et al. 1987] C. D. Masters et al., "World Resources of Crude Oil, Natural Gas, Natural Bitumen and Shale Oil," in *Proceedings of the Twelfth World Petroleum Congress* 5 (New York: Wiley).

[McDonald 1971] Stephen L. McDonald, *Petroleum Conservation in the United States: an Economic Analysis* (Baltimore: Johns Hopkins Press for Resources for the Future).

[Mead 1986] Walter J. Mead, "Competition in Outer Shelf Oil and Gas Lease Auctions," *National Resource Journal* 26 (Winter): 95–111.

[Mead and Sorenson 1980] ——— and Philip E. Sorenson, *Competition and Performance in OCS Oil and Gas Lease Sales and Development, 1954–1969*, report to U.S. Geological Survey (contract no. 14-08-0001-16552), March.

[Meyer 1977] Lorenzo Meyer, *Mexico and the United States in the Oil Controversy: 1917–1942* (Austin: University of Texas Press).

[Mikdashi 1982] Zuhayr Mikdashi, *The Community of Oil Exporting Countries* (London: George Allen & Unwin).

[Mikdashi Cleland and Seymour eds. 1970] Zuhayr Mikdashi, Sherril Cleland, and Ian Seymour, eds. *Continuity and Change in the World Oil Industry* (Beirut: Middle East Research and Publishing Center).

[Moran 1987] Theodore H. Moran, "Managing an Oligopoly of Would-be Sovereigns," *International Organization* 41, no. 4 (Autumn).

[Moroney 1993] John R. Moroney, "Energy, Capital, and Technological Change in the United States," *Resources and Energy* 14, no. 4:363–380.

[Nowell 1988] Gregory Patrick Nowell, "Realpolitik vs. Transnational Rent-seeking: French Mercantilism and the Development of the World Oil Cartel, 1860–1939" (Ph.D. diss. MIT).

[Odell 1987] Peter Odell, "Back to Cheap Oil?" *Lloyds Bank Review*, no. 165 (July).

[Ofurhie et al. 1991] M. A. Ofurhie, M. C. Amaechi, and A. O. Idowu, "How Nigeria Looked at Fiscal Terms, Prospects," *Oil and Gas Journal*, November 25.

[Penrose 1968] Edith Penrose, *The Large International Firm in Developing Countries: The International Petroleum Industry* (London: George Allen & Unwin).

[Perez 1969] Ruben Sader Perez, *The Venezuelan State Oil Reports to the People* (Caracas: Corporacion Venezolana del Petroleo, February).

[Petroleum Finance 1988] *World Petroleum Markets: A Framework for Reliable Projections,* World Bank Technical Paper 92 (Washington, D.C.: World Bank).

[PIW Swaps Report 1990] Amy Jaffe, Sarah Miller, and Thomas Wallin, *The Complete Guide to Oil Price Swaps* (New York: Petroleum Intelligence Weekly Special Report, October).

[Planck 1949] Max Planck [1858–1947], *Scientific Autobiography and Other Papers* (New York: Philosophical Library [Greenwood Press]).

[PPS 1966] "Crude Oil Production Costs," *Petroleum Press Service* 33, no. 5 (May):177–178.

[PPS 1968] "Cheapness with Security," *Petroleum Press Service* 35, no. 1 (January):2–4.

[Quandt 1977] William B. Quandt, *Decade of Decisions: American Policy Toward the Arab-Israeli Conflict, 1967–1976* (Berkeley: University of California Press).

[Radetzki et al. 1993] Marian Radetzki et al., *Energy and Economic Reform in the FSU* (New York: Macmillan).

[Repetto et al. 1989] Robert Repetto, William Magrath, Michael Wells, Christine Beer, Fabrizio Rossini, *Wasting Assets: natural resources in the national income accounts* (Washington: World Resources Institute).

[Richards and Waterbury] Alan Richards and John Waterbury, *A Political Economy of the Middle East* (Boulder, Colo.: Westview Press).

[Roberts 1993] John Roberts, "Middle East Developments," (paper presented at Twelfth CERI International Oil and Gas Markets Conference, September 26–28, Calgary).

[Robinson 1988] Jeffrey Robinson, *Yamani: The Inside Story* (London: Simon & Schuster Ltd.).

[Rockefeller 1978] *Working Paper on International Energy Supply* (New York: Rockefeller Foundation, March).

[Rostow 1980] W. W. Rostow, *How to Get from Here to There* (Austin: University of Texas Press).

[Saad 1969] Farid W. Saad, "France and Oil" (Ph.D. diss., MIT).

[Sadowski 1992] Yahya Sadowski, "Sandstorm with a Silver Lining?" *Brookings Review* (Summer).

[Samii Weiner and Wirl 1989] Massood V. Samii, Robert J. Weiner, and Franz Wirl, "Determinants of Crude Oil Prices: OPEC versus Speculators" (Cambridge, Mass.: Harvard Energy and Environmental Policy Center, June).

[Sampson 1975] Anthony Sampson, *The Seven Sisters: The Great Oil Companies and the World They Shaped* (New York: Viking Press).

[Schlesinger 1978] James R. Schlesinger, before the Joint Economic Committee, U.S. Congress, March 21.

[Schlesinger-Fahd 1978] "Meeting Between Secretary of Energy Schlesinger and Crown Prince Fahd on January 15, 1978." Telegram, Department of State, classified Secret. Sections later declassified, introduced as evidence in [IRS].

[Schneider 1983] Steven A. Schneider, *The Oil Price Revolution* (Baltimore: Johns Hopkins University Press).

[Schumpeter 1943] Joseph A. Schumpeter, *Capitalism, Socialism, and Democracy.* (New York: Oxford University Press).

[Schumpeter 1954] ———, *History of Economic Analysis* (New York: Oxford University Press).

[Schurr and Homan 1969] Sam H. Schurr and Paul T. Homan, *Middle East Oil and the Western World* (New York: American Elsevier).

[Schweitzer 1994] Peter Schweitzer, *Victory: the Reagan Administrations's Secret Strategy That astened the Collapse of the Soviet Union* (New York: Atlantic Monthly Press).

[Senate 1972] *Natural Gas Policy Issues,* Hearings before the Committee on Interior and Insular Affairs, U.S. Senate, 92d Cong., 2d sess., February 25–March 2.

[Senate 1973] *Oil and Gas Imports Issues.* Hearings before the Committee on Interior and Insular Affairs, U.S. Senate, 93d Cong., 1st sess.

[Shahanshah 1971] Shahanshah of Iran on Oil, transcript issued in London by Transorient Ltd., including statements and interviews in January and February.

[Shultz et al. 1970] Cabinet Task Force on Oil Import Control (Secretary George P. Shultz, chairman), *The Oil Import Question: A Report on the Relationship of Oil Imports to the National Security* (Washington, D.C.: Government Printing Office, February).

[Skeet 1988] Ian Skeet, *OPEC: Twenty-five Years of Prices and Politics* (Cambridge: Cambridge University Press).

[Smith 1987] James L. Smith, "International Petroleum Taxation: Reasons for Instability," *World Energy Markets. Proceedings of the Ninth International Conference, International Association of Energy Economists,* pp. 571–579.

[Smith 1992] ———, "The Fuel That Came in from the Cold," unpublished.

[Smith 1993] ———, rapporteur of *The Russian Oil Industry: Foreign Investment Opportunities,* conference convened by *Petroleum Intelligence Weekly,* Royal Institute of International Affairs, and Moscow Centre for Foreign Investment and Privatization, London, February 11 and 12.

[Solow 1974] Robert M. Solow, "The Economics of Resources or the Resources of Economic," *American Economic Review: Papers & Proceedings* 64:1–14.

[Solow 1992] Robert M. Solow, "An Almost Practical Step Toward Sustainability," published by Resources for the Future, October 8, 1992.

[Spiegel 1985] Steven L. Spiegel, *The Other Arab-Israeli Conflict* (Chicago: University of Chicago Press) .

[Springer 1977] E. H. Springer, "Discounts in Disguise: Price Shaving by Members of OPEC," State Department Bureau of Intelligence and Research, Report 831, July 22.

[Stiglitz 1976] Joseph E. Stiglitz, "Monopoly and the Rate of Extraction of Exhaustible Resources," *American Economic Review* 66 (September): 655–666.

[Taylor 1991] Alan R. Taylor, *The Superpowers and the Middle East* (Syracuse: Syracuse University Press).

[UN 1955] United Nations, Economic Commission for Europe, *The Price of Oil in Western Europe* (E/ECE/205, Geneva).

[Verleger 1982] Philip K. Verleger, Jr., *Oil Markets in Turmoil: An Economic Analysis* (Cambridge, Mass.: Ballinger).

[Verleger 1986] ———, "The Diversification of Risk: The Effect of Investment in Refining and Marketing in Western Europe on the Income of an Integrated Refiner," unpublished paper.

[Verleger 1987] ———, "The Evolution of Oil as a Commodity," in [Gordon Jacoby and Zimmerman 1987], pp. 161–186.

[Verleger 1993] ———, *Adjusting to Volatile Energy Prices* (Washington, D.C.: Institute for International Economics).

[Verleger-PEM] ———, *Petroleum Economics Monthly*.

[Viorst 1994] Milton Viorst, *Sandcastles: the Arabs in Search of the Modern World* (New York: Knopf).

[Watkins and Waverman 1986] G. C. Watkins and Leonard Waverman, "Oil Demand Elasticities," in David O. Wood, ed., *The Changing World Energy Economy: Proceedings of the Eighth Annual IAEE North American Conference* (Boston, November).

[Watt 1989] Donald Cameron Watt, *How War Came: The Immediate Origins of the Second World War* (New York: Pantheon Books).

[Weiner 1992] Robert J. Weiner, "The Oil Futures Market in the Gulf Crisis," Kennedy School Center for Business and Government, Working Paper 92-3.

[Wigel and Sandoval 1979] H. S. Wigel and D. A. Sandoval, *An Analysis of World Oil Markets, 1974–1979* (DOE/EIA-0184/9).

[Wilmott 1982] H. P. Wilmott, *Empires in the Balance: Japanese and Allied Pacific Strategies to April 1942* (Annapolis: U.S. Naval Institute Press).

[Wood 1990] David A. Wood, "Appraisal of Economic Performance of Global Exploration Contracts," *Oil and Gas Journal*, October 29, p. 48.

[Yamani 1978] Ahmad Zaki Yamani, "The Changing Pattern of World Oil Supplies," *Middle East Economic Survey*, July 17, p. S1.

[Yergin and Hillenbrand 1982] Daniel Yergin and Martin Hillenbrand, eds., *Global Insecurity: A Strategy for Energy & Economic Renewal* (Boston: Houghton Mifflin).

[Zannetos 1966] Zenon S. Zannetos, *The Theory of Oil Tankship Rates* (Cambridge, Mass.: MIT Press).

Index